T0296546

THE THEORY OF ELECTRICITY

THE THEORY OF ELECTRICITY

BY

G. H. LIVENS, M.A.

Sometime Fellow of Jesus College, Cambridge; Professor of Mathematics in the University College, Cardiff

SECOND EDITION

CAMBRIDGE
AT THE UNIVERSITY PRESS
1926

CAMBRIDGE
UNIVERSITY PRESS

University Printing House, Cambridge CB2 8BS, United Kingdom

Cambridge University Press is part of the University of Cambridge.

It furthers the University's mission by disseminating knowledge in the pursuit of education, learning and research at the highest international levels of excellence.

www.cambridge.org
Information on this title: www.cambridge.org/9781316626160

First edition 1918
Second edition 1926
First paperback edition 2016

A catalogue record for this publication is available from the British Library

ISBN 978-1-316-62616-0 Paperback

PREFACE

THIS book is a revised form of my *Theory of Electricity* published in 1918. Its somewhat smaller size is due to the omission of the applications of the theory in special problems—adequately dealt with in other books—and also of redundancies employed in the previous text to emphasise certain neglected aspects of the theory. The point of view of the older text has been entirely maintained but it is hoped that it is now presented in a somewhat more mature form.

As in the first edition the object has been to present a complete account of the purely theoretical side of the subject in the only form in which it appears to be satisfactory from the point of view both of mathematical consistency and of physical completeness. Particular attention has been given to the rigorous formulation of the underlying physical principles and to their translation into a mathematical theory. Various alternative points of view which have arisen in the development of the different aspects of the subject are, however, discussed when discrimination in the present stage of our knowledge seemed out of the question, but more emphasis has been laid on certain points where discrimination has been rather unjustifiably employed by other writers.

I have kept in the main to the usual and now almost classical nomenclature and symbolism of the subject; but one change—and a rather drastic one—I have been forced to make in the interests of physical consistency. The magnetic quantity denoted elsewhere and here also by **B** has been designated, not the *magnetic induction*, but the *magnetic force*, which is what it is in fact. The quantity usually denoted by **H** then appears as the derived or induced vector and is thus here called the *magnetic induction* (of mechanically effective force).

My thanks are due to a few encouraging critics of the first edition, to those whose writings on this subject have come into my hands since that manuscript was completed, and to the officials of the University Press for the manner in which they have again produced the book.

G. H. L.

University College, Cardiff
June 7th, 1926

CONTENTS

CHAP.		PAGE
I.	THE ELECTRIC FIELD	1
II.	DIELECTRIC THEORY	51
III.	ELECTRIC CURRENTS	110
IV.	THE MAGNETIC FIELD	153
V.	THE DYNAMICS OF THE MAGNETIC FIELD	178
VI.	MAXWELL'S ELECTROMAGNETIC THEORY	212
VII.	ELECTROMAGNETIC OSCILLATIONS AND WAVES	265
VIII.	THE ELECTRODYNAMICS OF MOVING MEDIA	329
APP.		
I.	ON THE MECHANISM OF MAGNETIC INDUCTION	390
II.	ON THE MECHANISM OF METALLIC CONDUCTION	404
INDEX		423

CHAPTER I

THE ELECTRIC FIELD

1. Electrification and electricity. A piece of glass and a piece of resin on being rubbed together will be found on separation to attract one another. Also if a second piece of glass be rubbed by a second piece of resin, and if the pieces be then separated and suspended in the neighbourhood of the former pieces of glass and resin, it will be observed that (i) the two pieces of glass repel each other, (ii) each piece of glass attracts each piece of resin, and (iii) the two pieces of resin repel one another. These phenomena of attraction and repulsion are called *electrical phenomena*, and the bodies which exhibit them are said to be *electrified* or *charged with electricity*. The electrical properties of the two pieces of glass are similar to one another but opposite to those of the two pieces of resin. If a body electrified in any manner behaves as glass does, that is, it repels the glass and attracts the resin, that body is said to be *positively* electrified, and if it attracts the glass and repels the resin it is said to be *negatively* electrified. No force either of attraction or repulsion can be observed between an electrified body and a body not electrified. When in any case bodies not previously electrified are observed to be acted on by an electrified body it is because they have been electrified by induction.

If a long metallic rod is suspended by silk threads so that one of its ends is near but does not touch a small body charged with electricity it will be found that this end of the rod has become charged with electricity opposite in sign to that on the small charged body, whilst the other end is found to be charged with electricity of the same sign as that on the given body. On the subsequent removal of the small electrified body it will be found that it has itself lost none of its charge, and that the rod has returned to its neutral state without electrification. If however the rod is so arranged that it can be divided into two parts we can separate the two parts before removing the inducing charge, and it will then be found that each part of the rod has actually retained its charge. The electrification of the metal rod which depends on the presence in its neighbourhood of an electrified body and which vanishes when that body is removed is called *electrification by induction.*

2. If two metallic bodies, one of which is charged with electricity, be suspended by silk threads at some distance apart and then connected by a thin metallic wire, it will be found that the electrification on the charged body has diminished and the uncharged body has acquired a charge of the same sign. In other words, the electrical condition has been transferred from the first to the second body by means of the wire. The wire is on this account called a *conductor* of electricity, and the second body is said to be *electrified by conduction.*

If a glass rod, a stick of resin or gutta-percha, or even a silk thread, had been used instead of the metal wire, no transfer of electricity would have taken place. These latter substances are therefore called *non-conductors of electricity* or *insulators*, and they are always used in electrical experiments as supports for electrified bodies. Although it is convenient in practice to draw this distinction between conductors and non-conductors, we shall find that in reality all substances resist the passage of electricity and all substances allow it to pass, although in very different degrees.

Of all substances the metals are by far the best conductors. Next come solutions of salts and acids, and lastly as very bad conductors, and therefore good insulators, come oils, waxes, silk, glass, and such substances as sealing-wax, resin, shellac, india-rubber, etc. Gases under ordinary conditions are good insulators. Flames however conduct well and, for reasons which will appear later, all gases become good conductors when in the presence of radium or other so-called radio-active substances. Distilled water is almost a perfect insulator, but any other sample of water will contain impurities which usually cause it to conduct tolerably well, so that a wet body is generally a bad insulator.

3. Previous to the researches of Faraday the generally accepted explanation of the phenomena of electrical action involved the fundamental concept of the two electric fluids: these two fluids were assumed to be composed of very small particles of a non-gravitative subtile matter of a more refined and penetrating kind than ordinary liquids and gases; and two particles of the same fluid repel each other whilst two particles of different fluids attract each other. It was then supposed that all bodies in their ordinary conditions contained equal amounts of both positive and negative electricity, and that in rubbing two bodies together as described above a difference in the quantities of each of the fluids in the two bodies is produced, so that the one has an excess of positive fluid and is positively charged and the other has the corresponding excess of negative fluid and is negatively charged. The theory thus effectively explains the phenomena of electrification by friction and also the forces of attraction and repulsion between electrified bodies: it would also account for the phenomena of conduction and induction if it is assumed that either or both the electric fluids are freely movable through good conducting media, being however more or less rigidly fixed to the elements of matter in good insulating media.

This view of the nature of electricity has remained to the present day with but little modification of its essential details. We know now that electricity is a fundamental constituent, if not the whole content, of all matter, although the two types occur in very different ways. The negative electricity always consists of *electrons* or indivisible atoms of electricity, which are apparently always of the same uniform size and with the same charge*. But the positive electricity behaves differently. There is no evidence of the existence of carriers of positive electricity of less than atomic mass, and generally the nature of the carriers varies according to the substance with which we are dealing. Evidence is accumulating however that these positive charges are themselves multiples of the smallest yet discovered, viz. the hydrogen atom, carrying a charge equal to that of one electron.

4. On this theory then every atom contains a certain number (roughly equal to half its atomic weight) of electrons grouped together in more or less stable congeries round a positive nucleus whose charge is just sufficient to neutralise the charge on the electrons. In every body there will be a continuous process of dissociation going on, the electronic configuration inside the atom being sufficiently unstable in many cases to be capable of breaking up on small provocation, with the consequent liberation of one or

* Their mass is $9{\cdot}042 \times 10^{-28}$ gm. and the charge on them is $4{\cdot}774 \times 10^{-10}$ electrostatic units.

more electrons and occasionally of positive elements as well. The result is that mixed up with the atoms of chemical matter comprising a body we have a greater or less percentage of negative electrons and perhaps also a few positive elements freely movable in the interstices between the atoms. An electrically charged body is one in which there is an excess (or deficit) of negative electrons. The distinction between insulators and conductors as regards the phenomena of induction and conduction depends essentially on the fact that in the conductors there is a large number of the free dissociated electrons which can be pulled about from one part of the medium to another under the action of forces exerted from other electrified bodies; whereas in insulators there is such an extremely small number of these free electrons that the phenomena depending on them can under most circumstances be neglected.

In the majority of our future discussions we shall have no special necessity to use this definite conception of electricity which now underlies practically the whole of modern physical theory, and we shall occasionally offer tentative illustrative explanations based on a less explicit conception. We shall however frequently find it conducive to clearness to adopt this concrete view and even to elaborate various details in the theory based upon it.

5. The law of action in electrical theory. The actual law for the interaction of electrified bodies was first experimentally determined by Coulomb, who measured the force by means of a torsion balance. The result he obtained may be stated in the following form.

If we suppose the dimensions of two bodies on which charges q and q' are placed to be small compared with the distance between them so that the result is not affected by any inequality of distribution of electrification on either body, and if also the bodies be supposed to be suspended in air at a considerable distance from other bodies, then the force between them is radial and of amount

$$\mathbf{F} = \gamma \frac{qq'}{r^2}$$

when they are at a distance r apart. The constant γ is a physical constant depending on the unit of distance chosen and also on the provisional unit adopted for measuring the charges q, q'.

This law of force is essentially an empirical one and no absolute proof is possible, although very strong theoretical evidence can be adduced in its favour. Experimental proofs, involving no direct measurement of the force as is done by Coulomb, have however been given and they show that the law must certainly be

$$\mathbf{F} = \gamma \frac{qq'}{r^{2+p}},$$

where p is less than 10^{-5}, so that for all practical purposes the evidence is conclusive.

6. We have said that the constant γ which enters into the equation

$$\mathbf{F} = \gamma \frac{qq'}{r^2}$$

is a physical constant depending on the units adopted. The reason is of course that the expression of a physical law must be independent of the units of measurement of the quantities involved and the dimensions of the quantities on the two sides of the above equation must be the same.

If we use $[m]$, $[l]$, $[t]$, $[q]$ to denote the respective dimensions of the chosen arbitrary units of mass, length, time and electric charge, the dimensional equation for the above law is

$$\left[\frac{ml}{t^2}\right] = [\gamma]\left[\frac{q^2}{l^2}\right],$$

so that

$$[\gamma] = \left[\frac{l^3 m}{q^2 t^2}\right]$$

defines the dimensions of γ. Now the definition of dimensions implies that the unit of any quantity is increased in the ratio of its dimensions when these are increased, so that the actual measure of a given quantity is reduced in the same ratio. Thus if we had chosen new arbitrary units for the fundamental quantities, which have respectively measures M, L, T, Q in terms of the former units, the new value Γ of the constant of the physical law of action between electric charges would be given by

$$\Gamma = \frac{\gamma}{\dfrac{L^3 M}{Q^2 T^2}},$$

and this relation defines completely the way in which this constant depends on the fundamental units.

7. We have so far assumed a knowledge of some arbitrary but definite unit of electric charge which does not depend on the other fundamental units employed. We can, however, simplify the matter by regarding the physical law of action as defining an unit of electric charge and then the constant γ can be chosen at will, provided that the electric charge is measured properly. For some purposes the simplest plan is to take, after Gauss, $\gamma = 1$ so that the equation for the law of action is

$$\mathbf{F} = \frac{qq'}{r^2}.$$

In this case the dimensions of an electric charge must clearly be

$$[q] = (l^{\frac{3}{2}} t^{-1} m^{\frac{1}{2}}),$$

and the unit charge is such that if condensed at unit distance *in vacuo* from a similar quantity it would exert unit force on it. This is Gauss' *absolute electrostatic unit* of charge. For purely theoretical purposes it appears, however, more convenient to take slightly different units of the same dimensions. In fact if we take $\gamma = \frac{1}{4}\pi$ we shall find that an otherwise frequently repeated factor 4π disappears entirely from our analysis.

The idea of the concentration of a charge at a point involved in the definition of our unit charge is of course merely a technical device introduced to simplify the mathematical expression of the law of action. We ought to say that the charges are on small bodies whose dimensions are infinitesimal in a physical sense compared with the distance at which the forces are investigated. In this sense we regard a 'point charge' in our theory much as we do a 'mass particle' in ordinary mechanics, and the proved real existence of the minute electron provides a direct physical picture for such a procedure.

8. The definition of the electric field of a system of point charges*. If an electrified body is brought into the space surrounding any system of charges it will in general produce a sensible disturbance in the electrification of the system by induction. But if the body is very small and its charge also very small the electrification of the other bodies will not sensibly be disturbed. The force acting on the body as a result of the action from the charges on

* J. Lagrange, *Par. sav.* (*étr.*) VII. (1773) (*Œuvres*, VI. p. 349).

the other bodies will then be proportional to its charge and will be reversed if the sign of the charge is reversed. That is if $\delta\mathbf{F}$ be the force and δq the charge, then when δq is an infinitesimal $\delta\mathbf{F}$ is proportional to δq or

$$\delta\mathbf{F} = \mathbf{E}\,\delta q,$$

where $\mathbf{E}$ is a vector function of the position of δq only, which is determinate when the system of charges is given. We thus may regard $\mathbf{E}$ as a property of the point.

The space in and around the given system of charges is called the *electric field* of those charges, and the vector $\mathbf{E}$, which represents the force 'per unit charge' at the point in the field, is called the *intensity of the electric force* at that point in the field.

In the neighbourhood of a single point charge q and at a distance r from it the electric force intensity is along the direction of r and of amount $\gamma q/r^2$. If we refer to ordinary rectangular coordinates with the point charge q_1 at the point (x_1, y_1, z_1), then the components of the force intensity along the coordinate axes at the point (x, y, z) are

$$\gamma q_1\left(\frac{x - x_1}{r_1^3},\ \frac{y - y_1}{r_1^3},\ \frac{z - z_1}{r_1^3}\right),$$

where $$r_1^2 \equiv (x - x_1)^2 + (y - y_1)^2 + (z - z_1)^2.$$

These are simply

$$-\left(\frac{\partial}{\partial x},\ \frac{\partial}{\partial y},\ \frac{\partial}{\partial z}\right)\frac{\gamma q_1}{r_1}.$$

In a similar manner it can be seen that if we have any system of point charges $q_1, q_2, \ldots q_n$ at the points $(x_1, y_1, z_1;\ x_2, y_2, z_2;\ \ldots;\ x_n, y_n, z_n)$ respectively, then the components of the total electric force at the point (x, y, z) of the field are

$$(\mathbf{E}_x, \mathbf{E}_y, \dot{\mathbf{E}}_z) = \gamma\sum_{s=1}^{n}\left[\frac{q_s(x - x_s)}{r_s^3},\ \frac{q_s(y - y_s)}{r_s^3},\ \frac{q_s(z - z_s)}{r_s^3}\right]$$
$$= -\left(\frac{\partial}{\partial x},\ \frac{\partial}{\partial y},\ \frac{\partial}{\partial z}\right)\left(\gamma\sum_{s=1}^{n}\frac{q_s}{r_s}\right),$$

where $$r_s^2 = (x - x_s)^2 + (y - y_s)^2 + (z - z_s)^2.$$

The function $$\phi = \gamma\sum_{s=1}^{n}\frac{q_s}{r_s},$$

from which the components of the force intensity at any point of the field are obtained by simple differentiation along the axes, is called the *potential** of the electric field at the point (x, y, z). It has an important physical significance which we shall discuss later: for the present it is merely defined so that

$$(\mathbf{E}_x, \mathbf{E}_y, \mathbf{E}_z) = -\left(\frac{\partial}{\partial x}, \frac{\partial}{\partial y}, \frac{\partial}{\partial z}\right)\phi,$$

and thus

$$\mathbf{E}_x \frac{dx}{ds} + \mathbf{E}_y \frac{dy}{ds} + \mathbf{E}_z \frac{dz}{ds} = -\frac{\partial\phi}{\partial x}\frac{dx}{ds} - \frac{\partial\phi}{\partial y}\frac{dy}{ds} - \frac{\partial\phi}{\partial z}\frac{dz}{ds}$$

$$= -\frac{d\phi}{ds}.$$

The component of the force intensity in any direction at a point is the space rate of fall or the negative gradient of the potential at that point and in that direction.

9. If we choose rectangular axes with the origin O conveniently near the system of charges, and if we write

$$r_0^2 \equiv x^2 + y^2 + z^2, \quad r_{s0}^2 = x_s^2 + y_s^2 + z_s^2,$$

and denote by θ_s the angle between the radii r_0 and r_{s0} from the origin, then

$$r_s^2 = r_0^2 + r_{s0}^2 - 2r_0 r_{s0} \cos\theta_s,$$

so that

$$\phi = \Sigma \frac{\gamma q_s}{\sqrt{r_0^2 + r_{s0}^2 - 2r_0 r_{s0}\cos\theta_s}},$$

and so also

$$r_0\phi = \Sigma \frac{\gamma q_s}{\sqrt{1 + \left(\frac{r_{s0}}{r_0}\right)^2 - 2\frac{r_{s0}}{r_0}\cos\theta_s}};$$

and similarly

$$r_0^2 \frac{\partial\phi}{\partial r_0} = -\Sigma \frac{\gamma q_s\left(1 - \frac{r_{s0}}{r_0}\cos\theta_s\right)}{\left[1 + \left(\frac{r_{s0}}{r_0}\right)^2 - 2\frac{r_{s0}}{r_0}\cos\theta_s\right]^{\frac{3}{2}}}$$

$$= -\Sigma\gamma q_s\left[1 + 2\frac{r_{s0}}{r_0}\cos\theta_s + 3\frac{r_{s0}^2}{r_0^2}\left(\frac{3\cos^2\theta_s - 1}{2}\right) + \dots\right].$$

* This function was used first in the theory of attractions by Laplace. The name potential was given to it by Green and independently by Gauss, *Allgemeine Lehrsätze über...Anziehungs- und Abstossungskräfte*, § 3 (*Collected Works*, v. p. 200).

There are now two particular cases of special importance to which we must refer.

(i) When Σq_s is not zero so that there is a definite resultant charge in the group, the potential at a great distance from the origin is to a first approximation equal to

$$\phi = \gamma \frac{\Sigma q_s}{r_0},$$

whilst the force is radial and of amount

$$-\frac{\partial \phi}{\partial r_0} = \gamma \frac{\Sigma q_s}{r_0^2}.$$

The second term in the approximation for both of these results vanishes if $\Sigma q_s r_{s0} \cos \theta_s = 0$, that is if the origin is at the centroid of the charges. We may thus conclude that to a second order of approximation the action at a distance of the series of point charges is just the same as if they were all collected at their mean centre.

(ii) If $\Sigma q_s = 0$ there is no mean electrical centre at a finite distance, but then the resultant charge is zero. In this case

$$\phi = \gamma \frac{\Sigma q_s r_{s0} \cos \theta_s}{r_0^2},$$

whilst for the radial force we have

$$-\frac{\partial \phi}{\partial r_0} = \gamma \frac{2\Sigma q_s r_{s0} \cos \theta_s}{r_0^3}.$$

In our main application of these results we shall treat the case where the point charges are confined within a very small volume. The above approximate results will then apply at distances from these volume elements which are large compared with the linear dimensions of the element.

10. The definition of the electric field at points outside a continuous distribution of charge*. The discussion of the previous paragraph applies only to a system of discrete point charges or electrons, and it is only in this sense that the analytical functions have any meaning. In actual practice however the distributions

* The points here briefly dealt with are discussed at length by Leathem, *Volume and Surface Integrals used in Physics* (1st ed. Cambridge, 1905). Cf. also J. Boussinesq, *Journ. de math.* [3] (1880), p. 89; H. Poincaré, *Amer. Journ. of Math.* 12 (1890), p. 284.

of charge with which we deal include such an enormous number of electrons that the expression of their field by functions of the type discussed above, even if it were possible, would be quite untractable. Fortunately however any such complete atomic analysis is useless in a physical theory, whose results can only be tested by observation and experiment on matter in bulk, for we are unable to take cognisance of the single molecule of matter, much less of the separate electrons inside it to which this analysis has regard. The development of the theory which is to be in line with experience must instead concern itself with an effective differential element of volume containing a crowd of molecules numerous enough to be expressible continuously, as regards their average relations as a volume density of matter.

Thus in any physical theory all that we are directly concerned with as regards the charge in any 'physically' small element of volume dv_1 is its total amount dq_1, and the ratio of these two magnitudes defines the density of the charge at the point, viz. ρ_1, where

$$dq_1 = \rho_1 dv_1.$$

The distinction here introduced between physically, as distinct from mathematically, small differential elements of volume is important and must be emphasised. In the speculations of pure mathematics there is no limit to the fineness of the subdivision of a region into volume elements, but in the physical theory there comes a limit when the element is so small that the number of elements of mass or charge in it is so small that the total mass included in the element depends appreciably upon its shape, so that the definition of density as the ratio of this total mass or charge to the volume ceases to have any meaning. The passage to the limit involved in a strict mathematical definition is thus not possible in a physical theory.

11. Now the element of charge $\rho_1 dv_1$ acts effectively at all points which are at a distance from it large compared with the linear dimensions of the element of volume dv_1 containing it, just like a charged particle, so that its field at such points is defined by the force vector $\delta\mathbf{E}$ and potential $\delta\phi$ which are determined by the relations

$$\delta\mathbf{E} = -\gamma\rho_1 dv_1 \operatorname{grad}\frac{1}{r_1}, \quad \delta\phi = \gamma\frac{\rho_1 dv_1}{r_1}.$$

In these expressions r_1 denotes the distance of the volume element dv_1 at the point (x_1, y_1, z_1) from the point (x, y, z) at which the functions are calculated. Thus for the whole system of charges grouped together in this way the field is defined by

$$\mathbf{E} = -\gamma \int \rho_1 dv_1 \operatorname{grad} \frac{1}{r_1},$$

whilst

$$\phi = \gamma \int \frac{\rho_1 dv_1}{r_1} *,$$

the integrals in each case being extended over the whole charge distribution.

The use of the definite integral expressions necessarily implies the possibility of endless subdivision in the strict mathematical sense of the electric charge, and attributes to the density ρ at any point the value obtained by passing to a limit in the usual way. This inconsistency is however removed by the simple device of replacing the actual distribution of electric charge by a hypothetical perfectly continuous distribution with the same density at each point and referring the integral expressions to this distribution. Such a continuous distribution is effectively the same as the actual one, at least as regards its effect at all points which are not too near the distribution, the actual distribution of charge in any physically small volume element being then quite irrelevant.

12. So far the field-point at which the force and potential are calculated is restricted to be at a distance from the nearest charge element which is large compared with the linear dimensions of the physically small element of the charge distribution; it is however easy to see that the definitions remain valid up to a distance comparable with the dimensions of the physically small element. Let us consider the potential integral: the difference between the sum $\Sigma \frac{q}{r}$ for the elements of charge in any physically small element of volume of linear dimensions l and the integral $\int \frac{\rho_1 dv_1}{r_1}$ taken throughout the same element will be of the same order of magnitude as either quantity separately so long as r for all points of the element is of the same order of magnitude as l, but that the difference will

* J. Lagrange, *Par. sav.* (*étr.*) VII. (1773) (*Œuvres*, VI. p. 45).

diminish to a quantity smaller in the ratio $\frac{r}{l}$ when r becomes great compared with l. Thus for purposes of estimating the order of magnitude it is reasonable to represent the difference between these two expressions for the element under consideration as

$$\int \frac{\alpha l \rho}{r^2}\, dv,$$

where α is a finite number. Thus the difference between the representation by a sum or by an integral of the potential at any point due to the charge between the spheres of radii l and a ($> l$) is

$$< 4\pi\alpha\rho' l \int_l^a dr,$$

$$< 4\pi\alpha\rho' l\,(a - l),$$

where ρ' is the maximum value of ρ in the region: this difference is negligibly small if a is not too big (say 1 cm.) on account of the smallness (physical) of l. A similar proof also applies to the other integral.

It thus appears that the definitions of the potential and force intensity by means of integrals extended throughout the hypothetical continuous distribution of charge, which replaces the actual or discrete one, are completely effective and valid without sensible error not only for points well outside the charge, but also for points whose distance from the nearest portion of charge is small of the order of the physically small length l. This includes the case when the point is so close to the apparent outer surface of the charged body as to be sensibly just not in contact with it, and also the case where the point is in a small but not imperceptibly small cavity of such a size that the piece excavated would have the properties of matter in bulk rather than the properties of a few molecules or electrons. Moreover the integrals involved in these definitions give rise to no mathematical difficulties. The subjects of integration are finite at all points of the region of integration, and the integrals themselves are finite and differentiable with respect to the coordinates (x, y, z) of the external point by the method known as differentiation under the sign of integration. It thus follows that we still have on the modified definitions

$$\mathbf{E} = -\operatorname{grad}\phi,$$

so that the electric force intensity at an external point in the field is still equal to the negative gradient of the potential at that point.

13. The definition of the electric field at points inside the continuous charge distribution. The generalised specification of the electric field of a continuous charge distribution given in the previous paragraph is perfectly definite, but applies only to points external to the charge distribution. As however we shall want to extend our analysis also to points inside the continuous charge distributions we must see whether the definitions still hold for such points*.

Let us first assume, without preliminary justification, that the hypothetical continuous distribution with which the real distribution of charge was replaced, effectively represents this actual charge at any point of the field however near to the actual charge it may be. The force and potential at a point inside the distribution will then be defined by the same integral expressions if these have any meaning at all: although the integrands in both cases become infinite the two integrals are however absolutely convergent in all cases if ρ is everywhere finite, so that there is a definite value to both integrals and there would be no difficulty in the application of these expressions in this case.

We can regard the matter physically in the following manner. Imagine a small volume cut out of the charge distribution around the internal point at which it is desired to calculate the functions. The potential and force due to the remaining distribution have then definite values which may however be large. The question is now: do the values of the functions at this point depend appreciably on the size and shape of the cavity, if it is made very small? If they do, the integrals given are either divergent or semi-convergent and no meaning can be attached to the functions they represent. If on the other hand, as is the case in the present instance, they do not depend on the shape or size of the cavity, if it is only made small enough, the integrals, although of the type called improper, have distinct values and the definitions remain.

* Cf. Gauss, *Allgemeine Lehrsätze etc.* § 6 (footnote 7) (*Works*, v. p. 202); O. Hölder, *Dissertation* (Tübingen, 1882), p. 6; J. Weingarten, *Acta math.* 10 (1887), p. 303; C. Neumann, *Leipz. Ber.* 42 (1890), p. 327.

Thus if we can assume that the continuous charge distribution effectively replaces the real one at all points of the field, the definitions of the force intensity and potential at an internal point are consistent and definite, and moreover the removal of a physically small portion of this charge round the point does not appreciably affect the values of the functions at the point, so that in their definition it is immaterial whether this small portion of the charge is present or not. But any attempt to justify the use of this effective distribution in calculating the field at a point whose distance from the nearest element of charge is of a higher order of smallness than a physically small differential length (l, of our previous analysis) can only result in failure. For now the single electron contributes to the potential, for example, a term q/r which in spite of the smallness of q may become very great as r diminishes, so that the presence of a few such electrons might easily become so important as to make the potential quite different from the value obtained from the continuous distribution and expressed by

$$\gamma\int\frac{\rho_1 dv_1}{r_1},$$

to which, as we have just mentioned, the part of the distribution near the point contributes only a negligible amount. But there is from the physical point of view no real motive for pursuing the enquiry further as we have in fact obtained a formulation of our field which is completely effective in a physical theory, and for the following reasons.

14. The real charge distribution may, as regards its action at any point inside the medium, be divided into two distinct portions by a physically small closed surface drawn round the point. The first part of the charge, viz. that outside the elementary surface, may be replaced by the continuous distribution as above which is, as regards its action at the point under consideration, effectively equivalent to it: to this we may also add, without appreciable modification, the continuation of this distribution throughout the interior of the small volume round the point. The second is the purely local distribution of charge elements inside the surface drawn. The contributions to the force and potential in the field at the internal point due to the former part of the charge are perfectly definite and are in fact expressed by the convergent integrals given above; but the local contribution from the elements of charge near the point is entirely unknown and may be continually changing. The only possible expressions for these functions are therefore the ones that omit altogether the contribution of these neighbouring elements.

If it were not possible thus to separate the physical functions into a molar and a molecular part a dependence would be involved between mechanical change and molecular structure, so that mechanical causes would alter the constitution of the medium and might even undermine its stability; whereas it is a postulate in ordinary mechanical theory that the physical properties of the medium are not affected by small forces*.

Thus for the purposes of a physical theory the force and potential at points inside the medium are properly defined as the corresponding quantities belonging to the field of the hypothetical distribution of charge which is thus concluded to be a completely effective representation of the real distribution. We have therefore both at external and internal points

$$\mathbf{E} = -\gamma \int \rho_1 dv_1 \operatorname{grad} \frac{1}{r_1},$$

and

$$\phi = \gamma \int \frac{\rho_1 dv_1}{r_1}.$$

Moreover since the integrals in these expressions are both absolutely convergent, it is legitimate in all cases to derive the former from the latter by the process of differentiation under the sign of integration so that we have

$$\mathbf{E} = -\operatorname{grad} \phi.$$

Thus the components of the force intensity at *any* point of the field in any direction is the space rate of fall of the potential in that direction.

* Cf. Larmor, *Aether and Matter* (particularly the footnote on p. 265), also *Phil. Trans.* 190 A (1897). "The principle of D'Alembert, which is the basis of the dynamics of finite material bodies, necessarily involves this order of ideas. That part of the aggregate forcive on the molecules in the element of volume which is spent in accelerating the motion of that element *as a whole* is written off; and the regular part of the remainder must mechanically equilibrate. But the wholly irregular parts of the molecular motions and forces are left to take care of themselves, which they are known to do for the simple reason that the constitution of the material body is observed to remain permanent."

15. On surface distributions and double sheets. We must now pass to the consideration of certain important types of discontinuity in the volume charge distribution with which we shall have to deal in our future work. Such cases actually occur in nature, and it seems necessary to consider what modifications are needed in the above definitions in order that they may apply to them.

The first example leads us to the idea of a surface density. If the volume density becomes very large in the neighbourhood of a surface f in the field, we may separate the comparatively infinite values from the rest by drawing two surfaces parallel and very close to f one on each side. The layer between these surfaces is then of very small thickness Δn, but the volume density ρ is so large that

$$\sigma = \int_0^{\Delta n} \rho \, dn$$

is finite when integrated across the common normal at any point of the surface. We then regard this part of the charge distribution as a surface distribution of density σ (i.e. amount per unit area) on the surface f. It is of course merely a big volume density concentrated in a shell of small thickness, but as we do not as a rule wish to be bothered about the constitution of the layer we treat it in this way.

The potential function associated with this part of the charge distribution is

$$\phi = \gamma \int_f df \int_0^{\Delta n} \frac{\rho \, dn}{r},$$

and since the shell is very thin this is practically

$$\phi = \gamma \int_f \frac{df}{r} \int_0^n \rho \, dn = \gamma \int_f \frac{\sigma \, df}{r} \text{*}$$

extended over the surface f, r denoting the distance of the element df from the point at which the potential is calculated.

The components of force are expressed in an analogous manner.

16. The second case of infinities in the volume density appears at first sight rather an artificial one, but as a matter of fact it actually exists in nature and the analysis associated with it is of

* G. Green, *Essay, etc.* Art. 4.

immense importance for other branches of the work. Imagine a surface f' placed parallel and infinitely near to a surface f so that the small normal distance between them is Δn. Now suppose the surface f' has a charge distribution of surface density σ' and the surface f one of density $-\sigma$. The potential of this distribution would be

$$\phi = \gamma \int_{f'} \frac{\sigma' df'}{r'} - \gamma \int_f \frac{\sigma df}{r};$$

and if we make $\sigma' df' = \sigma df$ so that there are equal and opposite charges on opposing elements of the surfaces and also put

$$\frac{1}{r'} = \frac{1}{r} + \frac{\partial}{\partial n}\left(\frac{1}{r}\right) \Delta n,$$

we get
$$\phi = \gamma \int_f \sigma \Delta n \frac{\partial}{\partial n}\left(\frac{1}{r}\right) df^*.$$

We are therefore no longer concerned with the surface distributions separately, but must treat them together. They form what is called a *double sheet* distribution. The quantity which mathematically specifies the sheet is the product $\sigma\Delta n$, which is called the *moment* of the sheet and is denoted by τ. We must have a very large σ to get a finite τ for $\tau = \Delta n \,.\, \sigma$: the surface densities involved are therefore large compared with the usual ones which occur separately.

The potential of this double sheet is

$$\phi = \gamma \int_f \tau \frac{\partial}{\partial n}\left(\frac{1}{r}\right) df,$$

and the components of force analogously.

These represent the only types of distribution with which we have to deal in our theories. Other types may occur in nature, but they are of little importance, if they occur at all†.

17. If there are surface densities and double sheets in the field the ordinary considerations as to convergence and continuity of the integrals expressing the force and potentials still apply provided the point under consideration does not lie on any of the surface infinities. If the point is on an ordinary distribution of surface density the potential is quite definite, the integral expressing it

* Helmholtz, *Ann. Phys. Chemie*, 89 (1853); *Collected Works*, I. p. 491.

† Cf. however W. Voigt, *Lehrbuch der Kristallphysik*, ch. III. (Leipzig, 1910).

being convergent, but there is a certain indefiniteness in the expression of the force, which is given by a conditionally convergent integral. The significance of these results in the physical theory is however obvious from our former discussions and need not now be further elaborated.

There are certain discontinuities introduced as we approach the surface distributions thus specified, but these are best attacked by the indirect method as discussed in the next sections.

18. Green's analysis of the electric field. We can now proceed to a discussion of the more important properties characteristic of the general electrostatic field in which the force and potential are defined by the integrals given in the preceding sections. The most direct method of approach is that provided by Green's analysis, an account of which must first be given.

Let ϕ, ψ be any two scalar functions. From them we can deduce a vector $\mathbf{A}$ of the form

$$\mathbf{A} = \phi\nabla\psi - \psi\nabla\phi$$

and thus

$$\operatorname{div} \mathbf{A} = \phi\nabla^2\psi - \psi\nabla^2\phi,$$

where, as always, we understand by ∇^2 the differential operator

$$\frac{\partial^2}{\partial x^2} + \frac{\partial^2}{\partial y^2} + \frac{\partial^2}{\partial z^2};$$

it is to be noticed that it is precisely the square of the Hamiltonian operator treated according to the ordinary rules.

Now suppose that the functions ϕ and ψ are continuous and have continuous derivatives inside the space v enclosed by the surface f. A simple application of Green's lemma then gives

$$\int_v (\phi\nabla^2\psi - \psi\nabla^2\phi)\, dv = + \int_f \left(\phi\frac{\partial\psi}{\partial n} - \psi\frac{\partial\phi}{\partial n}\right) df,$$

where we use $\frac{\partial\phi}{\partial n}$ as the component of the gradient of ϕ along the outward normal at the element df of the surface. This is Green's Theorem.

The more important forms of this theorem are however obtained by adopting a special form for one of the functions.

If P is a variable point in the region v with coordinates (x, y, z)

and P_1 any fixed point with coordinates (x_1, y_1, z_1), then if we use r_1 for the distance PP_1 we have

$$r^2 = (x - x_1)^2 + (y - y_1)^2 + (z - z_1)^2,$$

and consequently

$$\left(\frac{\partial}{\partial x}, \frac{\partial}{\partial y}, \frac{\partial}{\partial z}\right)\frac{1}{r} = -\frac{(x - x_1,\ y - y_1,\ z - z_1)}{r^3},$$

and therefore also

$$\frac{\partial^2}{\partial x^2}\left(\frac{1}{r}\right) = -\frac{1}{r^3} + \frac{3(x - x_1)^2}{r^5},$$

and similarly for $\frac{\partial^2}{\partial y^2}\left(\frac{1}{r}\right)$ and $\frac{\partial^2}{\partial z^2}\left(\frac{1}{r}\right)$. We therefore see that

$$\nabla^2\left(\frac{1}{r}\right) = 0.$$

If now the point P_1 lies outside the region v we can put

$$\psi = \frac{1}{r},$$

and the above theorem takes the form

$$\int_v \nabla^2\phi \frac{dv}{r} + \int_f \left\{\phi \frac{\partial}{\partial n}\left(\frac{1}{r}\right) - \frac{1}{r}\frac{\partial\phi}{\partial n}\right\} df = 0.$$

19. If however P_1 lies inside the region v, then $\frac{1}{r}$ regarded as a function of the position of P will be infinite at P_1, and if we wish to use our formula with $\psi = \frac{1}{r}$ we must exclude P_1 by putting a small surface round it. We shall do this by taking a small sphere of radius r_1 round the point P_1 as centre. The space between this and the surface f we call v'. We can now apply our theorem to this region v', so that

$$\int_v \nabla^2\phi \frac{dv}{r} + \int_{f'} \left\{\phi \frac{\partial}{\partial n}\left(\frac{1}{r}\right) - \frac{1}{r}\frac{\partial\phi}{\partial n}\right\} df' = 0,$$

where f' includes, in addition to the surface f, also the surface of the small sphere.

If now we denote by $d\omega$ the element of solid angle, the element of spherical surface of radius r is $r^2 d\omega$ and the element of volume is $r^2 dr\, d\omega$. It is also to be noticed that on the surface of the small sphere the normal n coincides with the direction of r (but is in the opposite sense) and so

$$\frac{\partial}{\partial n}\left(\frac{1}{r}\right) = +\frac{1}{r_1^2},$$

and therefore the part of the surface integral due to the surface of the sphere is

$$\int \phi \, d\omega - r_1 \int \frac{\partial \phi}{\partial r} \, d\omega,$$

where the integrations are over the surface of the unit sphere. If at the point P_1 the function ϕ and its differential coefficients are continuous, then $\frac{\partial \phi}{\partial r}$ is finite, and thus if we make the sphere infinitely small the second integral

$$r_1 \int \frac{\partial \phi}{\partial r} \, d\omega$$

tends to zero, and if the value of ϕ has the value ϕ_1 at P_1 the first integral tends to

$$4\pi\phi_1.$$

In the volume integral the part due to the inside of the sphere is excluded: but this is

$$\iint \nabla^2 \phi r \, dr \, d\omega,$$

and obviously vanishes with r_1 if $\nabla^2\phi$ is finite. We have thus in all

$$4\pi\phi_1 = -\int_v \nabla^2 \phi \frac{dv}{r} + \int_f \left\{\frac{1}{r}\frac{\partial \phi}{\partial n} - \phi \frac{\partial}{\partial n}\left(\frac{1}{r}\right)\right\} df',$$

where the integral with respect to v is now over the whole region inside f and the surface integral is over f only; all traces of the cavity drawn about the point P_1 have disappeared.

20. The analysis so far is limited to the case in which the functions involved are continuous over the whole region v. We can however immediately extend it to include the most important cases involving discontinuity. Supposing that ϕ and its first differential coefficients are discontinuous over the surface f' lying in this region, otherwise having determinate continuous values throughout the region. Draw a normal n at each point of the surface f' and regard directions in it as positive when in some definite chosen sense. The side of the surface f' on the side of increasing n we call the positive side and the other the negative side, and we distinguish the values of functions on the two sides by suffices + and −, e.g. ϕ_+ and ϕ_-.

We can then apply our previous formula if we include in the boundary of v the two sides of the surface f' as well as the surface f. On the positive side of f' dn is negative and on the negative side it is positive. The point P_1 is assumed not to lie on the surface f'. We thus get

$$4\pi\phi_1 = -\int_v \nabla^2\phi \frac{dv}{r} + \int_f \left\{\frac{1}{r}\frac{\partial\phi}{\partial n} - \phi\frac{\partial}{\partial n}\left(\frac{1}{r}\right)\right\} df'$$
$$-\int_{f'}\left[\left(\frac{\partial\phi}{\partial n}\right)_+ - \left(\frac{\partial\phi}{\partial n}\right)_-\right]\frac{df'}{r} + \int_{f'}(\phi_+ - \phi_-)\frac{\partial}{\partial n}\left(\frac{1}{r}\right) df',$$

a formula which will hold even if f' consists of several separate surfaces.

This is the general result. In applications however one often has to apply it to indefinitely extended fields from which certain finite spaces are excluded. In such cases a detailed discussion of the behaviour of the infinite integrals becomes necessary, and each case must be treated on its merits. The following general result is however easily deduced if the theorem is applied as though the field were bounded by a very large enclosing surface which is ultimately extended indefinitely in all directions, and it provides a sufficient criterion in most cases.

21. Suppose the function ϕ, now given throughout all space, is such that

$$\operatorname*{Lt}_{r\to\infty} \phi = 0$$

and

$$\operatorname{Lt} r^2 \frac{\partial\phi}{\partial n} \text{ is finite,}$$

then the part of the surface integral corresponding to the infinite boundary will be zero in the limit and the formula can be written as

$$4\pi\phi_1 = -\int_v \nabla^2\phi \frac{dv}{r} + \int_f \left\{\frac{1}{r}\frac{\partial\phi}{\partial n} - \phi\frac{\partial}{\partial n}\left(\frac{1}{r}\right)\right\} df'$$
$$-\int_{f'}\left[\left(\frac{\partial\phi}{\partial n}\right)_+ - \left(\frac{\partial\phi}{\partial n}\right)_-\right]\frac{df'}{r} + \int_{f'}(\phi_+ - \phi_-)\frac{\partial}{\partial n}\left(\frac{1}{r}\right) df',$$

where the volume integral is extended over the whole of space outside certain specified regions, the first surface integral over the boundaries of these regions and the second over all discontinuity surfaces in the region investigated.

A function ϕ limited by the usual conditions of continuity as well as the above conditions at infinity will be said to be regular in the space investigated.

This general result shows that a function ϕ which fulfils the conditions, and which apart from the surfaces f' is with its derivatives continuous in the whole of space, is uniquely determined in the whole region if the values of $\nabla^2\phi$ are given at each point and also the discontinuities in ϕ and its normal gradient on all the surfaces f'.

The continuity of $\nabla^2\phi$ is not involved.

If we write
$$\nabla^2\phi = -4\pi\rho,$$
$$\left(\frac{\partial\phi}{\partial n}\right)_+ - \left(\frac{\partial\phi}{\partial n}\right)_- = -4\pi\sigma,$$
and
$$\phi_+ - \phi_- = 4\pi\tau,$$
then our formula shows that
$$\phi = \int\frac{\rho\, dv}{r} + \int_{f'}\frac{\sigma df'}{r} + \int_{f'}\tau\frac{\partial}{\partial n}\left(\frac{1}{r}\right) df',$$
where we now consider the whole of space without any excluded regions.

If we consider the values of ρ, σ and τ to be those as defined above, then this formula is merely the expression of an identity. It contains an expression by definite integrals of a general function ϕ subject merely to the specified continuity conditions.

22. If however we regard the question from the other point of view and consider the quantities ρ, σ and τ as given *a priori*, then we want to know whether the function ϕ defined in the same way satisfies the same conditions. It is easily proved that it does.

We define ϕ at any point P by the relation
$$\phi = \int\frac{\rho\, dv}{r} + \int\frac{\sigma df'}{r} + \int\tau\frac{\partial}{\partial n}\left(\frac{1}{r}\right) df',$$
the first integral being taken over the whole of space and the second and third over those surfaces on which σ and τ have finite values.

If the point P is at an external point, i.e. at a point in space in the immediate neighbourhood of which $\rho = 0$, then we can differentiate each of these integrals with respect to the coordinates of P under the sign of integration. We get

$$\nabla^2\phi = \int \rho\nabla^2\left(\frac{1}{r}\right) dv + \int \sigma\nabla^2\left(\frac{1}{r}\right) df' + \int \tau \frac{\partial}{\partial n}\nabla^2\left(\frac{1}{r}\right) df',$$

and since $\nabla^2\left(\frac{1}{r}\right) = 0$ we have

$$\nabla^2\phi = 0.$$

If however P is at any other point where the value of ρ is not zero, we must proceed in a different manner. Notice that it is only possible for discontinuities in ϕ or its derivatives to occur near one of the surfaces f'; we may therefore from our previous theorem write

$$4\pi\phi = -\int \nabla^2\phi \frac{dv}{r} - \int_{f'}\left[\left(\frac{\partial\phi}{\partial n}\right)_+ - \left(\frac{\partial\phi}{\partial n}\right)_-\right]\frac{df'}{r} + \int_{f'}(\phi_+ - \phi_-)\frac{\partial}{\partial n}\left(\frac{1}{r}\right) df',$$

and thus on elimination of ϕ we get

$$\int(\nabla^2\phi + 4\pi\rho)\frac{dv}{r} - \int\left[\left(\frac{\partial\phi}{\partial n}\right)_+ - \left(\frac{\partial\phi}{\partial n}\right)_- + 4\pi\sigma\right]\frac{df'}{r} + \int(\phi_+ - \phi_- - 4\pi\tau)\frac{\partial}{\partial n}\left(\frac{1}{r}\right) df' = 0.$$

Thus we see that, on account of the arbitrariness of the position of the point P, we must have*

$$\nabla^2\phi + 4\pi\rho = 0,$$

$$\left(\frac{\partial\phi}{\partial n}\right)_+ - \left(\frac{\partial\phi}{\partial n}\right)_- + 4\pi\sigma = 0,$$

and

$$\phi_+ - \phi_- - 4\pi\tau = 0,$$

which are precisely the same as the previous conditions.

23. The physical application of these purely analytical results is now obvious and we may directly conclude that the potential function of any electric field defined in the manner previously specified must be subject to the following conditions.

* At any point of the field we can choose an infinite number of variations of the position of P such that along them any two of the integrals together are constant.

(i) It must be such that

$$\nabla^2\phi + 4\pi\gamma\rho = 0$$

at all points of the field where there is a finite volume density ρ: and where $\rho = 0$

$$\nabla^2\phi = 0.$$

The first of these equations contains Poisson's extension* of the second, which is Laplace's equation†. It expresses the general characteristic property of the potential function in its differential form and provides us with a test that any stated function is a potential function.

(ii) At any point on the surface distributions of charge

$$\left(\frac{\partial\phi}{\partial n}\right)_+ - \left(\frac{\partial\phi}{\partial n}\right)_- + 4\pi\gamma\sigma = 0‡,$$

and

$$\phi_+ - \phi_- - 4\pi\gamma\tau = 0§,$$

where σ is the density of the surface distribution and τ the moment of the double sheet at the point of the surfaces.

These two conditions are in reality merely the particular forms which the general property expressed by Poisson's equation assumes when applied to the respective limiting forms of distribution. Notice that in crossing a simple surface distribution σ the normal force only is discontinuous, the potential and tangential forces being continuous, whereas in crossing a double sheet the potential and tangential forces are discontinuous, but the normal force is continuous.

(iii) ϕ must be an otherwise continuous function regular everywhere in the field; and also if the charge distribution is a finite one it satisfies the conditions that the limiting values of

$$R\phi \text{ and } R^2\frac{\partial\phi}{\partial R}\|$$

* *Nouveau Bulletin des Sciences par la Société Philomathique de Paris*, 3 (1813), p. 388. Gauss gave the first correct proof, *Allgemeine Lehrsätze etc.*, §§ 9, 10.

† *Par. Hist.* [85], pp. 135, 252 (1782). (*Œuvres*, x. pp. 302, 278.) Cf. also *Mécanique Céleste*, t. II.

‡ Poisson, *Par. Mém.* [XII.], p. 30 (1811). Cauchy, *Bull. Soc. Phil.* (1815), p. 53. Green, *Essay, etc.* § 4.

§ Helmholtz, *Ges. Abh.* I. p. 489.

‖ These conditions are quoted by Lejeune-Dirichlet, *Journ. f. Math.* 32 (1846), p. 80 (*Werke*, II. p. 40).

remain finite when R, the distance of the field-point from the finite origin of coordinates, increases indefinitely.

We might sum this last condition up by saying that the function ϕ is regular everywhere and at infinity*.

Moreover the analysis shows that there is *only one* solution of these conditions and that is *the one* we have found.

24. If therefore we can by any means obtain a solution of these conditions for given values of ρ, σ and τ, then we have a complete specification of the field of the given charges. The inverse problem in electrostatics is the determination of such solutions. It is easily seen that the solution required assumes the form

$$\phi = \gamma \int \frac{\rho dv}{r} + \Phi,$$

where Φ is an appropriate solution of the differential equation of Laplace,

$$\nabla^2 \Phi = 0,$$

which must be so chosen that ϕ satisfies the specified boundary conditions at the surface infinities in the charge distribution.

Now it appears that the first part of the complete solution thus obtained is a perfectly continuous function of position with continuous first derivatives. The complementary function Φ has therefore to take full account of the discontinuities in ϕ, and it is in fact the potential of the surface distributions which give rise to these discontinuities.

The problem thus resolves itself into a determination of the function Φ which satisfies the equation

$$\nabla^2 \Phi = 0$$

at all points of the field, and

$$\frac{\partial \Phi}{\partial n_+} - \frac{\partial \Phi}{\partial n_-} + 4\pi\gamma\sigma = 0,$$

$$\Phi_+ - \Phi_- - 4\pi\gamma\tau = 0$$

at the surface distributions.

* I am reminded that this last expression is not a usual one: it appears however a useful and concise method of expressing a definite property of such functions and saves detailed repetition of the conditions in every case where they occur.

The nature of the solution required will of course depend essentially on the type of surface or surfaces on which the charge infinities exist, and it is in fact only in the cases where these surfaces are of the simplest geometrical form that a solution can be obtained at all*. It is moreover clear that any desired solution will be obtained in its simplest form in terms of those coordinates in which the equation to the surface or surfaces to which it is to be appropriate is in its simplest form. Thus in dealing with any particular type of surface it is desirable to begin by transferring the fundamental differential equation to coordinates suitable for that surface, and then to tabulate the different types of solution. We can then choose the particular solution appropriate to the problem in hand by bringing in the surface conditions.

25. Gauss' analysis of the electric field. So far our basis is purely analytical, and as a consequence the physical significance of the results is not very clear. In order to obtain a better insight into this other side of the matter we shall proceed from a different standpoint and along more elementary lines.

The chief characteristic property of the electric field contained in Poisson's equation is expressed in an integral form by a theorem usually ascribed to Gauss† but which was probably first stated by Faraday in a physical manner.

If any closed surface f is taken in the electric field, and if $\mathbf{E}_n$ denote the component of the electric force intensity at any point of the surface in the direction of the outward normal dn, then

$$\int_{f'} \mathbf{E}_n df = 4\pi\gamma Q,$$

where the integration extends over the whole of the surface and Q is the total charge enclosed by it.

This theorem is a mere mathematical verification for a single point charge q: for at any point of the closed surface distant r from q

$$\mathbf{E}_n = \frac{\gamma q}{r^2} \cos(\widehat{nr}),$$

* Cf. Heine, *Handbuch der Kugelfunktionen* (2nd ed. Berlin, 1878); Byerly, *Fourier's Series and Spherical, Cylindrical and Ellipsoidal Harmonics* (Boston (U.S.A.), 1893); Whittaker, *Modern Analysis* (2nd ed. Cambridge, 1915).

† *Allg. Lehrsätze.* The theorem was also given by Kelvin in 1842; cf. *Papers on Electricity and Magnetism.* The present demonstration is due to Stokes.

where $(\widehat{nr})$ denotes the angle between the positive directions of n and r; but if $d\omega$ is the element of solid angle subtended by the surface element df at the point charge q, then

$$df \cos(\widehat{nr}) = r^2 d\omega,$$

and thus

$$\mathbf{E}_n df = \gamma q \, d\omega,$$

so that

$$\int_f \mathbf{E}_n df = \gamma q \int d\omega$$

$$= 4\pi\gamma q \text{ if } q \text{ is inside } f$$

$$= 0 \quad \text{ if } q \text{ is outside.}$$

If there are any number of charges $q_1, q_2, \ldots q_n$ present in the field, then

$$\mathbf{E}_n = \mathbf{E}_{1n} + \mathbf{E}_{2n} + \ldots$$

is the sum of the normal components of force due to the separate point charges, and thus we get by simple addition

$$\int_f \mathbf{E}_n df = 4\pi\gamma \text{ (total charge inside } f\text{).}$$

Moreover, although we have proved this theorem for a system of point charges it remains valid when the charges are merged into continuous volume or surface distributions as the following analysis, which exhibits the theorem as an immediate consequence of Poisson's equation, proves. If we consider the integral

$$\int_v \operatorname{div} \mathbf{E} \, dv$$

taken throughout the space v inside the surface f, its value is

$$4\pi\gamma \int \rho \, dv;$$

but by Green's lemma it also consists of

$$\int_f \mathbf{E}_n df,$$

together with the surface integrals arising at the surface distributions of charge. These are simply

$$\int_{f'} (\mathbf{E}_{n+} - \mathbf{E}_{n-}) \, df',$$

f' referring to the surfaces on which the charge infinities are distributed. But

$$\mathbf{E}_{n_+} - \mathbf{E}_{n_-} = -4\pi\gamma\sigma,$$

so that this latter integral is

$$-4\pi\gamma\int_{f'}\sigma\,df',$$

and we have
$$4\pi\gamma\left[\int_v \rho\,dv + \int_{f'}\sigma\,df'\right] = \int_f \mathbf{E}_n\,df,$$

which is precisely Gauss' theorem. If we use

$$\mathbf{E}_n = -\frac{\partial\phi}{\partial n},$$

then the equation may be written in the form

$$-\int_f \frac{\partial\phi}{\partial n}\,df = 4\pi\gamma Q,$$

which exhibits it as the integral form of the characteristic property of the potential function of an electrostatic field.

26. The integral $\int \mathbf{E}_n\,df$ over any closed surface is defined as the total *normal induction* through that surface. By the above theorem this is equal to $4\pi\gamma$ times the total charge inside the surface.

We see here the great simplicity which would follow the adoption of a value for γ which would make

$$4\pi\gamma = 1.$$

That is, if we take
$$\gamma = \frac{1}{4\pi},$$

then the normal induction over any such closed surface would be equal to the charge inside.

On account of this fact we shall in future assume this value for γ and then the potential function of a volume distribution of density ρ will be

$$\phi = \frac{1}{4\pi}\int\frac{\rho\,dv}{r},$$

and it will satisfy the equation

$$\nabla^2\phi + \rho = 0.$$

Taking this value for γ implies increasing the unit of electricity in the ratio $\sqrt{4\pi} : 1$; the value of the force between two charges of q and q' units respectively, which is now

$$\frac{1}{4\pi}\frac{qq'}{r^2},$$

remaining of necessity the same value in all units.

27. Lines of force and equipotential surfaces*. The electric field due to any static system of charges is completely specified if we know the magnitude and direction of the electric force intensity at any point of the field. The direction of the force at any point is understood to be the direction in which a small point charge δq would be displaced if put there. If we follow this direction from point to point we obtain curves which are called *lines of force.*

A line of force in an electric field is a curve such that along it the force is always tangential. It follows that the positive direction of the line is always that of decreasing potential. Hence a line of force cannot return into itself, but must have a beginning and an end. We shall prove presently that it can only begin on a positively charged surface and end on a negatively charged one.

If $d\mathbf{s}$ is a vector element of arc of the line of force which passes through the point P (x, y, z) in the field, its direction must be the same as that of the electric force $\mathbf{E}$ at the point. This implies that

$$d\mathbf{s} = \lambda \mathbf{E},$$

where λ is a scalar quantity, the ratio of the measures of $d\mathbf{s}$ and $\mathbf{E}$. This vector equation is equivalent to the three cartesian equations

$$\frac{dx}{\mathbf{E}_x} = \frac{dy}{\mathbf{E}_y} = \frac{dz}{\mathbf{E}_z},$$

which are the usual differential equations of the lines in these coordinates.

Now suppose an electric field to be given and all the lines of force drawn in it. At any position P in the field where there is no electricity we place a small surface element df, perpendicular to the lines of force at that place. Then all the lines of force which pass through the edges of this surface element will form the sides of a tube called a *tube of force.*

* Faraday, *Roy. Soc. Trans.* 141 (1831), p. 2.

The inside of a tube of force is also filled with lines of force, i.e. through each point in its interior we can draw one line of force but only one. If there were several they would cut at this point, and this is impossible because at each point the line of force gives the direction of the resultant force intensity of the field, and this is uniquely determinate unless the force is zero, an exceptional case which is reserved for future consideration. Thus as a general rule lines of force cannot cut one another and none of them can pass through the sides of a tube of force.

At another point P' of our tube of force let us draw another surface element df' also perpendicular to the direction of the lines of force there. Now apply Gauss' theorem to the portion of the tube PP'. In summing up the induction over the surface we can neglect the curved sides of the tube, since the electric force intensity is at each point tangential and thus has no normal component. Let now $\mathbf{E}$, $\mathbf{E}'$ be the force intensities at P and P' respectively: their directions are along tangents to a mean axis in the tube and may therefore be considered normal to the end surfaces df, df', both of which are very small. A simple application of Gauss' theorem thus gives

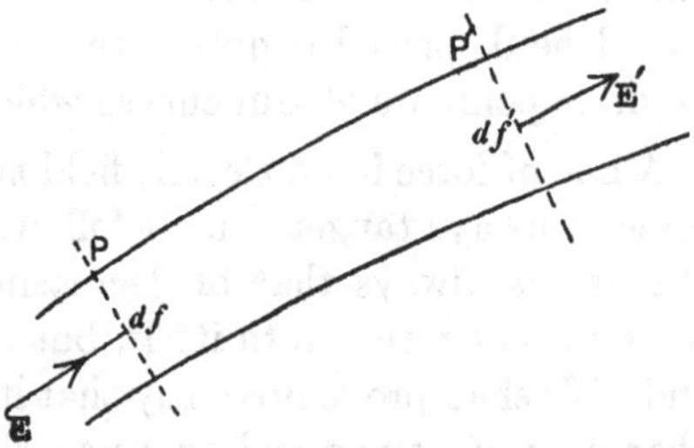

Fig. 1

$$\mathbf{E}df = \mathbf{E}'df',$$

provided there is no electricity in the part of the tube between P and P'.

The intensity of force at every point of the tube is inversely proportional to the cross-section of the tube at the point. We have thus a convenient method of graphically representing the intensity of a field of electric force. We fill up the space of the field by drawing tubes of force, choosing their cross-sections so that the constant value of $\mathbf{E}df$ along each is the same for all. The density of the tubes, or merely their thickness, would then give a graphical measure of the strength of the field. The tubes are thickest in the positions of small intensity.

28. A still simpler method of representing the field is obtained by considering the potential function*. The force intensity is a vector with three components at each point of the field and it is much easier if we notice that this vector is always the gradient of the scalar quantity, which we have called the potential, so that instead of having three things to consider we have only one. The field of force is completely specified by this one function ϕ. We could therefore map out the field by plotting the function ϕ, the simplest method being to draw the surfaces over which ϕ is constant: such surfaces are called *equi-potential surfaces*, *level surfaces* or simply *equi-potentials*.

Since the potential is as a general rule a one-valued function of position we see that two equi-potentials cannot in general cut one another. Now let us choose two adjacent equi-potentials whose potential difference is $\Delta\phi$ and let Δs be a small line drawn from a point on the surface of higher potential to a point on the other surface. We know then that $\frac{\Delta\phi}{\Delta s}$ is in the limit the force intensity in the direction of Δs; but $\frac{\Delta\phi}{\Delta s}$ is a maximum when Δs is a minimum. Since we may regard the two adjacent surfaces as parallel, at least when we confine ourselves to small opposing regions on them, Δs will be a minimum when it is the normal distance Δn between the surfaces. Thus the resultant force intensity is

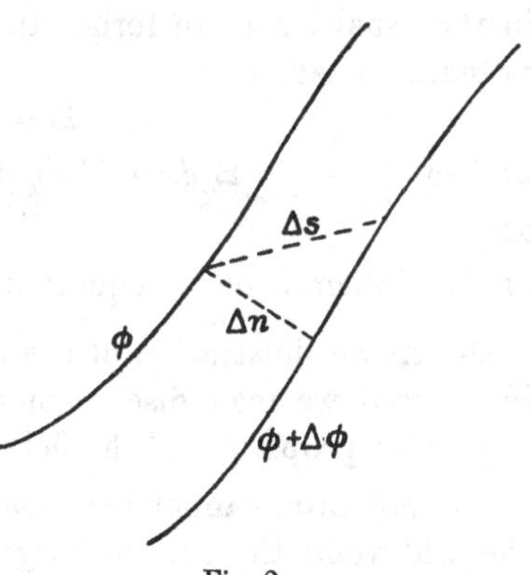

Fig. 2

$$\mathbf{E} = \frac{\Delta\phi}{\Delta n},$$

and is normal to the equi-potential surfaces.

The lines of force are therefore everywhere normal to the equi-potentials. Conversely, of course, we may conclude that if we can find a surface everywhere normal to the lines of force, it must be a level surface.

* Maclaurin, *Treatise on Fluxions*, § 640 (Edinb. 1742). Clairaut, *Figure de la terre* (Par. 1743).

Thus if we draw all the level surfaces in the field so that the potential difference for any two succeeding ones is the same, the density of the surfaces so drawn is everywhere proportional to the force intensity in the field. We thus obtain another very real representation of the field. If we draw the tubes of force as well we notice that the cross-sections of the tubes are proportional to the distances between the level surfaces. This would provide a good test as to whether the field had been correctly mapped. The tubes are thickest where the surfaces are widest apart.

The equi-potential surfaces are mathematically the integral surfaces of the total differential equation

$$\mathbf{E}_x dx + \mathbf{E}_y dy + \mathbf{E}_z dz = 0$$

in the usual cartesian form. In fact in the statical field where a potential exists we have

$$\mathbf{E} = -\nabla\phi,$$

and so $$\mathbf{E}_x dx + \mathbf{E}_y dy + \mathbf{E}_z dz = -d\phi,$$

and so $$\phi = c$$

are the integrals of the equation.

29. As an illustration of the usefulness of the conceptions here introduced we may discuss in terms of them some of the more important properties of the fields in question*.

The potential cannot be a maximum or minimum at a point of the field where there is no charge. For supposing that at any point P there is a true maximum value of ϕ, then at all other points in the immediate neighbourhood of P the value of ϕ is less than its value at P. Hence P will be surrounded by a series of closed equipotential surfaces, each outside the one before it, and at all points of any one of these surfaces the electrical force will be directed outwards so that the total normal induction through the surface will be positive and cannot be zero: it follows then that there must be a positive charge inside the surface, and since we may take the surface as near to P as we please, there is a positive charge at P.

In the same way we may prove that if ϕ is a minimum at P there must be a negative charge at P.

* Gauss, *Allg. Lehrsätze etc.* (1840). Stokes, *Camb. and Dublin Math. Journal*, IV. (1849). See also papers by Lord Kelvin in the same journal (1842–3) and a memoir by Chasles, *Connaissances des Temps* (1845).

This enables us also to complete the proof of the statement made above that lines of force must start from a place where there is positive electricity and end at a place where there is negative electricity, for it can only begin at a position of maximum potential and end at a position of minimum potential.

It is of course possible for a line of force to begin on a positive charge and go to infinity, the potential decreasing all the way, in which case the line of force has, strictly speaking, no second end at all. So also a line may come from infinity and end on a negative charge.

30. The field of a system of conductors. A very important class of electrostatic problem is concerned with the field of a system of charged metallic conductors, as we may gather from the experiments described at the beginning of this chapter. The field surrounding such a system possesses certain characteristic properties which we can very easily interpret in terms of lines of force and equi-potential surfaces. Metallic conductors as we know them are chiefly characterised by the presence in them of an exceedingly large number of electrons (about 10^{23} per c.c.) which are freely movable in the space between the atoms. Thus as long as any electric force acts on a free electron inside a conductor equilibrium is not possible.

Thus if such a body be brought into an electric field the electric force will act in opposite directions on the positive electricity (in the atoms of the body) and the negative electrons at each point of the metal and separate them so that they no longer cancel one another's effects. A number of the negative electrons in any volume element A will be driven by the electric force into a neighbouring element B; the consequence of this is that the element A becomes positively charged whilst the element B is negatively charged, the total quantity of electricity remaining the same.

Between the two neighbouring elements A and B a new field will arise due to their charge. This new field is superposed on the old one, but is in the opposite direction to it, as lines of force always go from a positive to a negative charge. In this way the field in the interior of the metal will tend to annul itself, and moreover this process will go on until the field in the whole of the body is zero.

We may now enquire as to the whereabouts of the charges which arise from the separation in each element and which give rise to the field which when superposed on the original field makes the resultant intensity zero throughout the whole of the interior of the conductor. Now since the force intensity at every point of the interior of the conductor is zero, the total normal induction, in Gauss' sense, through any closed surface entirely in the metal must also be zero and therefore this surface can contain no electricity. There is therefore no charge in the interior of the conductor. The charge is concentrated on its surface. Thus any conductor introduced into a field of intensity $\mathbf{E}_0$ will have *induced* on its surface a charge which gives rise to a field of intensity $\mathbf{E}_1$ which must be such that at every point in the conductor

$$\mathbf{E}_0 + \mathbf{E}_1 = 0.$$

Moreover at a point on the surface of the conductor the total electric force $\mathbf{E}_0 + \mathbf{E}_1$ can have no component tangential to the surface, as otherwise an unending separation of charge would take place in the surface itself. The electric force intensity of the total field just outside the conductor is therefore entirely normal to the surface.

Even if we introduce a charge to the metallic body from any source, this charge must distribute itself over the surface of the body so that the above conditions are still satisfied.

31. In the interior of the conductor the force intensity of the total field is everywhere zero. Therefore the potential of the field must be constant throughout the interior: this is the *potential of the conductor*. The surface of the conductor is therefore an equipotential surface in the field. This is another reason why the lines of force just outside a conductor are normal to its surface and it also shows the great importance of the potential function introduced on general lines as above.

Since there are no lines of force inside a conductor, the tubes of force of the external field must end on the conductors. We can now show quite simply that they must begin on positive electricity and end on negative and also that the quantities of electricity on the two ends of the tube are equal and opposite. Consider the portion of a small tube of force from a conductor up to the cross-section

df of it and apply Gauss' theorem to the tube closed by a slight extension into the metallic conductor. We see at once that

$$\mathbf{E}\,df = \delta q,$$

where δq is the charge on the portion of the conductor included in the tube. Similarly from the other part of the tube

$$-\mathbf{E}\,df = \delta q'.$$

Thus $\delta q = -\delta q' = \mathbf{E}\,df.$

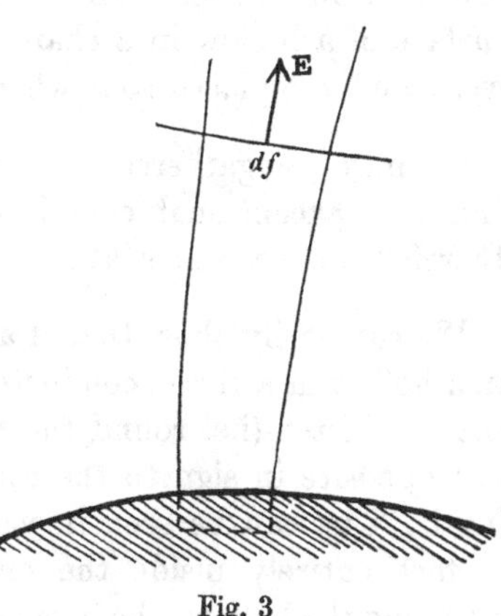

Fig. 3

We may thus regard an electrified system as consisting always of positive and negative electricity in equal amounts, each element of electricity being associated with an equal and opposite element and connected with it by a tube of force running through the dielectric medium.

Moreover since along a tube of force

$$\mathbf{E}\,df = \text{const.},$$

we see that a line of force is always a line of ascending or of descending potential throughout its length; the sign of **E** cannot change along any one line. From this we conclude that the potential cannot be an absolute maximum or minimum in free space. The greatest potential in the field must occur on a conductor (among the conductors we must include the earth, or infinity, as we say, if necessary). Moreover this conductor must have its charge all of the positive kind because if there were negative electricity at any point of its surface lines of ascending potential would pass from the conductor and this is impossible.

Similarly the least potential must occur on a conductor on which the electricity is wholly negative.

32. We also see that if the potential in a field is constant over any surface not enclosing any charge, it must be the same constant value throughout the interior of that surface, because if it were not we could draw lines of force from one point of the surface to another and we should then have an ascending and descending potential (the initial and final values at the ends are the same) in the same line. There can therefore be no charge on the interior surface of a hollow in a charged conductor unless of course there were a charge placed somewhere inside the hollow.

As may be gathered from the experiments described above, this last exceptional case is of great practical importance and therefore deserves special examination.

We can easily show that if any number of charged bodies exist in a hollow in a closed conductor the charge on the inner surface of the conductor (i.e. round the hollow) will be equal in magnitude but opposite in sign to the total charge on the system of bodies inside. This can be seen in various ways: we can draw a closed surface entirely inside the material of the conductor and surrounding the hollow; the normal force at every point of this surface being a point in the interior of the metal must be zero so that the total normal induction through the surface is zero; the total charge inside the surface must therefore also be zero. The total charge is, however, the sum of the charge induced on the inner surface of the conductor and the charges on the bodies in the hollow: these must therefore be equal and opposite.

But in the experiments mentioned the total charge of the metallic vessel into which the charged bodies were inserted was zero; so that the charge on the outer surface of the vessel must be equal and opposite to that on the inner surface: it will therefore be equal to the total charge on the bodies inserted in the interior. This is the result deduced experimentally.

Referring back again to the result just established, that the quantity of electricity at the end of the tube of force is numerically equal to the induction along the tube, we may conclude that the density of the charge at any point of a conductor in any field is given by

$$\sigma = \mathbf{E}_n = -\frac{\partial \phi}{\partial n},$$

where $\mathbf{E}_n = -\frac{\partial \phi}{\partial n}$ determines the normal component of the electric force at a point in the field just outside the conductor near the point where the density is examined. This is of course a particular case of the more general result established above that

$$\left(\frac{\partial\phi}{\partial n}\right)_+ - \left(\frac{\partial\phi}{\partial n}\right)_- + \sigma = 0,$$

because $\left(\frac{\partial\phi}{\partial n}\right)_-$ is zero, the potential inside the conductor being constant.

The total charge on the conductor is

$$-\int_f \frac{\partial\phi}{\partial n}\, df$$

taken over its surface.

33. On the nature of the potential. The important analytical function called the potential has a fundamental physical significance which we must now examine.

Lagrange* and von Helmholtz† have given mathematical proofs of the existence of the potential function in electrostatic and allied theories, associating it with the potential energy function of any ordinary dynamical system, on the assumption that the action forces involved consist merely of attractions or repulsions in the direct lines of the particles and according to some function of the distance. There is however no reason at all why all mutual physical actions of this kind should be built up of direct attractions, and one case of a different kind is known, the action of a magnetic pole on a current filament discovered by Oersted, where the forces are certainly not radial attractions‡.

The difficulties may however be avoided and at the same time a far wider idea of the meaning of the potential obtained by connecting it with the doctrine of the conservation of energy. Let us examine a general problem from this point of view.

Suppose we have a number of physical systems M_1, M_2, ... acting upon one another across space and suppose also that we have a small element δm of a similar system.

* *Par. sav.* (*étr.*) VII. (1773) (*Œuvres*, VI. p. 349).

† *Über die Erhaltung der Kraft* (Berlin, 1847). Cf. Planck, *Das Prinzip der Erhaltung der Energie* (Leipzig, 1913).

‡ See chapter IV, where this particular case is discussed in detail.

We shall first suppose either that δm is so small that it does not disturb the finite systems when moved about in their neighbourhood, or that these finite systems are held rigid during the motion of δm, so that they are not thereby disturbed. When the element δm receives a small displacement δs work from some external agency is required of amount $T_s \delta s$, where T_s is the component of the force exerted on δm in the direction of its displacement which is just sufficient to balance the action of the systems M_1, M_2,.... The work done in any finite displacement from the initial position 1 to a second position 2 is

$$\sum_1^2 T_s \delta s = \int_1^2 T_s ds.$$

But this work ought to be the same for all paths, if the general path is reversible, i.e. if the work done by external agency in any displacement is equal to the work done on the external agent by δm in the same displacement taken the reverse way. If the paths are all reversible and the work not equal for all of them we could take the element δm down one path and bring it back along another, so that the work lost in taking it down is less than that gained in bringing it back and thus on the whole there would be a gain of work. Where could this work come from? M_1, M_2, ... are all effectively rigid and so the work must have been created from nothing! This might be so but for the fact that we could repeat the process as often as we please and thus get an unlimited quantity of work, and all out of nothing. This is reasonably taken to be incredible as it involves the idea of perpetual motion: the essence of the matter is the unlimited extent.

34. Thus the argument from perpetual motion shows that for any natural law of action across space

$$\int_1^2 T_s ds$$

is independent of the path described in going from position 1 to position 2. The only other assumption involved is that of the reversibility of each path. In general this condition is satisfied.

The above argument is restricted to the case in which the systems M_1, M_2, ... were supposed to be uninfluenced by the motion of δm. We can however easily remove this restriction and consider for example the case where M_1, M_2, ... are systems of charged conductors, when the moving of a small charge δq about alters the distribution on the conductors by influence. However, even in this case, if we take δq round a closed path so that at the end the distribution is everywhere the same as the original one, the work done in the complete cycle must still be zero. There cannot be any loss of work, for if it is reversible we could promptly turn it into a gain.

Thus in the most general case

$$\int_1^2 T_s ds$$

is independent of the path from position 1 to 2. That is, there must be a function Φ of the position of δm such that

$$\Phi_2 - \Phi_1 = \int_1^2 T_s ds$$

is true for any two positions of the points 1 and 2. If we can find this function Φ, then

$$T_s = \frac{\partial \Phi}{\partial s}.$$

The function Φ represents for a definite position of the element δm the work done on it by external agency against the action of the systems M_1, M_2, ... in bringing the element from a standard position assumed to correspond to the value $\Phi = 0$, and this work may be regained by the external agents for any ulterior purpose that may be desired if the reverse path is traversed. This is interpreted in the usual way by saying that the element δm possesses a store of potential energy in virtue of its position relative to the systems M_1, M_2, ..., the amount in the typical position being Φ more than that in the standard position for which $\Phi = 0$.

35. If we now regard the physical systems as the system or systems of charged bodies acting across space in the manner specified, or in fact in any manner, and if δm is a small charged body carrying the quantity δq of electricity, then since T_s is the force which balances the electrical action on δm we know that

$$T_s = -\delta q \frac{\partial \phi}{\partial s},$$

and thus

$$\Phi = -\phi \delta q + \text{const.}$$

and the existence of the potential energy function Φ implies the existence of the analytical potential function ϕ and *vice versa*.

This proof of the existence of the potential function of our analytical theory rests on a physical basis. It assumes nothing about the method of transmission of the force, but merely that the effects are reversible and that perpetual motion does not exist. If there were friction effects or if the charge δq were moved about rapidly, electric currents and perhaps also electromagnetic waves would be excited, which would result in heat production, and the essential condition of reversibility would then be lost.

In the whole of this discussion we have neglected altogether the store of energy possessed by every system in the form of heat. Why should not work come out of this store of heat? Such a question is easily answered by an appeal to Carnot's principle. The Carnot Axiom states that if we have systems in thermal equilibrium then it is impossible for work to be done at the expense of the heat they contain. The test of thermal equilibrium is equality of temperature. Thus to make the above argument correct we must put in the criterion of equal temperatures. If two bodies were at different temperatures they could be used as a heat engine from which work could be obtained.

36. The energy in the electrostatic field. We have found that in order to establish an electric field by bringing the charges which define it into position (by friction, conduction, etc.) a certain amount of work has to be done but that when once the field is established its maintenance requires the expenditure of no work, provided of course there is no leakage to be counteracted. Thus there is a certain amount of work associated with each electric field, and this amount must be independent of the method of establishing the field in order that the energy principle may be verified: it measures the amount of energy in the electrostatic field relative to the same group of masses in their uncharged state, which would be transformed into other forms of energy if the charges were removed or cancelled.

A theory of the present type regards the electrical conditions in any field as characteristic of the electric charges in the field, the distribution of these charges being the most essential thing required for the specification of the system. In such a theory we therefore require a definition of the electric energy which makes it depend on the charges and potentials. This is readily obtained in the following way.

The work required to bring up a small charge δq to any place where the potential is ϕ is, by the generalised definition of the potential function, equal to

$$\phi\,\delta q.$$

This statement is valid and consistent whatever the complexity of the field.

Thus the work required to increase the density of the volume charge at any point of the field by $\delta\rho$ and the density of any surface charge by $\delta\sigma$ is

$$\delta W_1 = \int \phi\,\delta\rho\,dv + \int_f \phi\,\delta\sigma\,df,$$

the first integral being extended to all points of space where there is a volume charge ρ and the second over all surfaces f on which there is a surface charge σ*.

This is the fundamental differential equation of the subject, representing as it does, in a differential form, the characteristic equation of energy for the system. If we can by any process succeed in integrating this equation we shall be in a position to know the complete mechanical circumstances of the system.

But in bringing up these charges the potential at each point of the field is increased by $\delta\phi$ so that the potential energy of the charges already existing in the field is increased by the amount δW_2 where

$$\delta W_2 = \int \rho\,\delta\phi\,dv + \int_f \sigma\,\delta\phi\,df.$$

But if the mechanical process of establishing the charges in the field is a reversible one so that all operations involved in it can be reversed—and this is essential to the existence of a potential function—the work which is done in charging the system must all

* Double sheets are excluded as they are of relatively small importance in the present aspects of the theory. It is however quite easy to generalise the discussion to include them.

be stored up as potential energy of a purely electrical nature in the field so that

$$\delta W_1 = \delta W_2,$$

and thus each of these is equal to the half of their sum or

$$\begin{aligned}\delta W_1 = \delta W_2 &= \tfrac{1}{2}(\delta W_1 + \delta W_2)\\ &= \tfrac{1}{2}\delta\left\{\int (\rho\,\delta\phi + \phi\,\delta\rho)\,dv + \int_f (\phi\,\delta\sigma + \sigma\,\delta\phi)\,df\right\}\\ &= \tfrac{1}{2}\delta\left\{\int \phi\rho\,dv + \int_f \phi\sigma\,df\right\}.\end{aligned}$$

We have therefore in such a case

$$W_1 = W_2 = \tfrac{1}{2}\int \rho\phi\,dv + \tfrac{1}{2}\int_f \sigma\phi\,df\,*.$$

Thus if we multiply each element of charge by half the potential at its position and add up over the whole distribution we get the general value for the potential energy of the electrical system referred to the state in which $\rho = \sigma = 0$ everywhere as the zero state.

37. The result here deduced in a physical manner can also be obtained analytically as follows. We know that

$$4\pi\phi = \int \frac{\rho\,dv}{r} + \int_f \frac{\sigma\,df}{r}.$$

Thus if we use dashed letters to denote these integrals in ϕ we see that

$$4\pi\delta W_1 = \iint \frac{\rho'\delta\rho\,dv\,dv'}{r'} + \int_f\int \frac{\sigma'\delta\rho\,dv\,df'}{r'} + \iint_f \frac{\rho'\delta\sigma\,df\,dv'}{r'} + \int_f\int_f \frac{\sigma'\delta\sigma\,df\,df'}{r'},$$

where r' denotes the distance between the typical elements of the double integrations over the volume of the field and the surfaces of discontinuity in it.

It follows immediately from this form of the expression, by integrating with respect to the undashed elements first—a process that is fully justified in view of the absolute convergence of the integrals concerned—that

$$\delta W_1 = \int \rho'\delta\phi'dv' + \int_f \sigma'\delta\phi'df',$$

* Kelvin, *Glasgow Phil. Soc. Proc.* 3 (1853); Helmholtz, *l.c.* p. 55; Clausius, *Die Potential Funktion*, §§ 63 and 64 (Leipzig, 1859).

where $$4\pi\delta\phi = \int \frac{\delta\rho\, dv}{r} + \int_f \frac{\delta\sigma\, df}{r}$$

is the increment of the potential at the typical field-point consequent on the addition of the charges $\delta\rho$ and $\delta\sigma$. We thus see that

$$\delta W_1 = \delta W_2$$

and the result deduced above now follows immediately by the same argument.

38. We have thus succeeded in establishing the existence of a mechanical potential energy function associated with our electrically charged system of a type similar to that with which we are familiar in ordinary mechanics and the usual properties of this function now follow as a matter of course. A system of charges on a rigid system of masses will adjust themselves in such a way as to make the electrical potential energy function W_1 stationary subject only to the constancy of the total charge and the conditions implied by the natural restraints of the system. In fact the mutual forces exerted between the various charge elements tend to produce motion of these elements and if the energy of such motions can be obtained at the expense of the internal store of potential energy they are almost certain to take place. Thus equilibrium is possible only in those configurations in which the potential energy has a stationary value as regards small displacements from the configuration, because it is only then that the small initial displacement from the configuration results in no appreciable change in the store of potential energy from which the kinetic energy of any further motions that take place must be derived.

Moreover it appears that if any configuration of the system is a configuration of stable equilibrium as regards the charge distribution throughout it, the stationary value of the potential energy must be an absolute minimum value, for it is only in such a case that the system will not, if slightly disturbed, depart widely from the configuration by the action of its own internal forces. Thus it is only when the natural restraints* of the system prevent any further running down of the electrical potential energy that equilibrium among the charges is permanently possible.

* The insulation of a conductor acts as a 'restraint' to prevent the charge on it getting across to another conductor.

39. If we apply this condition for the equilibrium of a system of charges, some or all of the elements in which are capable of free movement within certain limited spaces, we see that the potential function ϕ must be constant throughout any space in which the charges are freely movable. This follows at once because the condition for equilibrium is that the variation of the potential energy consequent on a slight rearrangement of the charges, which is

$$\delta W_1 = \int \phi \, \delta\rho \, dv + \int_f \phi \, \delta\sigma \, df,$$

must vanish subject to the condition that the total charge in each partial space must be constant, or that

$$\int \delta\rho \, dv + \int_f \delta\sigma \, df = 0,$$

where the volume integral is taken throughout the partial space and the surface integral over the boundary of that space and any surfaces of discontinuity inside it. This leads to the result that ϕ has a constant value throughout the partial space.

40. The importance of this last result lies in its application to the field of a number of charged conductors. The elements of a charge on a conductor are, effectively speaking, freely movable throughout the material of the conductor but not beyond its outer surface. Thus in order that the charge distribution may be one of stable equilibrium it must be such that the potential ϕ is constant throughout the volume of each conductor. We have already seen that this means that the charge on the conductor exists entirely on its outer surface; this of course also results from the fact that the greatest dispersion then exists in the group of elements constituting the charge and the least value of the potential energy is attained.

41. There is an important reciprocal theorem due to Gauss* which is worth quoting at this stage: we shall first give it in terms of discrete charge elements and then indicate its extension to the more general case.

Let q be an element of charge in any distribution and ϕ the

* *Allgemeine Lehrsätze über...Anziehungs- und Abstossungskräfte,* § 19 (*Collected Works,* v. p. 200).

potential of that distribution at any point and in a second system let these be q' and ϕ'; we have then

$$\Sigma q\phi' = \Sigma q'\phi,$$

where in $\Sigma q\phi'$ every element of the first distribution is multiplied by the potential of the second distribution at the position of the element, and similarly in $\Sigma q'\phi$.

The usual proof of this theorem in the present case is that each sum is equal to the double sum

$$\frac{1}{4\pi}\Sigma\frac{qq'}{r},$$

wherein the summation extends to each element of charge of the one system with each element of the other system and r is the distance between them.

A more fundamental interpretation of the relation is obtained by noticing that

$$\Sigma q\phi'$$

is the work required to bring up the first charged system supposed rigidly fixed in its final relative configuration into its position relative to the second, and that therefore this must be equal to the work required to bring up the second charge system into its position relative to the first which is expressed by the second sum $\Sigma q'\phi$; these are in fact merely two different ways of establishing the combined field.

In this sense we see that it must also be true in the more general case with effectively continuous charge distributions, or in other words if the two systems of charges are specified by their charge densities ρ and ρ' at the typical point of space and if the potentials of their respective fields are ϕ and ϕ' then

$$\int \rho\phi' dv = \int \rho'\phi\, dv,$$

the integrals in each case being taken throughout the entire charge distribution.

If the second system of charges is the first increased by very small increments $\delta\rho$, then ϕ' differs from ϕ only by a differential amount $\delta\phi$ and thus we may put

$$\rho' = \rho + \delta\rho, \quad \phi' = \phi + \delta\phi,$$

and on substituting these values in the above relation we see at once that

$$\int \rho\,\delta\phi\,dv = \int \phi\,\delta\rho\,dv,$$

a relation deduced above from the energy principle.

42. The mechanical forces on the matter in the field. The analytical theory up to the present stage has been concerned merely with the electromotive forces of the field, the question of the ponderomotive forces has so far not arisen: we have only explained how charges are separated and not how electrified bodies attract one another.

It is however at once evident that such forces must exist. The *electromotive* force acting on a charge connected with a material body and in equilibrium must be counterbalanced by an equal and opposite force resulting from the action of the material body on the same charge: and the reaction to this latter force will be an equal and opposite force exerted from the charge on the material medium with which it is rigidly connected. Thus any material body carrying a charge and in equilibrium will be acted upon by a force equal to and in the same direction as the resultant electromotive force on the system of charges contained in it. Thus if ρ denote the density of the charge distribution at the point (x, y, z) in the body where the intensity of the electric force is $\mathbf{E}$, then there will be a ponderomotive force on the body determined by

$$\mathbf{F}_1 = \int \rho\mathbf{E}\,dv,$$

the integral being extended throughout the volume of the body at no point of which is ρ infinite.

This is the general form; but it is convenient to have the specialisation of it applicable when there are surface distributions associated with the body. This is easily obtained by considering such a surface distribution as the limiting case of a volume density concentrated in a thin layer. On this view the force on the small element df of the surface is practically

$$\int \mathbf{E}\rho\,dv,$$

this integral being taken throughout the small volume of the sheet standing on df. Now $\mathbf{E}$ varies continuously throughout the sheet from a value $\mathbf{E}_1$ on one side to a value $\mathbf{E}_2$ on the other, and thus the average value throughout the sheet is $\frac{1}{2}(\mathbf{E}_1 + \mathbf{E}_2)$* and with this the force on df is

$$\mathbf{F}_2 df = \tfrac{1}{2}(\mathbf{E}_1 + \mathbf{E}_2)\int \rho\, dv = \tfrac{1}{2}(\mathbf{E}_1 + \mathbf{E}_2)\,\sigma df,$$

and thus for the whole surface charge

$$\mathbf{F}_2 = \tfrac{1}{2}\int (\mathbf{E}_1 + \mathbf{E}_2)\,\sigma df,$$

and a proper combination of the forces $\mathbf{F}_1$ and $\mathbf{F}_2$ will give the correct form of the force on any electrified body or at least the linear components of the resultant force. The angular components can be obtained in an analogous manner or they may be calculated in a general way from the results deduced below where the question of these forces is regarded from another and more general point of view.

43. The mere existence of mechanical forces on the matter in the field is involved in the idea of the energy of the charged system. When two charged bodies are moved relative to each other, the total electrical energy in the field is altered and if the charges are kept constant the loss (or gain) of energy is due to some other system linked with the electrical one. It reappears in fact as a gain (or loss) in the mechanical energy of the charged bodies, which determines the mechanical forcive between them. Thus to obtain the forces we need only give the bodies small virtual displacements and include the virtual work in these displacements in the general expression of the work done on the system during a general virtual change in its configuration.

If the positions of the material bodies of the system are determined by the generalised coordinates $\theta_1, \theta_2, \ldots$ in the usual

* This statement and the whole proof depending on it is perhaps not as rigorous as might be desired. The usual argument divides the force close up to the surface into a local and a general part; the local part is due to the small portion of the surface charge near the point and the general part to the remainder. For two near points equidistant from the surface but on opposite sides the local parts are equal in magnitude but opposite in direction whilst the general parts are the same at both points. The general part of the force which alone is mechanically effective is then equal to the mean of the total forces at the two points.

Lagrangian sense and if the force components applied by external agency corresponding to the coordinates are $\Theta_1, \Theta_2, \ldots$ respectively, then the work done by external agency during a displacement is

$$\Theta_1\delta\theta_1 + \Theta_2\delta\theta_2 + \ldots.$$

This is the generalised Lagrangian definition of a force component in statics: it is defined so that the force multiplied into the small change in the coordinate is the work done, provided none of the other coordinates vary.

The work done in increasing the charge distribution of the system is

$$\int \phi\,\delta\rho\,dv + \int_f \phi\,\delta\sigma\,df,$$

if the bodies are fixed; if however in addition the matter receives a small virtual displacement as above we must add the work done against the forces acting in these displacements. The general form for the work done on the system during the most general virtual change in its configuration (a complete definition of a configuration involving a knowledge of the charge distribution and the positions of the material bodies) is thus

$$\delta W = \int \phi\,\delta\rho\,dv + \int_f \phi\,\delta\sigma\,df + \Theta_1\delta\theta_1 + \Theta_2\delta\theta_2 + \ldots;$$

and again the usual argument based on the assumption of reversibility and the negation of perpetual motion requires that δW should be a complete differential of some function W which ultimately measures relative to some standard configuration the potential energy which the electrified system possesses in virtue of its charge. In other words W is the electrical potential energy of the system.

We know however from the discussions of the previous section that the electrical potential energy of the system can be expressed in the form

$$W = \tfrac{1}{2}\int \rho\phi\,dv + \tfrac{1}{2}\int_f \sigma\phi\,df,$$

so that as soon as these integrals can be effected W is known and the complete mechanical relations of the system are theoretically determinate.

44. If the charge distribution on the system of masses is maintained constant during the slight displacement of the system, the first part of the total expression for δW does not occur and the increase in the electrical potential is simply given by

$$\delta W_c = \Theta_1 \delta\theta_1 + \Theta_2 \delta\theta_2 + \ldots,$$

where W_c denotes exactly the same quantity as W above but the suffix implies that the charge distribution throughout each body is maintained constant during a displacement of that body. In this case the work of the external forces, viz.

$$\Theta_1 \delta\theta_1 + \Theta_2 \delta\theta_2 + \ldots,$$

which is extracted from a machine outside the system, is added to the store of internal energy which the system of masses possesses in virtue of the charges rigidly attached to them.

We have also in this case

$$\frac{\partial W_c}{\partial \theta_s} = \Theta_s,$$

so that the force in any one of the material coordinates under the specified conditions is determinate.

45. In some important cases the total charge and its distribution are altered in such a way as to maintain the potential distribution throughout the various masses constant. In this case the work done on the system during a small virtual displacement will involve the complete expression given above, but this is not now in a convenient form as it requires a knowledge of the distribution of $\delta\rho$ and $\delta\sigma$ necessary to secure the maintenance of the potential distribution. To obtain a more suitable form for such cases we have only to rewrite it in the form

$$\delta\left(\int \rho\phi\, dv + \int_f \sigma\phi\, df - W\right) = \int \rho\,\delta\phi\, dv + \int_f \sigma\,\delta\phi\, df - \Theta_1 \delta\theta_1 - \ldots,$$

and then notice that since in the present instance

$$\tfrac{1}{2}\int \rho\phi\, dv + \tfrac{1}{2}\int_f \sigma\phi\, df = W,$$

the term on the left is still δW, so that

$$\delta W = \int \rho\,\delta\phi\, dv + \int_f \sigma\,\delta\phi\, df - \Theta_1 \delta\theta_1 - \ldots.$$

We now see immediately that when the potential distribution

throughout the various masses is maintained constant throughout the displacement the change of the internal potential energy is

$$\delta W_p = -\Theta_1\delta\theta_1 - \Theta_2\delta\theta_2 - \ldots,$$

where we have used W_p to denote the value of W in which the potential distribution in the various masses is maintained constant.

Again, as above,

$$\Theta_1\delta\theta_1 + \Theta_2\delta\theta_2 + \ldots$$

represents the energy expended by the external system in the mechanical work of shifting the masses and

$$\delta W_p$$

is the increase in the internal potential energy of the system of charged masses. These quantities are now of opposite sign so that some other outside source must be involved in the action. The only other available source of energy is, generally speaking, that which is used to maintain the potential distribution, and the amount of energy derived from it must be double the amount of the mechanical work gained from the system; the other half of the total supply goes to increase the internal potential energy of the charge distribution.

We have also in this case

$$\frac{\partial W_p}{\partial \theta_s} = -\Theta_s, \qquad s = 1, 2, \ldots\ldots$$

so that

$$\frac{\partial W_p}{\partial \theta_s} = -\frac{\partial W_c}{\partial \theta_s}.$$

Thus a variation of the configuration of the system which increases the internal energy when the charge distribution is maintained constant decreases this energy when the potential distribution is maintained constant.

CHAPTER II

DIELECTRIC THEORY

46. Introduction*. The fact that certain bodies after being rubbed appear to attract other bodies at a distance from them was known to the ancients. In modern times a great variety of other phenomena have been observed and found to be related to these phenomena of attraction at a distance. The first definite formulation of this distance action between two bodies was given by Newton in his law of gravitational attraction, which states that one piece of matter attracts another according to a simple law, not affected by any intervening matter. This law of gravitation proved so successful in astronomy that it was made the pattern for the solution of more abstract problems in physics†, especially after it was discovered that the laws of electric and magnetic attraction were of precisely the same type, and attempts were made to model the whole of natural philosophy on this one principle, by expressing all kinds of material interaction in terms of laws of direct mechanical attraction across space. Of course if material systems are constituted of discrete atoms, separated from each other by many times the diameter of any one of them, this simple plan of exhibiting their interactions in terms of direct forces between them would probably be exact enough to apply to a wide range of questions, provided we could be certain that the laws of force depended only on the positions and not also on the motions of the atoms.

The doctrine of action at a distance, which thus expresses the view that our knowledge in such cases *may* be completely represented by means of laws of action at a distance expressible in terms of the positions (and possible motions) of the interacting bodies without taking any heed of the intervening space, was specially

* Cf. the article "Aether" by Sir J. Larmor in the *Encyclopaedia Britannica* (1911). Also the book *Aether and Matter* by the same author.

† The most successful of these applications has been in the theory of capillary action elaborated by Laplace, though even here it appears that the definite results attainable by the hypothesis of mutual atomic attractions really reposed on much wider and less special principles—those, namely, connected with the modern doctrine of energy.

favoured by the French and German scientific schools and in W. Weber's hands an almost complete electric theory was built up on it*. The doctrine was however strongly repudiated by Newton himself and hardly ever became influential in the English school of abstract physicists.

The modern view of these things, according to which the hypothesis of direct transmission of physical influences expresses only part of the facts, is that all space is filled with physical activity and that while an influence is passing across from a body A, to another body B, there is some dynamical process in action in the intervening region, though it appears to the senses to be mere empty space. The problem is of course whether we can represent the facts more simply by supposing the intervening space to be occupied by a medium which transmits physical actions, after the manner that a continuous material medium, solid or liquid, transmits mechanical disturbances. The object of the following pages is to answer this question in the affirmative along the lines laid down by Faraday in his experimental researches.

47. Faraday's† ideas on the nature of the electric action between charged bodies. The theory of electric actions developed in the previous chapter is essentially a distance action theory, inasmuch as it depends merely on the concept of the electric charge with a definite law of action between point charges. No reference whatever is made to the medium between the charges and in fact our theory is true only if one single homogeneous medium pervades the whole of space. Such is in general not the case and the theory thus needs generalisation on the lines suggested by Faraday and worked out mathematically by Maxwell. Faraday firmly believed that the action between two charged bodies was transmitted through and by the medium between them. To test his ideas experimentally he tried to alter the medium by interposing between the charges different dielectric substances. The procedure actually adopted to obtain the exact effect was to use two equal spherical condensers, one with an air dielectric and the other with some other

* An historical account of the developments of electrical theory on this basis is given with full references by Reiff and Sommerfeld in *Ency. der math. Wissensch.* v. 12, pp. 1–62.

† Cf. *Experimental Researches*, especially I. 1231, 1613–16; III. 3070–3299.

substance, such as sulphur. By connecting the two inner spheres of the condensers together and the outer ones to earth any charge is divided between them. If the condensers are of exactly the same size the charge should be divided equally if the presence of the sulphur makes no difference. Faraday found however that they were not equally divided; but were in a definite ratio $\epsilon : 1$; the presence of the sulphur increases the capacity of the one condenser ϵ times. This constant ϵ was found to be typical of the substance used in the second condenser and is therefore called the *specific inductive capacity* (S.I.C.) of the substance or simply its *dielectric constant**.

The value assumed by the dielectric constant for certain standard substances will appear from the list appended where approximate values are given:

Glass (crown)	6	Water	81
„ (flint) ...	8·5	Benzene	2·3
Ice	93·9	Petroleum ...	2·1
Rubber	2·2	Alcohol (methyl)	35·4
Paper	2·2	„ (ethyl)	26·8
Ebonite	2·8	Glycerine	39·1

Shellac, oils, rubber and ebonite have values of the constant between 2 and 3; glass and the micas have a value between 6 and 9. Many liquids are much higher, water being about 80 at ordinary temperatures; but at low temperatures the value seems to become constant at about 2 or 3 for a variety of liquids which are widely different at ordinary temperatures. The value varies also largely with the purity and constitution generally of the specimen under examination.

48. In attempting to find some cause or theory of this new effect Faraday was induced to a closer investigation of the electric field in the cases when the otherwise simple circumstances are complicated by the presence of some dielectric material, his method being to trace out experimentally the lines of force and to form them into tubes of force in the manner indicated in the previous chapter. He intuitively got the idea that the amount of electricity at one end of the tube was the same as that at the other and he

* Dielectric constants were independently determined by Cavendish in 1773. Cf. Maxwell, *Treatise*, I. p. 54

then succeeded in verifying this experimentally even in the case when dielectric media of great complexity were present in the field. Moreover he discovered that the other important property of tubes of force, viz. that the product of the force intensity at any point in a tube by the cross-section at that point is constant all along, still held in the general case.

Such obviously fundamental results naturally led Faraday to think that there must be some physical cause for this equality and he put it down to some physical action in the tube. He imagined the tube to be full of an incompressible fluid so that if the liquid were pushed along the tube, whatever excess is produced at one end has an equivalent diminution at the other, both being equal to the amount crossing any section of the tube during the establishment of the displacement.

The essential point in this idea of Faraday's is that an electric field arises through an electric displacement or induction along the curved lines of force, resulting in an accumulation of positive charge at one end and negative at the other. The term 'electric displacement' thus introduced is however not to be taken too literally in this sense. The idea that survives is that it is some vector of the same nature as the displacement of a fluid which is related in the usual manner to the lines of force which experiment maps out.

49. In the mathematical formulation of this scheme it is first necessary to define the electric force independently of the idea of simple attraction. We define the electric force intensity $\mathbf{E}$ at a point in the field so that $\mathbf{E}\delta q$ is the force on a small charge δq placed there. Before however we put the charge there we must make room for it, we must remove the matter inside a small cavity surrounding it. The question as to whether this removal affects the force at the point is reserved for future consideration, and for the present we shall assume that the force intensity thus defined is a mathematically definite vector quantity.

The force $\mathbf{E}$ as thus defined is still the gradient of a potential function ϕ. A simple justification of this statement could be based, in the manner previously indicated, on the perpetual motion idea, the argument being that, under steady conditions, carrying a charge round a closed circuit ought to involve no work on the whole, assuming that the conditions are the same at the end as at the beginning. We shall therefore assume the existence of a potential function ϕ at each point of the field.

We must next define the electric displacement or something akin to an electric displacement. The intensity of the displacement $\mathbf{D}_s$ in any direction at a point is such that the total amount displaced across any small area df_s perpendicular to the direction during the establishment of the field is

$$\mathbf{D}_s df_s.$$

Suppose now we consider that the field is established by slowly increasing all charges proportionally: the configuration of the field at each instant as regards the lines of force and displacement will then be similar to the final one. Now consider a tube formed by the lines of displacement drawn through any small area df, which is perpendicular to its mean axis, and let df_s be any adjacent slanting cross-section making an angle θ with df. This tube is a tube of displacement at each instant during the establishment of the field and thus the totality of displacement across any section of it is the same all along: that is

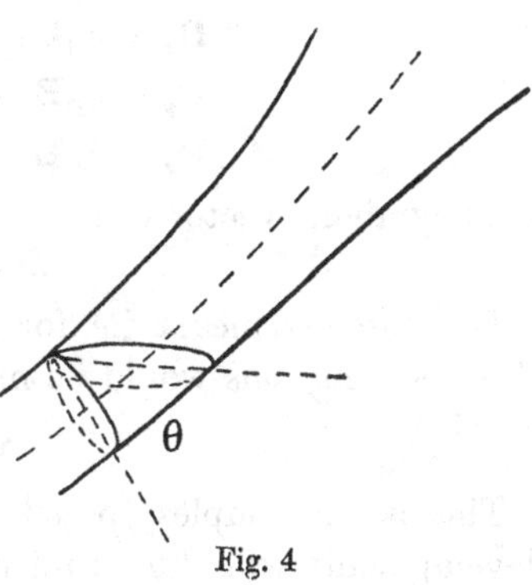

Fig. 4

$$\mathbf{D}_s df_s = \mathbf{D} df;$$

but $df_s \cos\theta = df$ and therefore

$$\mathbf{D}_s = \mathbf{D}\cos\theta,$$

and thus the quantity $\mathbf{D}$ so defined is an ordinary vector, its direction being at each point along the line of displacement through that point.

It is easily verified that the final result of any more complicated method of establishing the field would be expressible by the same quantity $\mathbf{D}$.

50. Thus in our electric field we have at each point two kinds of vectors:

(i) the force intensity **E** which is the gradient of a potential function, and

(ii) the flux intensity **D**.

Now our theory implies that **E** is the cause of **D**; the displacement at a point is conditioned by the force intensity there. The simplest possible law of causality we can have is that in which there is a simple proportionality, so that if we double the cause we double the effect. In this case the components $(\mathbf{D}_x, \mathbf{D}_y, \mathbf{D}_z)$ of the displacement would be linear functions of the components $(\mathbf{E}_x, \mathbf{E}_y, \mathbf{E}_z)$ of the electric force, i.e.

$$\mathbf{D}_x = \epsilon_{11}\mathbf{E}_x + \epsilon_{12}\mathbf{E}_y + \epsilon_{13}\mathbf{E}_z,$$
$$\mathbf{D}_y = \epsilon_{21}\mathbf{E}_x + \epsilon_{22}\mathbf{E}_y + \epsilon_{23}\mathbf{E}_z,$$
$$\mathbf{D}_z = \epsilon_{31}\mathbf{E}_x + \epsilon_{32}\mathbf{E}_y + \epsilon_{33}\mathbf{E}_z,$$

or in symbolic vector form

$$\mathbf{D} = (\epsilon)\,\mathbf{E}.$$

For isotropic media, i.e. for those having the same properties in all directions, this would be more simply expressed by the vector relation

$$\mathbf{D} = \epsilon\mathbf{E}.$$

This is the simplest possible form of the theory. Subsequent developments will show that it is very approximately the correct one. Let us now follow it to some of its more important consequences, confining ourselves however to the case of isotropic media.

51. Now suppose all the lines of displacement drawn out in the field and formed into tubes. The total displacement across any cross-section of a tube is constant along the tube. Moreover the displacement at one end of a tube where it abuts on a charge is equal to that along the tube and on the simplest assumption is measured by the charge there. If we are dealing only with conductors carrying charges and if a tube ends at a place where the surface density is σ and the cross-section there is df_1, then if df is the cross-section at any other part of the tube where the flux density is **D**

$$\mathbf{D}\,df = \sigma\,df_1.$$

On this idea the surface density, or charge at the ends of a displacement tube, is merely the terminal aspect of the displacement in the tubes. The way the displacement reveals itself is by piling up surface density.

In this theory the ordinary Gaussian surface integral theorem has a distinct physical significance. Our notion of the displacement means that if we take a closed surface of any kind in the field and integrate the normal displacement over it, then the total thus obtained must be equal to the total charge inside. Thus if $\mathbf{D}_n$ is the normal component of the displacement at the position of the element df of this surface

$$\int_f \mathbf{D}_n df = Q,$$

where Q is the total charge inside the surface f. This means that if the charge inside is a volume charge of finite density ρ, then

$$\int_f \mathbf{D}_n df = Q = \int_f \rho\, dv,$$

which by Green's lemma reduces to

$$\int_f \operatorname{div} \mathbf{D}\, dv = \int_f \rho\, dv,$$

and as this is true for any volume f we must have

$$\operatorname{div} \mathbf{D} = \rho$$

at each point of space: if $\rho = 0$, as is usually the case in all electrostatic problems, then

$$\operatorname{div} \mathbf{D} = 0,$$

and as much is displaced out of any region as flows into it, i.e. the displacement is like the flux of an incompressible fluid, as assumed by Faraday and Maxwell; **D** then satisfies the conditions for a *stream vector*.

52. We have already seen that the electric force intensity at each point of the field is a vector whose components are

$$(\mathbf{E}_x, \mathbf{E}_y, \mathbf{E}_z) = -\left(\frac{\partial}{\partial x}, \frac{\partial}{\partial y}, \frac{\partial}{\partial z}\right)\phi;$$

moreover our assumptions imply that the displacement **D** is conditioned by the electric force **E** and the relation adopted was

$$\mathbf{D} = (\epsilon)\, \mathbf{E} = -(\epsilon) \operatorname{grad} \phi;$$

hence we have the characteristic equation of the field in the form

$$\operatorname{div} ((\epsilon) \operatorname{grad} \phi) = -\rho,$$

which is the equation that replaces Poisson's equation for our theory. It is the characteristic equation of the potential function on the new generalised method of procedure invented by Faraday.

If the dielectric medium is isotropic this equation reduces to

$$\frac{\partial}{\partial x}\left(\epsilon\frac{\partial\phi}{\partial x}\right)+\frac{\partial}{\partial y}\left(\epsilon\frac{\partial\phi}{\partial y}\right)+\frac{\partial}{\partial z}\left(\epsilon\frac{\partial\phi}{\partial z}\right)=-\rho,$$

and if it is homogeneous as well it becomes

$$\nabla^2\phi=-\frac{\rho}{\epsilon},$$

a modified form of Poisson's equation. This result stated in words implies that the result of having the dielectric throughout the region is that the same distribution of potential requires a distribution of charge ϵ times larger than in free space. Thus the constant ϵ of homogeneous isotropic medium which has been here introduced merely as the physical constant in the relation between the electric force and displacement, is identical with Faraday's dieleotric constant ϵ.

53. As in the previous method of analysis of the electric field it is essential that we consider the alterations in the above analytical procedure necessitated by singularities in the distribution of charge, so that we may know how to deal with them when they turn up in any applications. The most important case is that in which ρ is infinite along a surface so that there is a surface distribution of density σ; we shall confine our attention to this example.

In the first place it is obvious that discontinuities can arise only in the neighbourhood of the surface distribution. Moreover ϕ must be continuous as we cross the surface as otherwise its gradient across would be infinite. If we now apply our generalised Gauss' theorem to a small flat cylinder enclosing a part of the surface of area df we can conclude at once that the normal component of the displacement across the surface is discontinuous by an amount $+\sigma$, i.e. if $\mathbf{D}_n$ is the component of $\mathbf{D}$ normal to the surface on which the surface density is situated and if also suffices 1 and 2 denote the values of the functions at near points on the same normal one at each side of the surface

$$\mathbf{D}_{n_1}-\mathbf{D}_{n_2}=\sigma;$$

but, as above, in the case of isotropic media

$$\mathbf{D}_{n_1} = -\epsilon_1 \frac{\partial \phi_1}{\partial n},$$

$$\mathbf{D}_{n_2} = -\epsilon_2 \frac{\partial \phi_2}{\partial n},$$

where we have also included for generality the possibility of different values of ϵ for the substance on the two sides of the surface. Thus

$$\epsilon_1 \frac{\partial \phi_1}{\partial n} - \epsilon_2 \frac{\partial \phi_2}{\partial n} = -\sigma.$$

If the surface is one of discontinuity in the isotropic dielectric medium, without any charge, $\sigma = 0$ and thus

$$\epsilon_1 \frac{\partial \phi_1}{\partial n} - \epsilon_2 \frac{\partial \phi_2}{\partial n} = 0,$$

and these equations give the form of the boundary conditions to which ϕ is subject in the present form of the theory.

54. It must now be noticed that the theory here given agrees perfectly with that deduced from the old attraction ideas for a vacuum. We have only to put $\epsilon = 1$ to get all the results of our previous theory. We have however gone deeper into the matter; instead of talking of mere attractions we have attempted to see what is going on and have generalised the theory to include the properties of the medium conveying the action. Instead of a simple theory of attractions we have now a theory of flux stimulated by electric force. The exposition is of course merely of the nature of an explanation; no proof can be given that it is the correct view of the affair. We have merely invented a consistent scheme whose continued existence depends only on the test of its reality.

55. Electric displacement. The whole of the present scheme turns on the electric displacement. What is this displacement? Why is it different when a dielectric substance is present? The first question has not yet been satisfactorily answered, but the second question was at least explained when Kelvin applied to such media Poisson's analysis for magnetically polarised media in combination with Faraday's idea of dielectric polarisation.

If there be brought near to a charged body A a rod composed of some dielectric or conducting material, the usual phenomena of electric induction are observed: the ends of the rod near to and remote from the charged body behave just as if they carried respectively charges of the opposite and of the same kind as A. If the rod is made of conducting material it can be charged permanently in this way: on cutting the rod at any point between its two ends and removing it from the neighbourhood of A the separated fragments are found to retain the charges which they appeared to carry under the influence of the charge on A. But if the rod is made of insulating material the separated fragments will be without charge at whatever point the rod be cut. The old-fashioned explanation of this fact is that the neutral rod is supposed to contain in each of its elements or particles equal quantities (comparatively large) of electricity of opposite sign which normally counterbalance one another. When the body is brought into the neighbourhood of the charged body A the electricity on that body attracts that one of the two charges in each element of the neutral body which is of the opposite sign and repels the other, the result being a separation of the charges. In a conductor the charges are quite free to move about as they like and the displacement may be of finite extent. In a dielectric on the other hand the charges appear to be bound to the molecules of the body by restraining forces of some kind*, quasi-elastic forces we may say, and the displacement is thereby limited to very small molecular dimensions, the charges each settling down into an equilibrium position where the electric forces of the field balance these quasi-elastic forces.

This theory has been completely substantiated by the discovery of the atomic structure of electricity and the consequent developments of experimental science in the elucidation of the 'electron theory' to which this discovery gave rise. According to this theory every atom contains as an essential element of its constitution a certain number of electrons more or less tightly bound in it, in addition to the necessary positive charge to make it neutral. The application of an electric field will then as above displace the negative electrons relative to the positive ones and thus render each atom bi-polar in the above sense.

* Mossotti assumes that the molecules are like small conductors insulated from one another. Cf. *Sur les forces qui régissent la constitution intime des corps* (Turin, 1836).

56. The theory of polarised dielectric media is a molecular one; the polarisation is supposed to belong to the individual molecules of the dielectric substances or perhaps to molecular groups; the essence of the affair is that the physical element (the smallest thing we are concerned with) is a bi-polar one, i.e. it has two equal and opposite poles of electric charge. Each molecule of a substance contains one or more elements of electrical charge of each sign, which are originally practically coincident; the application of an electric field would then tend to pull them about until some elastic resilience (supposed for the present to be proportional to the displacement) balances the electric pull.

There is another theory of polarised media which assumes that the separate molecules are permanently polarised (i.e. the charges never coincide) but are usually arranged in all sorts of directions. It is owing to the fortuitous distributions of the directions that the total statistical effect is null in the normal condition. The application of an electric field would then turn each molecule round, all towards a definite direction, against elastic resilience and their separate fields would no longer cancel.

Either of these theories would do for the present purposes but we prefer the first for reasons which will afterwards appear. The statistical view of the two is the same, and this is, after all, all that we are concerned with at present. There are certain facts which suggest that there is, at least in some cases, a certain justification for the second view but these will be dealt with separately.

The essential fact in any such theory is that all electric excitement is accompanied by electric separation. In a conductor the opposite charges present at each point will be pulled right apart and as the supply is practically unlimited the separation will go on while the electric force exists; in an insulator or dielectric substance on the other hand the extent of the separation is limited by certain elastic resiliences tending to hold the separate charge elements to the matter.

57. Mathematical formulation of the theory. We start by analysing the electric field of a polarised dielectric mass. The element of the analysis is the simple bi-pole consisting of two point charges ($+q$, $-q$) placed at a small distance apart. The law of inverse squares is assumed for each constituent. If the pole $+q$ is at A and $-q$ at B, then the potential due to this element at any point P in the field is ϕ where

$$4\pi\phi = \frac{q}{PA} - \frac{q}{PB}.$$

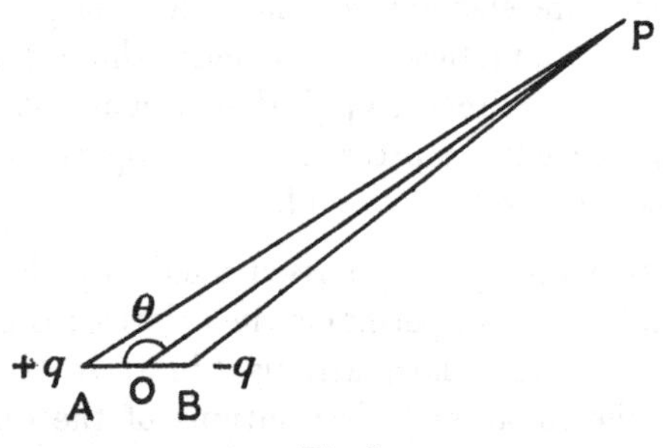

Fig. 5

Let O be the mid-point of AB and θ be the angle AOP; also put $AB = \delta s$, $OP = r$, then since δs is very small we have

$$PA = r - \frac{\delta s}{2}\cos\theta, \quad PB = r + \frac{\delta s}{2}\cos\theta,$$

so that
$$4\pi\phi = q\left(\frac{1}{r - \frac{\delta s}{2}\cos\theta} - \frac{1}{r + \frac{\delta s}{2}\cos\theta}\right)$$

$$= \frac{q\,\delta s\cos\theta}{r^2} \text{ practically.}$$

In dealing with elements of this nature we do not as a rule know either q or δs, only the product $q\,\delta s$; but this is all we are concerned with in the investigation. We therefore use a special name for it and call it the *moment* of the element; the line AB is called the axis of the bi-pole and since the direction of this axis is an essential part of the specification of the bi-pole we shall find it convenient to treat the moment as a vector quantity, for then it will take account of the direction as well as the magnitude of the electric moment. We use the symbol $\mathbf{e}$ to denote such a moment*.

* The mode of presentation here adopted is due to Larmor. Cf. *Aether and Matter*, App. A, "On the principles of the theory of magnetic and electric polarisation."

If we put the bi-pole at the origin of coordinates, then the potential of the element at the (x, y, z) point is

$$\phi = \frac{(\mathbf{er})}{4\pi r^3}$$
$$= \frac{1}{4\pi}\left(\mathbf{e}_x \frac{x}{r^3} + \mathbf{e}_y \frac{y}{r^3} + \mathbf{e}_z \frac{z}{r^3}\right)$$

so that the potential of the element is the sum of the potentials of its separate components.

58. The molecule or element of a dielectric medium may consist of a whole system of simple doublets of the kind here examined. The molecule may contain a whole lot of positive and negative elements and if we group them in pairs (positive and negative) we have a system of bi-polar elements. We could find the resultant of the moments of all these separate elements and we should then define this as the moment of the molecule. If we are considering their effects at ordinary distances the different positions of the centres of all these bi-poles in the molecule will not concern us; for all practical purposes we can regard the centres as coincident on account of the extreme smallness of the molecule. It thus appears that we can treat each molecule, however complicated it may be, just as if it were a simple bi-pole with a definite moment, obtained perhaps by considering the positive charges as practically equivalent to a positive charge at its mean centre and similarly with the negative. The essential point is thus that we can treat the molecule as a simple element and we need not for the present trouble ourselves with details of how it is built up.

Now suppose we have a whole lot of these bi-polar molecules forming a finite mass. We must then treat the thing as a whole and take as the element of our analysis a volume element δv. The aggregate of the moments of all the bi-poles in this element is again obtained by combining them all vectorially without regard to their different positions in the element. The resulting moment must be proportional to δv if that volume is small*; suppose it is $\mathbf{P}\delta v$, where of course $\mathbf{P}$ is a vector. This quantity $\mathbf{P}$ expresses the way in which the body is polarised; it is the polarisation *per* unit volume at the position, and in the language of physics is called

* 'Physically small.'

the *intensity* of the polarisation. This vector expresses all that we can know or recognise experimentally about the polarisation of the body.

If the bi-polar elements or molecules are distributed anyhow in all different directions, $\mathbf{P} = 0$; but if there is any degree of convergence of their axes to a definite direction, then $\mathbf{P}$ has a definite value different from zero.

59. Having now defined this quantity $\mathbf{P}$ which completely specifies the electric state of the polarised body, we can by its means determine the electric field in the neighbourhood of the body, without at present stopping to consider the actual method by which this polarisation is produced.

Each volume element δv_1 of the body is like a little bi-pole of moment $\mathbf{P}_1 \delta v_1$ and thus its potential at the point P is evidently

$$\phi = \frac{(\mathbf{P}\mathbf{r}_1)}{4\pi r^2}\,\delta v_1,$$

$\mathbf{r}_1$ denoting the unit vector along the direction of the radius r_1 from the position of the element δv_1 to P. The potential of the whole body at any point P is therefore

$$\phi = \frac{1}{4\pi}\,\Sigma\,\frac{(\mathbf{P}_1\mathbf{r}_1)}{r_1^2}\,\delta v_1,$$

wherein Σ denotes a sum taken over all the elements δv of the body. This may also be written in the form of an integral

$$\phi = \frac{1}{4\pi}\int \frac{(\mathbf{P}_1\mathbf{r}_1)}{r_1^2}\,dv_1,$$

which is the same as

$$-\frac{1}{4\pi}\int (\mathbf{P}_1\nabla)\,\frac{dv_1}{r_1},$$

the vectorial operator ∇ involving differentials with respect to the coordinates of P.

The intensity of force at the point P in the field can now be written down in an analogous manner. It appears as a vector—the negative gradient of the potential ϕ

$$\mathbf{E} = -\operatorname{grad}\phi.$$

One proviso, and an important one, has been omitted in the above statement: *the point P must be well outside the dielectric substance.*

60. The integral definitions here given necessarily involve some sort of continuity in the distribution of the polarisation intensity, and rather more than is actually the case in a real medium constituted of discrete molecules. To give them a definite sense in a strict mathematical theory we can however replace as in the first chapter the real medium by a perfectly continuous distribution of polarisation with the proper intensity **P** at each point. This hypothetical distribution is effectively equivalent to the real one at all points of space which are not too near it.

We may next enquire as to how close to the distribution does this representation of the force and potential by integrals remain valid. In this connection it must first be noticed that since we have assumed the nucleus of the bi-pole which forms the basis of this theory to be entirely confined within a molecule or atom it may reasonably be supposed that the law of its action as defined above remains valid up to within a physically small differential distance from the molecule, which is a length defined so as to include a large number of molecular diameters; in other words the effective combination of positive and negative elements of charge in the molecule into doublets in the manner specified above is valid up to within this small distance from the molecule. But the substitution of an effectively continuous charge distribution for the distribution of both positive and negative elements thus combined is valid to the same extent, so that we may conclude that the hypothetical distribution of polarisation effectively replaces the actual discrete one as regards its field up to within a physically small differential distance from the polarised body, this being the distance at which the actual distribution of the charge in any small volume element ceases to be irrelevant.

61. The next process in the development of the mathematical theory is to specify the electric field at points in the interior of the polarised medium. Waiving for the present any other difficulties involved in the extension, let us assume that the hypothetical continuous distribution of polarisation effectively replaces the old one at all points of the field however near to the medium it may be and let us examine the force and potential at the internal point. We might jump to the conclusion that in this case the force and potential are correctly represented by the integrals as given above for external points; but any attempt to use these definitions for internal points for which the corresponding integrands both become infinite must be preceded by a justification of their convergence at such points. If the integrals representing them are convergent at internal points, then the force and potential so defined will have definite meanings at such points.

From a physical point of view we see that the contribution to the values of these functions at any point from adjacent parts of the medium involves a large factor in the integral and we want to know whether their aggregate is comparable with that from the rest of the body. If this is the case the integrals are at best semi-convergent and the definitions are almost useless, because we do not know anything about the local configuration of the elements; it may be anything for all we know.

But we have already seen that the integral expressing the potential is absolutely convergent, so that on the present basis, the local contribution due to the continuous distribution of polarity near the point under investigation is negligible: strictly speaking the effects of these adjacent parts involve the dimensions of their volume linearly and thus in the aggregate their effect is negligible compared with that of the rest of the body. The physical way of stating this is, as we had it before, that the scooping out of a vanishingly small cavity round P makes no difference to the integral; the shape of the cavity does not matter so long as it is indefinitely small.

It is however otherwise with the integral for the force at P. The x-component of this force is in fact represented by the integral

$$-\int \operatorname{grad}_x (\mathbf{P}_1\nabla)\frac{dv_1}{r_1}$$

which, if we use (x_1, y_1, z_1) as the coordinates of the elements dv_1 and (x, y, z) as those of P, can be written in the form

$$\int dv \left[-\frac{\mathbf{P}_x}{r_1^{\,3}} + 3\mathbf{P}_x \frac{\overline{x - x_1}^2}{r_1^{\,5}} + \frac{3\,(x - x_1)}{r_1^{\,5}}\,(\overline{y - y_1}\mathbf{P}_y + \overline{z - z_1}\mathbf{P}_z)\right],$$

which is precisely of the type which is known to be semi-convergent. The local parts of the polarisation even in the hypothetical continuous medium thus have an effective influence on the value of the force. Thus our definitions of the force, at least, in the internal field by means of the effectively continuous distribution of polarisation breaks down! This does not however seriously disturb our formulation of the theory because we can proceed as in the first chapter to modify the definitions to make them more consistent with a physical theory.

62. The necessary modification will come better if we first obtain Poisson's transformation of the potential integral given above. We shall assume that the point P at which the field is investigated is well outside the dielectric substance, so that there is no doubt about an application of the above definitions and the consequent effectiveness of the continuous distribution of polarisation. If P is actually inside the medium we just remove a part of the medium inside a physically small cavity round P, so that it is still in free space. We can then adopt without any further hesitation the above definitions for the force and potential at P. The potential is in fact such that

$$4\pi\phi = -\int (\mathbf{P}_1\nabla)\frac{dv_1}{r_1},$$

but since

$$\nabla\frac{1}{r_1} = -\nabla_1\frac{1}{r_1},$$

where ∇_1 denotes the same vector differential operator as ∇ but taken with respect to the coordinates (x_1, y_1, z_1) of dv_1 instead of (x, y, z) as above, this is also

$$4\pi\phi = \int (\mathbf{P}\nabla_1)\frac{dv_1}{r_1},$$

the integrals in each case being taken throughout the volume of the polarised medium, excluding the part removed from the small cavity about P if it is made. A simple transformation by Green's theorem then shows that

$$\begin{aligned} 4\pi\phi &= \int dv_1 (\mathbf{P}\nabla_1)\frac{1}{r_1} \\ &= -\int \frac{\operatorname{div}\mathbf{P}}{r_1}\,dv_1 + \int \frac{\mathbf{P}_n}{r_1}\,df_1, \end{aligned}$$

where the surface integral is extended over the surface of the body (including the walls of the cavity if P is inside) and the volume integral over the volume of the body.

This means that the potential of this polarised body is the same as the gravitational potential of the mass distribution specified as:

(i) a volume density

$$\rho = - \operatorname{div} \mathbf{P}$$

throughout the body excluding the cavity if made; although, as a matter of fact, the cavity may be filled in with the continuation of this distribution throughout its volume without making any appreciable difference to the integral for ϕ which is convergent at the point;

(ii) a surface density $\sigma = \mathbf{P}_n$*

over the surface of the body and cavity.

All this applies only to a point outside the substance of the polarised medium. If the point is right outside the medium the values of the force and potential due to such a mass distribution are quite definite and are in fact identical with those already given on the more direct definition; this distribution of attracting charges or masses effectively replaces the distribution of polarisation as regards its action at all external points. It is however in the analysis of the field at internal points that this mode of treatment helps us.

63. If the point P is inside the polarised medium we can draw round it a small surface whose linear dimensions are physically small. The distribution of polarisation in the medium outside this surface can then as regards its action at P be effectively replaced by the continuous distribution of attracting masses just described. We thus see that the total field at P can be separated into distinct components. Firstly the volume distribution ρ outside the surface gives a definite force and potential at P, no matter what size or shape the cavity may be, provided only that it is physically very small. Similar remarks apply to the surface distribution σ on the outer surface of the body. But the distribution σ on the walls of and the distribution of molecules inside this cavity give a potential and force at P which depend entirely on the shape and size of the cavity even if it is very small and will in general be comparable with the other parts. This latter component of the field is however

* At an interface between two different dielectric media there is a surface distribution of density

$$\sigma = P_{n_1} - P_{n_2}.$$

a purely local part depending entirely on the molecular configuration round the point and the conditions of polarisation existing in them: as we do not know the local molecular configuration, which may be changing rapidly, we cannot know what this local part really amounts to; but we have succeeded in separating it from the main part of the action due to the rest of the body.

We now adopt the arbitrary course of simply neglecting this local molecular part of the field, so that we can confine our discussions entirely to that definite part of the force which is due to the medium as a whole, i.e. the molar part. This is merely following a usual method in physics and involves but a simple extension of the ideas underlying the Young-Poisson principle of the mutual compensation of molecular forcives employed in their theory of capillary actions*. It requires that such local forcives shall set up a purely local physical disturbance of the molecular configuration in the material, until it is thereby balanced. Another example of this principle is provided in the ordinary theory of elasticity where in addition to the local strain forces in an elastic medium there are the comparatively very powerful cohesive forces, which are however presumed to form an equilibrating system and not to affect the phenomena as a whole. It is fortunate that we can in this way eliminate the influence of the neighbouring elements.

We can therefore define the electric field inside the body as that field when the effect of the local part is omitted, and this definition will apply quite consistently to outside points as well. We have been able to separate the local part from the total, and the field of force of our subsequent discussion is that due to the rest. On this definition the integrals expressing the force and potential are always convergent and apply to inside as well as to outside points as they are then just like the corresponding functions of our former theory involving only ordinary volume and surface distributions of charge. The electric force is then always the gradient of the potential.

* Cf. Larmor, *Aether and Matter*, App. A; Young, "On the Cohesion of Fluids," *Phil. Trans.* (1805); Poisson, *Nouvelle Théorie de l'Action Capillaire* (Paris, 1831); Rayleigh, "On the theory of surface forces," *Phil. Mag.* 1883, 1890, 1892; especially 1892 (1), pp. 209–220; Van der Waals, *Essay on the continuity of the liquid and gaseous states.*

64. On these definitions the electric force vector **E** is still the gradient of the potential ϕ. Also since ϕ is due to the specified volume and surface distribution

$$\begin{aligned}\nabla^2\phi &= -\rho\\ &= +\operatorname{div}\mathbf{P},\end{aligned}$$

at each point of the field and at the boundary of the dielectric in air

$$\frac{\partial\phi_1}{\partial n_1} - \frac{\partial\phi_2}{\partial n_2} = -\sigma = -\mathbf{P}_n;$$

but $$\mathbf{E} = -\operatorname{grad}\phi = -\nabla\phi,$$

so that $$\operatorname{div}\mathbf{P} = \nabla^2\phi = -\operatorname{div}\nabla\phi = -\operatorname{div}\mathbf{E},$$

or $$\operatorname{div}(\mathbf{E}+\mathbf{P}) = 0.$$

The vector $(\mathbf{E}+\mathbf{P})$ is therefore a streaming vector, it satisfies the usual equation of continuity of incompressible fluid flow. We call it the vector of *electric displacement** and denote it by **D**. This electric displacement is the important vector of the theory: its importance lies in the fact that the flux or displacement through any surface only depends on its boundary so that we can take the flux as estimated as so much through a circuit. For we have

$$\operatorname{div}\mathbf{D} = 0,$$

so that if we take any closed surface f in the field we get

$$\int(\operatorname{div}\mathbf{D})\,dv = 0$$

taken throughout the region bounded by f. But by Green's theorem this consists of

$$\int \mathbf{D}_n\,df$$

together with the surface integrals arising from discontinuities when we pass into the polarised medium; these are the integrals of

$$-(\mathbf{D}_{n_1} - \mathbf{D}_{n_2})$$

over the parts of the surfaces concerned which are inside f, or

$$-\frac{1}{4\pi}(\mathbf{E}_{n_1} - \mathbf{E}_{n_2} + \mathbf{P}_{n_1} - \mathbf{P}_{n_2}),$$

which is zero: we thus have

$$\int \mathbf{D}_n\,df = 0.$$

* Maxwell, *Treatise*, I. p. 64. The vector D is the displacement proper.

Now suppose we have an unclosed surface f abutting on the closed curve s, then we can take another surface f' with the same curve s as boundary and the two surfaces f and f' together form a closed surface. Thus if n denote the normal to the element of either surface in a definite sense through the circuit, the above equation gives

$$\int_f \mathbf{D}_n df - \int_{f'} \mathbf{D}_n df' = 0,$$

and the result is as stated.

The real significance of these results does not however appear until we discuss the subject of electromagnetism; it is hidden by the more general circumstances under which the present theory has to be developed.

65. The general problem. We have so far merely discussed the fields of polarised media without any reference to the way the polarisation is created. As a rule however we cannot have dielectrics polarised unless they exist in an external electric field, i.e. in an imposed field due to an extraneous electrical system. The introduction of the dielectric into the field results in each element of it being turned into a little bi-pole in the manner above indicated and the total field of all these bi-poles has alone been under investigation, although as a matter of fact it merely represents the addition to the original field brought about by the introduction of the dielectric substance into it. For the general case therefore we must superpose on the field above investigated that original field which existed before the dielectrics were introduced and which we shall for the present suppose to be due to certain volume and surface densities ρ_0 and σ_0. The above discussion of the convergence of the force and potential integrals is not hereby affected, since the additional parts of these functions due to such a distribution are already known to have definite values at all points of space. The electric force is therefore still the gradient of a potential function. We now use ϕ_1 for the potential of the field above investigated, ϕ_0 for the potential of the original field, and ϕ for that of the total field; a similar suffix-notation is also adopted for the other quantities involved. We have now

$$\phi = \phi_0 + \phi_1,$$

and

$$\mathbf{E} = -\operatorname{grad}\phi = -\nabla\phi.$$

Thus $$\nabla^2\phi = \nabla^2\phi_0 + \nabla^2\phi_1,$$
and $$\nabla^2\phi_0 = -\rho_0, \quad \nabla^2\phi_1 = -\rho_1 = -\operatorname{div}\mathbf{P},$$
so that we now have $$\operatorname{div}(\mathbf{E}+\mathbf{P}) = \rho_0.$$
The induction or displacement vector is no longer a stream or solenoidal as we say; if we again use **D** to represent it we have
$$\operatorname{div}\mathbf{D} = \rho_0.$$

66. At a boundary surface of the dielectric medium which also carries a surface charge of density σ_0 we have
$$\frac{\partial\phi}{\partial n_1} - \frac{\partial\phi}{\partial n_2} = \left(\frac{\partial\phi_1}{\partial n_1} - \frac{\partial\phi_1}{\partial n_2}\right) + \left(\frac{\partial\phi_0}{\partial n_1} - \frac{\partial\phi_0}{\partial n_2}\right)$$
$$= -\sigma_1 - \sigma_0,$$
or $$-\mathbf{E}_{n_1} + \mathbf{E}_{n_2} = -\mathbf{P}_n - \sigma_0,$$
or $$-\mathbf{E}_{n_1} + (\mathbf{E}_{n_2} + \mathbf{P}_n) = -\sigma_0.$$
In free space, i.e. on side 1 of the general surface of discontinuity,
$$\mathbf{D} \equiv \mathbf{E};$$
in the dielectric medium however
$$\mathbf{D} = \mathbf{E} + \mathbf{P},$$
so that the above surface condition can be written as
$$\mathbf{D}_{n_1} - \mathbf{D}_{n_2} = \sigma_0.$$
The normal induction is discontinuous across the surface charge by an amount σ_0; if there is no surface charge
$$\mathbf{D}_{n_1} = \mathbf{D}_{n_2}.$$
Similar conditions are also found to hold at charged or uncharged surfaces of discontinuity in the dielectric medium itself, i.e. surfaces separating not the dielectric medium from a vacuum, but one medium from a second different one.

We have thus a complete specification of the field in terms of the electric force **E**, the electric induction or displacement **D** and the intensity of polarisation **P** where we know that at each point of the field
$$\mathbf{D} = \mathbf{E} + \mathbf{P}$$
in the vector sense. These three vectors give us the distribution of force, induction and polarisation; but any two of them are sufficient as the third is determined when the other two are known.

Although the vector **P** is perhaps the more fundamental physical one, we shall regard the first two vectors as the independent variables; they turn out to be the more significant ones of the theory.

67. We have now examined the electrical field under the conditions of the dielectric being present and having induced in it a polarisation of intensity **P** at each place. But how do we know what polarisation will be induced in a given dielectric substance and of what use is the above analysis? We evidently want an additional physical principle to complete the scheme.

The electric force at any point in the dielectric medium is **E** (neglecting the local part) and the electric displacement induced is **D**, and as we have pictured the affair the polarisation and therefore the displacement is conditioned by the electric field. Thus if there is to be any law about the matter at all one of these quantities is a function of the other. The simplest possible relation we could have is a simple proportionality so that if we double the cause we double the effect. Expressed mathematically a relation of this kind means that the components of the displacement are linear functions of the components of the electric force

$$\mathbf{D}_x = \epsilon_{11}\mathbf{E}_x + \epsilon_{12}\mathbf{E}_y + \epsilon_{13}\mathbf{E}_z,$$
$$\mathbf{D}_y = \epsilon_{21}\mathbf{E}_x + \epsilon_{22}\mathbf{E}_y + \epsilon_{23}\mathbf{E}_z,$$
$$\mathbf{D}_z = \epsilon_{31}\mathbf{E}_x + \epsilon_{32}\mathbf{E}_y + \epsilon_{33}\mathbf{E}_z,$$

or expressed shortly by a vector equation

$$\mathbf{D} = (\epsilon)\,\mathbf{E}.$$

For homogeneous media this relation would assume the simpler form

$$\mathbf{D} = \epsilon\mathbf{E}.$$

We might of course assume more generally that

$$\mathbf{D} = \epsilon\mathbf{E} + \epsilon_1\mathbf{E}^2 + \ldots,$$

but we presume that if ϵ is small the other terms beyond the first are negligible and we find that it fits the facts. In any case the simpler form is right for very small fields and anything more complicated is mathematically unworkable.

It might be thought that it would be better to take the polarisation as proportional to the total electric force *including* the local part. The local influences have however been regarded as equal and opposite actions and reactions occurring in and between the molecules concerned and cannot add anything to the total result in any definite direction. The presumption is that these local effects are erratic and cannot influence a vector effect at the place.

68. We can now complete our formulation of the scheme: we have in the vector sense

$$\mathbf{D} = (\epsilon)\,\mathbf{E}$$

as our physical relation; but

$$\mathbf{D} = \mathbf{E} + \mathbf{P},$$

and thus

$$((\epsilon) - 1)\,\mathbf{E} = \mathbf{P},$$

and **P** the electric polarisation intensity is thus also proportional to **E**.

Also on this theory

$$\mathbf{D} = (\epsilon)\,\mathbf{E} = -(\epsilon)\,\mathrm{grad}\,\phi,$$

and since

$$\mathrm{div}\,\mathbf{D} = \rho,$$

we have

$$\mathrm{div}\,\{(\epsilon)\,\mathrm{grad}\,\phi\} = -\rho_0,$$

which is the characteristic equation of the theory; the potential function ϕ must always satisfy this equation when there are dielectrics about.

Notice that for isotropic media this becomes

$$\frac{\partial}{\partial x}\left(\epsilon\frac{\partial\phi}{\partial x}\right) + \frac{\partial}{\partial y}\left(\epsilon\frac{\partial\phi}{\partial y}\right) + \frac{\partial}{\partial z}\left(\epsilon\frac{\partial\phi}{\partial z}\right) = -\rho_0.$$

The boundary conditions to apply at a surface of discontinuity in the medium on which there is also a surface charge of density σ_0 are obtained in the usual way; they are*

$$\phi_1 = \phi_2,$$

$$(\epsilon_1)\frac{\partial\phi_1}{\partial n} - (\epsilon_2)\frac{\partial\phi_2}{\partial n} = \sigma_0,$$

or if $\sigma_0 = 0$ simply

$$(\epsilon_1)\frac{\partial\phi_1}{\partial n} = (\epsilon_2)\frac{\partial\phi_2}{\partial n}*.$$

We are thus enabled to express everything in terms of the potential function of the field which must satisfy the above characteristic relations at each point of the field. We notice that these

* These conditions are perhaps not so happily expressed as they might have been. In the general case of anisotropic media the expression of the normal component of the induction vector in terms of the potential gradient requires great care.

relations are identical in form with those obtained on the less explicit basis of the Faraday-Maxwell theory, the constant ϵ being the same as there adopted. The electrical conditions are the same on either theory, the only difference is the more physical basis for the mode of action of the dielectric medium in modifying the electric field by which it is surrounded.

69. The electric displacement $\mathbf{D}$ is now a composite vector,

$$\mathbf{D} = \mathbf{E} + \mathbf{P},$$

containing two essentially distinct parts, both of which are presumably caused by the electric force intensity $\mathbf{E}$. The first part, the *aethereal displacement*, is present at each point of the field even if there are no dielectric media there, but the second one, the *material polarisation*, arises solely on account of, and in, these dielectrics. The second part is a true displacement of electricity, but all we can say of the first is that, somehow or other, whatever its ultimate nature may prove to be, it nevertheless behaves just as though it were a true displacement of electricity.

We must notice that the direction of the displacement $\mathbf{D}$ is in the positive direction of the electric force and is therefore in a direction away from the positive charge; further the vector $\mathbf{D}$ is a totality of displacement. If the rate of change of the displacement at any time t be denoted by the vector $\mathbf{R}$, then the amount displaced across any surface element df during the small time dt is $\mathbf{R}_n dt\,df$; during the finite interval since the initial instant t_0, required in setting up the field, the total displacement across the surface element is therefore

$$\mathbf{D}_n df = df \int_{t_0}^{t} \mathbf{R}_n dt$$

and this defines the vector $\mathbf{D}$ of our theory.

70. The mechanical relations of the dielectric field. In the previous chapter we found that the total energy in the electrostatic field was expressible in terms of the charge distribution in such a way that the amount of energy required to increase the volume density of charge at any point by $\delta\rho$ and the surface density by $\delta\sigma$ was given in the form

$$\delta W = \int \phi\, \delta\rho\, dv + \int_f \phi\, \delta\sigma\, df,$$

the first integral being taken throughout the whole of the field and the second over all those surfaces where there is a surface density.

Although the previous discussions have entirely ignored the dielectric medium and its effect on the electrical conditions of the system, the result just quoted is perfectly correct whatever the complexity of this dielectric medium; this results from the general definition of the potential function ϕ which we have given above.

Now the simple hypothesis of action through a medium regards the electric charges merely as manifestations of a varied condition in the whole of the medium throughout the field. In such a theory therefore the energy of the electrical system is distributed throughout the space of the field so that each element of volume furnishes a part to the total amount, which part depends solely on the electrical conditions existing in the element. Now on the Faraday-Maxwell form of the theory the essential specification of the conditions at any point of the field is involved in the vector $\mathbf{D}$, the electric displacement, so that the alteration of the energy produced in the system by a small arbitrary charge in its specification should be expressible in terms of the alteration of the conditions of the medium, viz. by $\delta\mathbf{D}$, at each point. But in the form of the theory adopted, this increase of displacement $\delta\mathbf{D}$ at each point of the field is effected by the agency of the electric force $\mathbf{E}$, which produces it and is alone effective in altering it, and if the analogy with material phenomena implied in our choice of names is valid the work done by the force intensity $\mathbf{E}$ in producing an additional virtual displacement $\delta\mathbf{D}$ throughout the small volume element dv would be

$$(\mathbf{E} \cdot \delta\mathbf{D})\, dv,$$

so that the total increase in the energy of the system should be expressible by the integral

$$\int (\mathbf{E} \cdot \delta\mathbf{D})\, dv$$

taken throughout the whole field.

71. Owing however to the great indefiniteness in our knowledge of the true nature of the vector $\mathbf{D}$, this deduction of an expression for the energy cannot be regarded as anything more than a mere analogy. It can however be shown that the result obtained on this analogy agrees exactly in the total amount with the former estimate. In fact we have at each point of space

$$\operatorname{div} \mathbf{D} = \rho,$$

and at points on surfaces on which there is a charge density σ

$$\mathbf{D}_{1n} - \mathbf{D}_{2n} = \sigma,$$

and thus the variations $\delta\mathbf{D}$ are connected with the variations $\delta\rho$ and $\delta\sigma$ by the conditions

$$\operatorname{div} \delta\mathbf{D} = \delta\rho$$

at each point of space and

$$\delta\mathbf{D}_{1n} - \delta\mathbf{D}_{2n} = \delta\sigma$$

at each surface distribution. Thus the first estimate of W is equivalent to

$$\int \phi \operatorname{div} \delta\mathbf{D}\, dv + \int_f \phi\, (\delta\mathbf{D}_{1n} - \delta\mathbf{D}_{2n})\, df,$$

and a simple transformation of the first integral by integration by parts shows that this is equal to

$$\delta W = -\int (\delta\mathbf{D} \,.\, \nabla)\, \phi\, dv$$

$$= \int (\mathbf{E} \,.\, \delta\mathbf{D})\, dv^*.$$

Thus if the change in the total energy of the whole system is distributed throughout the field with a density $(\mathbf{E} \,.\, \delta\mathbf{D})$ at each point, the total amount is consistent with our other theories. For the above-mentioned reason we have however no really definite theoretical basis for regarding this distribution of the energy as the actual one. Maxwell assumed that it was† and there are certainly many points in favour of his view. It is the simplest distribution which suits the case, and for this reason alone it might be regarded as the correct one. More we cannot say except that future developments confirm the assumption.

72. We have however seen that the vector $\mathbf{D}$ occurring in the above expressions is a composite vector containing a part due to

* To make this argument quite rigorous it is necessary to include a bounding surface in the field at a great distance from the origin and to examine the integral over it which results in the integration by parts. In any real case however the field will be always regular at infinity and this surface integral tends to vanish.

† *Treatise*, I. p. 167.

the actual electric displacements in the polarised dielectric media: we found in fact that

$$\mathbf{D} = \mathbf{E} + \mathbf{P},$$

so that the integral for the energy change becomes

$$\int (\mathbf{E},\ \delta\mathbf{E} + \delta\mathbf{P})\, dv = \tfrac{1}{2}\int \delta\mathbf{E}^2 dv + \int (\mathbf{E}\delta\mathbf{P})\, dv.$$

The first part of this expression represents the aethereal part of the energy which would be there even if there were no dielectric media in the field. But the second part is definitely associated with these media, and arises on account of the quasi-elastic connections between the electric particles constituting the polarisation and the material atoms, so that its complete analysis involves a detailed discussion of the mechanics of molecularly constituted media.

73. In dealing with matter containing a large number of particles, atoms of matter or their contained electrons. we must realise that each particle is in the general case subject to the action of certain forces which can however be divided into different types.

(i) *External forces.* These are the definitely controllable forces exerted directly by external systems: these would include the electric forces exerted by the field on those of the particles which are charged with electricity.

(ii) *Internal forces.* These are the forces exerted on the individual particle by all the other particles in the system.

(iii) *Reaction forces.* These are the forces of constraint exerted indirectly as a consequence of some condition or conditions restricting the motion of the system. These forces are characterised by the fact that they do no work on the particles on which they act so long as the restricting conditions which give rise to them are not violated.

Thus if in any small interval of time the total kinetic energy of the system of particles constituting the matter changes by δT we shall have by the principle of energy that

$$\delta T = \delta W_0 + \delta W_i,$$

where δW_0 is the work done during that time by the external forces and δW_i is the work done by the internal forces.

74. The kinetic energy T of the particles consists in the main of the energy (T_0) of the average drifts or orientations of the molecules; but there is also the average residue of energy (T_i) concerned with the part of the motion of the molecules which is devoid of any regularity (which is super-imposed on the regular motion sorted out) and of which we know nothing except its total quantity; this latter part is the thermal energy and is a function only of the temperature of the body; it is of course the only part that exists if there is no visible motion of the medium as a whole $(T_0 = 0)$.

Again the energy of the interactions of the various particles (δW_i) is also necessarily presented to us divided into different groups. There is a part involving the interaction with the particle under consideration of the other particles of the system at a finite distance from it, which integrates into an energy function of the applied mechanical forces exerted between the various elements of the system. Of the remainder of this energy a regular or organised part can be separated out which represents the energy of elastic stress (molecular and intra-molecular) and is a function of the deformation of the volume element treated as a whole (polarisation constituting also electric deformation): this stress arising from the immediate surroundings in part compensates for the element of mass under consideration, the applied mechanical forces aforesaid. The remaining usually wholly irregular part may be considered as compensated on the spot by other such forces of different origin that are not at present under review.

75. Using then T_0 for the kinetic energy of the organised motions, T_i for the internal heat energy, W_m for the work of the mechanical forces between different elements of the system, W_s for the work involved in the elastic stress, then the equation above can be written as

$$\delta T_0 + \delta T_i = \delta W_0 + \delta W_m + \delta W_s.$$

If we limit ourselves to the statical case with the dielectrics as a whole at rest, then $T_0 = 0$. Further, $\delta W_0 + \delta W_m$ represents the work done by all the directly observable mechanical forces on the different particles and may be treated as a single quantity δW_a so that in this case

$$\delta T_i = \delta W_a + \delta W_s,$$

or using V_s as the potential of the internal elastic forces we have, since $\delta W_s = -\delta V_s$,

$$\delta W_a = \delta T_i + \delta V_s,$$

so that the work done by the external forces goes partly to increase the kinetic energy of the internal motions and partly to increase the internal potential energy of elastic or dielectric strain: including these two under the single symbol E_i as internal energy we have

$$\delta W_a = \delta E_i.$$

76. In applying these considerations to the case of a mass of some dielectric substance in an electric field we shall find it convenient to separate the applied forces represented in δW_a into two parts: the mechanical forces applied generally as pressures on the outer boundaries of the media and the electrical forces exerted on the charges rigidly connected therewith. Denoting the work in these two parts by δW and δW_e respectively we shall have

$$\delta W_a = \delta W + \delta W_e,$$

so that $$\delta W_e = \delta E_i - \delta W.$$

This equation can of course be applied to any mass of the dielectric and in the sequel we shall in fact usually apply it to a small element of volume δv, which is of course assumed to be large enough to enclose a sufficient number of the particles to enable the statistical method to be applied to it.

77. But δW_e is easily calculated: in fact the energy required to establish any constituent bi-polar element at any point in the medium can be regarded as mathematically equivalent to the work required to separate the positive pole $+q$ from coincidence with the negative pole $-q$. If the intensity of force at the point is $\mathbf{E}$, supposed uniform in the neighbourhood of the element, and $\mathbf{e}$ represents the vector moment of the doublet finally established, this work is easily seen to be equal to

$$(\mathbf{Ee}).$$

Thus summing for all the doublets in the element of volume δv we have the total work done in polarising the element equal to

$$\delta W_e = \Sigma (\mathbf{Ee}).$$

In the present theory of polarised media we saw that the force $\mathbf{E}$ at any point internal to the medium consists of a definite molar and an irregular molecular part which we succeeded in separating by means of the ideal volume and surface densities of Poisson; the method consisting essentially in computing the force by combining opposed poles of neighbouring elements, instead of taking the single polarised element as the unit. It appears that the adjacent poles nearly compensate each other except as regards a simple volume density whose attraction has no local or molecular part and a surface density partly at the outer surface and partly at the surface of the cavity which contains the point under consideration. The effect of the latter surface density depending as it does wholly on the immediate surroundings is the molecular or cohesive part of the average forcive. It is the irregular part of the forcive on the contained element of the dielectric which arises from the excitation of neighbouring molecules and is expressed in terms of them alone. It is not transmitted by a material stress; but forms a balance on the spot with cognate internal molecular forcives of other types. Thus in seeking for the mechanical relations for the dielectric as a whole we shall be justified in neglecting this local part of the total force and its associated energy. We can thus use **E** as the electric force as defined above, omitting the local part; and it is then clear that **E** will be practically constant throughout the small volume element δv and thus

$$\delta W_e = (\mathbf{E}, \Sigma\mathbf{e});$$

but

$$\Sigma\mathbf{e} = \mathbf{P}\delta v,$$

and thus the work done in polarising the element δv to intensity **P** is

$$\delta W_e = (\mathbf{EP})\,\delta v,$$

and for the whole medium the work done is

$$W_e = \int (\mathbf{EP})\,dv.$$

This is the energy required by the system as a whole on account of the polarisation induced in it.

78. As explained above this energy is to be regarded as consisting partly in the mechanical potential energy of the polarisations of the elements of volume and partly in mechanical work done against internal quasi-elastic forces preventing displacement of the elementary charges ultimately constituting its polarisation. To effect a separation of the two parts thus involved we proceed by the method, usual in such problems, of varying the configuration of the system generally and calculating the coefficients of each part of the variation in the general expression for the virtual work thus obtained.

An infinitesimal displacement of the volume δv from a place where the field is $\mathbf{E}$ to where it is $\mathbf{E} + \delta\mathbf{E}$ involves a total change in W_e equal to

$$\delta W_e = [(\mathbf{E}\delta\mathbf{P}) + (\mathbf{P}\delta\mathbf{E})]\,\delta v$$

and then it is at once obvious that the second part with its sign reversed, viz.

$$-(\mathbf{P}\delta\mathbf{E})\,\delta v,$$

is the virtual work δW of the mechanical forces performed during the shifting of the element, for it is the part of δW_e which remains when the polarisation of the element is held rigid during the displacement so that no work is done in the internal degrees of freedom corresponding to the displacements involved in it.

The other part of the total energy δW_e expended during the displacement of the volume element δv corresponds to δE_i and thus

$$\delta E_i = (\mathbf{E}\delta\mathbf{P})\,\delta v,$$

or integrated throughout the system for the total. This represents work done against the quasi-reactions to the setting up of the polarisation. It has nothing whatever to do with the mechanical forces on the dielectric as a whole, but is stored up in the polarisation of the medium as internal energy of intra-molecular strain*.

79. We may now draw some important conclusions respecting the relations between $\mathbf{P}$ and $\mathbf{E}$, which follow directly from a simple application of the energy principle with the forms for the energy above determined. Confining ourselves to the element δv we see that the work supplied by it to outside systems, which it drives, in traversing any path is

$$-\int \delta W = dv \int (\mathbf{P}\delta\mathbf{E}),$$

the integrals being taken along the path. If $\mathbf{P}$ is a function of $\mathbf{E}$ so that the operation is reversible, this work must vanish for any

* The argument here employed is given implicitly by Larmor, *Phil. Trans.* A, 190 (1897); and in more detail for the cognate magnetic problem in *Proc. R. S.* 71 (1903), pp. 236–239.

closed cycle, otherwise energy would inevitably be created, either in the direct path or else in the reversed one, for the complete system of which dv is an element. The negation of perpetual motion therefore requires in this case that

$$(\mathbf{P}\,\delta\mathbf{E}) = d\phi$$

is a complete differential of some function of **E**. Moreover this function ϕ can only involve even powers of $(\mathbf{E}_x, \mathbf{E}_y, \mathbf{E}_z)$ for it must naturally remain unaltered if the direction of the electric force is reversed, and for weak fields it is practically quadratic. Implying this restriction, which proves sufficient in practice, we may therefore write

$$\phi = \tfrac{1}{2}\,[\epsilon_{11}'\mathbf{E}_x^2 + \epsilon_{22}'\mathbf{E}_y^2 + \epsilon_{33}'\mathbf{E}_z^2 + 2\epsilon_{12}'\mathbf{E}_x\mathbf{E}_y + 2\epsilon_{23}'\mathbf{E}_y\mathbf{E}_z + 2\epsilon_{31}'\mathbf{E}_z\mathbf{E}_x],$$

so that

$$\mathbf{P}_x = \epsilon_{11}'\mathbf{E}_x + \epsilon_{12}'\mathbf{E}_y + \epsilon_{13}'\mathbf{E}_z,$$
$$\mathbf{P}_y = \epsilon_{21}'\mathbf{E}_x + \epsilon_{22}'\mathbf{E}_y + \epsilon_{23}'\mathbf{E}_z,$$
$$\mathbf{P}_z = \epsilon_{31}'\mathbf{E}_x + \epsilon_{32}'\mathbf{E}_y + \epsilon_{33}'\mathbf{E}_z,$$

in agreement with the results of our previous speculations. But now we see that the doctrine of energy requires that there should be no rotational quality in the polarisation or that

$$\epsilon_{12}' = \epsilon_{21}', \quad \epsilon_{13}' = \epsilon_{31}', \quad \epsilon_{23}' = \epsilon_{32}',$$

relations not necessarily implied by our previous argument.

We can simplify the quantity ϕ by taking a proper choice of axes, and thereby reduce it to the form

$$\phi = \tfrac{1}{2}\,(\epsilon_1'\mathbf{E}_x^2 + \epsilon_2'\mathbf{E}_y^2 + \epsilon_3'\mathbf{E}_z^2).$$

This determines the principal electrocrystalline axes of the substance at each point: the law of polarisation is now given by

$$\mathbf{P}_x = \epsilon_1'\mathbf{E}_x, \quad \mathbf{P}_y = \epsilon_2'\mathbf{E}_y, \quad \mathbf{P}_z = \epsilon_3'\mathbf{E}_z,$$

which for isotropic media reduces simply to

$$\mathbf{P} = \epsilon'\mathbf{E}.$$

80. The total mechanical work done in establishing the field may now be calculated by building up the charge distribution ρ gradually in the presence of the dielectric media, the induced polarity simultaneously taking the appropriate value at each stage of the process. By the general definition of potential, the work done in bringing up small increments of charge $\delta\rho$ at each point of the field in which the potential is ϕ is

$$\delta W = \int \phi\, \delta\rho\, dv,$$

but generally $$\rho = \operatorname{div} \mathbf{D},$$

so that the small variation of $\mathbf{D}$ at any point of the field induced by the above change in ρ is defined by

$$\delta\rho = \operatorname{div} \delta\mathbf{D},$$

whence $$\delta W = \int \phi \operatorname{div} \delta\mathbf{D}\, dv,$$

and by Green's lemma this is

$$= -\int (\delta\mathbf{D} \,.\, \nabla)\, \phi\, dv + \int \phi\, \delta\mathbf{D}_n\, df,$$

the latter integral being taken over an indefinitely extended surface bounding the field. For a finite charge system this integral vanishes and thus

$$\delta W = -\int (\delta\mathbf{D} \,.\, \nabla)\, \phi\, dv;$$

but $$\mathbf{E} = -\nabla\phi,$$

so that $$\delta W = \int (\mathbf{E} \,.\, \delta\mathbf{D})\, dv,$$

as in the previous theory of Maxwell-Faraday. But now $\mathbf{D}$ is a composite function

$$\mathbf{D} = \mathbf{E} + \mathbf{P},$$

so that $$\delta\mathbf{D} = \delta\mathbf{E} + \delta\mathbf{P},$$

and thus $$\delta W = \int (\mathbf{E}\delta\mathbf{E})\, dv + \int (\mathbf{E}\delta\mathbf{P})\, dv,$$

and therefore the total work done in establishing the field can be calculated in the form

$$W = \int_0 \delta W = \int dv \int_0^E (\mathbf{E}\delta\mathbf{E}) + \int dv \int_0^P (\mathbf{E}\delta\mathbf{P})$$

$$= \tfrac{1}{2}\int \mathbf{E}^2 dv + \int dv \int_0^P (\mathbf{E}\delta\mathbf{P}).$$

The first term in this expression represents the electrical potential energy stored up in the electrical field on account of the distribution of electricity involved in the charges and polarisations. It may be regarded, as in the ordinary theory, as the potential function of the mechanical forces equivalent to the electrical attractions between the various charged and polarised elements of matter in the field. The second part represents the energy stored up in the dielectric media as a consequence of the strained condition involved in its polarisation.

For a linear law of polarisation we have

$$\mathbf{P} = (\epsilon')\,\mathbf{E},$$

so that the total energy is

$$W = \tfrac{1}{2}\int \mathbf{E}^2 dv + \tfrac{1}{2}\int (\epsilon_{11}'\mathbf{E}_x{}^2 + \epsilon_{22}'\mathbf{E}_y{}^2 + \epsilon_{33}'\mathbf{E}_z{}^2 + 2\epsilon_{12}'\mathbf{E}_x\mathbf{E}_y + \ldots)\,dv$$

$$= \tfrac{1}{2}\int (\epsilon_{11}\mathbf{E}_x{}^2 + \ldots + 2\epsilon_{12}\mathbf{E}_x\mathbf{E}_y + \ldots)\,dv,$$

where $\epsilon_{rr} = 1 + \epsilon_{rr}'$, $\epsilon_{rs} = \epsilon_{rs}'$ are the constants of the relations in paragraph 79. But this is

$$W = \tfrac{1}{2}\int (\mathbf{ED})\,dv$$

and as before we may regard it as distributed continuously throughout the field with a density at any point equal to

$$\tfrac{1}{2}\,(\mathbf{ED}),$$

the part

$$\tfrac{1}{2}\mathbf{E}^2$$

belonging to the aether and the rest to the matter.

81. A simple application of Green's analysis will show that in this case the total energy of the system is given also by

$$W = \tfrac{1}{2}\int \rho\phi\,dv + \tfrac{1}{2}\int_f \sigma\phi\,df,$$

which is the form suitable for applications based on the distance action theory. In fact in such a case

$$W = \tfrac{1}{2}\int (\mathbf{ED})\,dv$$

$$= -\tfrac{1}{2}\int (\mathbf{D}\nabla)\,\phi\,dv,$$

and this transforms by the analytical theorem to

$$W = -\tfrac{1}{2}\int \phi\,\operatorname{div}\mathbf{D}\,dv - \tfrac{1}{2}\int_f \phi\,(\mathbf{D}_{n_1} - \mathbf{D}_{n_2})\,df,$$

the first integral being taken throughout the whole of the field and the second over those surfaces where **D** is discontinuous as regards its normal component. The integral over the infinite boundary as usual tends to zero and is neglected altogether. But

$$\operatorname{div} \mathbf{D} = -\rho,$$

and
$$\mathbf{D}_{n_1} - \mathbf{D}_{r_2} = -\sigma,$$

so that
$$W = +\tfrac{1}{2}\int \rho\phi\, dv + \tfrac{1}{2}\int_f \sigma\phi\, df,$$

which is identical with the form obtained in the more restricted theory given in the first chapter. The present analysis indicates the necessary restrictions which must be placed on the method of argument used on the former occasion*.

82. The existence of the mechanical potential for a dielectric in any field necessarily implies of course the existence of mechanical forces on these media.

To deduce these forces from their potential energy function we have only to vary this function with regard to the geometrically possible displacements of the medium as a whole. In any such displacement however the polarisation must be kept constant, for it is determined wholly by the internal degrees of freedom of the medium. We see at once by giving the medium a small linear displacement that the force acting on it is the vector

$$-\operatorname{grad} W = \int \operatorname{grad} (\mathbf{P} \cdot \mathbf{E})\, dv,$$

the differentials however not operating on **P**.

This is the same as

$$\int \nabla (\mathbf{P} \cdot \mathbf{E})\, dv,$$

the operator ∇ not affecting **P**.

This determines the linear components of the force; there may also be a torque. To obtain its components give the body a small vectorial rotation $\delta\omega$: in this displacement the element dv goes

* It must however be emphasised that the restrictions are of theoretical interest only. The simple linear relation satisfies all the requirements of actual fact.

into a position where $\mathbf{E}$ has the value $\mathbf{E} + [\mathbf{E} . \delta\omega]$ and so the variation of the energy is

$$\delta W = \int (\mathbf{P} . [\mathbf{E} . \delta\omega])\, dv = -\int (\delta\omega . [\mathbf{E} . \mathbf{P}])\, dv,$$

and thus the couple is

$$-\int [\mathbf{EP}]\, dv.$$

83. There is however also a simple alternative method of deducing these results from the ideas involved in the theory of the polarisation in the medium. The mechanical force acting on a single doublet of moment $\mathbf{e}$ at a point in the field where the electric force intensity is $\mathbf{E}$ is represented by the vector

$$(\mathbf{e} . \nabla)\, \mathbf{E}.$$

Thus by simple addition over all the doublets in the volume element dv of the polarised body, we obtain that the linear force on the element of the medium is equal to

$$\begin{aligned} \Sigma\, (\mathbf{e} . \nabla)\, \mathbf{E} &= (\Sigma \mathbf{e} . \nabla)\, \mathbf{E} \\ &= (\mathbf{P} . \nabla)\, \mathbf{E} dv \end{aligned}$$

or $(\mathbf{P} . \nabla)\, \mathbf{E}$ per unit volume.

In this calculation the value taken for $\mathbf{E}$ excludes the local part of the total forcive which acts on any pole at the place. In such cases when we are dealing with a summation throughout the element of volume, the local actions in each charge element, which really arise from the other elements in the volume, must all be cancelled by complementary reactions, so that in the aggregate such terms will not occur.

This expression for the linear component of the mechanical force per unit volume on the medium is slightly different from that obtained from the energy, the difference being actually

$$(\mathbf{P} . \operatorname{curl} \mathbf{E}),$$

which is however zero if

$$\operatorname{curl} \mathbf{E} = 0,$$

i.e. if the electrical forces have a potential which is true in the present statical theory: the difference should however be noticed for future reference.

The elementary theory also shows that the simple bi-pole above mentioned is subject to a torque of amount

$$[\mathbf{e} \,.\, \mathbf{E}]$$

and thus we obtain for the torque on the element of volume of the polarised medium

$$[\mathbf{P} \,.\, \mathbf{E}]\, dv,$$

or in all a vector $[\mathbf{P} \,.\, \mathbf{E}]$ per unit volume. This is the same as the previous result.

84. On the transmission of force through the dielectric medium. The transmission hypothesis underlying the Faraday-Maxwell theory of electric action here under discussion regards all electrical phenomena merely as the result of a certain state of affairs established in the surrounding dielectric field. It would therefore be necessary on such a theory to regard the ponderomotive forces resulting from the attractions and repulsions between the charges as the terminal aspects of some state of stress in the medium between. We must now enquire as to a possible representation of the manner in which these forces are transmitted across the space between the bodies. The problem reduces itself to finding the state of stress in the elastic medium which agrees with the known boundary values. Unfortunately we do not know the nature of the elasticity of the medium, so we can only proceed tentatively. Faraday divined a very simple scheme which has high claims on account of its simplicity, but it is not general enough for our purposes.

After experimentally investigating the nature of the electric fields around conductors, Faraday came to the conclusion that the forces between them could be accounted for by a pull along the tubes of force, i.e. as if they were tending to contract like stretched elastic bands. This would obviously account for the attraction, but with it alone the elements of the transmitting medium could not be in equilibrium. He then saw somehow that in addition there must be an equal pressure in all directions perpendicular to the tubes: the tube tends not only to contract itself along its length but also to expand against a normal pressure all round. Under such a stress system the medium would be in equilibrium as regards its own parts but would transmit the force from one body to another.

Although Faraday invented this scheme he was unable to prove its reality: it was Maxwell who formulated it mathematically and put it in a very precise form.

85. As we are here going rather deeply into this question it might be as well to indicate how the stress in any medium is analysed*. Consider the medium in the neighbourhood of any point separated over a small interface there. Forces would then be required to be applied at each of the exposed surfaces to hold the medium in equilibrium: the same forces would be required for each of the two interfaces since they represent the action and reaction which were exerted between the parts of the medium on the two sides of the slit before they were separated. We shall consider these forces measured as so much per unit area. If for the slit made of area δf we know that the force required to hold the medium on either side of the interface in equilibrium is $\mathbf{T}\delta f$, then the vector $\mathbf{T}$ defines the stress for this particular direction of df. If we knew so much for every direction of the slit we should have a complete knowledge of the state of the stress at the point. It can however be shown that it is quite sufficient to know it for three directions only. Moreover if these three directions are the directions of a convenient set of coordinate rectangular axes and the components of the stress for the three directions of the elementary area perpendicular to these directions are

$$\begin{array}{l} T_{xx},\ T_{xy},\ T_{xz}, \\ T_{yx},\ T_{yy},\ T_{yz}, \\ T_{zx},\ T_{zy},\ T_{zz}, \end{array}$$

then the equations of equilibrium of the medium are

$$\frac{\partial T_{xx}}{\partial x} + \frac{\partial T_{xy}}{\partial y} + \frac{\partial T_{xz}}{\partial z} = \mathbf{F}_x,$$

$$\frac{\partial T_{yx}}{\partial x} + \frac{\partial T_{yy}}{\partial y} + \frac{\partial T_{yz}}{\partial z} = \mathbf{F}_y,$$

$$\frac{\partial T_{zx}}{\partial x} + \frac{\partial T_{zy}}{\partial y} + \frac{\partial T_{zz}}{\partial z} = \mathbf{F}_z,$$

$$T_{yz} - T_{zy} = \mathbf{G}_x,\quad T_{zx} - T_{xz} = \mathbf{G}_y,\quad T_{xy} - T_{yx} = \mathbf{G}_z,$$

* The general theory is discussed by Love, *Mathematical Theory of Elasticity* (2nd ed. Cambridge, 1906).

where **F** and **G** are the force and couple per unit volume exerted on the medium from outside.

The application of these results in the present theory is made by a reversal of the usual line of argument. If the applied forcive **F**, here supposed known, can be expressed in such a way that

$$\mathbf{F}_x = \frac{\partial X_x}{\partial x} + \frac{\partial X_y}{\partial y} + \frac{\partial X_z}{\partial z},$$

and similar forms are obtained for the other components, then we can say that the bodily forces on that part of the system enclosed in any surface f can be represented as the result of an elastic stress traction over the surface f: in fact

$$\int \mathbf{F}_x dv = \int (lX_x + mX_y + nX_z)\, df,$$

and thus we can at once write

$$T_{xx} = X_x, \quad T_{xy} = X_y, \quad T_{xz} = X_z,$$

as providing a sufficient solution of the problem.

86. Returning now to the problem of the stresses in the dielectric field, and assuming that for the present at least there are no surfaces of discontinuity*, the bodily forcive per unit volume acting on the charges and polarisations in the part of the system enclosed in any surface f is

$$\mathbf{F} = \mathbf{E}\rho + \nabla\,(\mathbf{EP}),$$

the differential operator in the second term affecting only the **E** function. But $\rho = \operatorname{div} \mathbf{D}$ so that

$$\mathbf{F} = \mathbf{E} \operatorname{div} \mathbf{D} + \nabla\,(\mathbf{EP}).$$

Considering now the x-component of this force alone we have in the general case

$$\begin{aligned} \mathbf{F}_x &= \nabla_x (\mathbf{PE}) + \mathbf{E}_x \operatorname{div} \mathbf{D} \\ &= \mathbf{P}_x \frac{\partial \mathbf{E}_x}{\partial x} + \mathbf{P}_y \frac{\partial \mathbf{E}_y}{\partial x} + \mathbf{P}_z \frac{\partial \mathbf{E}_z}{\partial x} + \mathbf{E}_x \frac{\partial \mathbf{D}_x}{\partial x} + \mathbf{E}_x \frac{\partial \mathbf{D}_y}{\partial y} + \mathbf{E}_x \frac{\partial \mathbf{D}_z}{\partial z}. \end{aligned}$$

Now remembering that **E** is the gradient of a potential and using the substitution

$$\mathbf{P} = \mathbf{D} - \mathbf{E},$$

* Any such may of course be replaced in the usual way by a continuous rapid transition layer.

we find that

$$\mathbf{F}_x = \mathbf{D}_x \frac{\partial \mathbf{E}_x}{\partial x} + \mathbf{D}_y \frac{\partial \mathbf{E}_x}{\partial y} + \mathbf{D}_z \frac{\partial \mathbf{E}_x}{\partial z} - \frac{1}{2}\frac{\partial}{\partial x}(\mathbf{E}_x{}^2 + \mathbf{E}_y{}^2 + \mathbf{E}_z{}^2) + \mathbf{E}_x \operatorname{div} \mathbf{D}$$

$$= \frac{\partial}{\partial x}(\mathbf{E}_x\mathbf{D}_x - \tfrac{1}{2}\mathbf{E}^2) + \frac{\partial}{\partial y}(\mathbf{E}_x\mathbf{D}_y) + \frac{\partial}{\partial z}(\mathbf{E}_x\mathbf{D}_z),$$

and similar expressions are obtained for the other components. We have therefore succeeded in our aim and may therefore conclude that the mechanical force acting on the part of any ordinary electrical system inside any closed surface f drawn in the field can be represented in the main by a system of interfacial tractions over the surface. These surface tractions are themselves the representatives of a stress system specified at any point by nine components

$$\begin{aligned}
T_{xx} &= \mathbf{E}_x\mathbf{D}_x - \tfrac{1}{2}\mathbf{E}^2, \\
T_{yy} &= \mathbf{E}_y\mathbf{D}_y - \tfrac{1}{2}\mathbf{E}^2, \\
T_{zz} &= \mathbf{E}_z\mathbf{D}_z - \tfrac{1}{2}\mathbf{E}^2, \\
T_{xy} &= \mathbf{E}_x\mathbf{D}_y, \qquad T_{yx} = \mathbf{E}_y\mathbf{D}_x, \\
T_{yz} &= \mathbf{E}_y\mathbf{D}_z, \qquad T_{zy} = \mathbf{E}_z\mathbf{D}_y, \\
T_{zx} &= \mathbf{E}_z\mathbf{D}_x, \qquad T_{xz} = \mathbf{E}_x\mathbf{D}_z,
\end{aligned}$$

or in matrix form

$$\left\| \begin{matrix} \mathbf{E}_x\mathbf{D}_x - \tfrac{1}{2}\mathbf{E}^2, & \mathbf{E}_x\mathbf{D}_y, & \mathbf{E}_x\mathbf{D}_z \\ \mathbf{E}_y\mathbf{D}_x, & \mathbf{E}_y\mathbf{D}_y - \tfrac{1}{2}\mathbf{E}^2, & \mathbf{E}_y\mathbf{D}_z \\ \mathbf{E}_z\mathbf{D}_x, & \mathbf{E}_z\mathbf{D}_y, & \mathbf{E}_z\mathbf{D}_z - \tfrac{1}{2}\mathbf{E}^2 \end{matrix} \right\|.$$

87. We must notice that in the general case of crystalline media for which

$$\mathbf{D}_x : \mathbf{D}_y : \mathbf{D}_z \neq \mathbf{E}_x : \mathbf{E}_y : \mathbf{E}_z$$

the stress obtained is not self-conjugate. It cannot therefore represent a stress system of simple mechanical type which is always self-conjugate. There is however no contradiction in principle; our material dielectric is now polarised and will thus be more than a mere medium of transmission as regards the mechanical force; it is subject to a definite additional type of strain (a polar torque) the reactions to which are sufficient to account for the more general form of stress found necessary to transmit the electrical actions. The inequality in the diagonal terms means that there is in fact a bodily torque whose x-component is for example equal to

$$\begin{aligned} T_{zy} - T_{yz} &= \mathbf{E}_y \mathbf{D}_z - \mathbf{E}_z \mathbf{D}_y \\ &= \mathbf{E}_y \mathbf{P}_z - \mathbf{E}_z \mathbf{P}_y \\ &= [\mathbf{EP}]_x \end{aligned}$$

per unit volume and this is precisely the torque which was determined from elementary principles.

88. To obtain a simpler expression* of this stress let us choose convenient axes. Choose the (x, y) plane as the plane of D and E and the x-axis as the internal bisector of the angle 2θ between these two vectors. The z-axis is chosen to form the usual right-handed system. Now

$$\begin{aligned} (\mathbf{D}_x, \mathbf{D}_y, \mathbf{D}_z) &= (D \cos\theta,\ D \sin\theta,\ 0), \\ (\mathbf{E}_x, \mathbf{E}_y, \mathbf{E}_z) &= (E \cos\theta,\ - E \sin\theta,\ 0), \end{aligned}$$

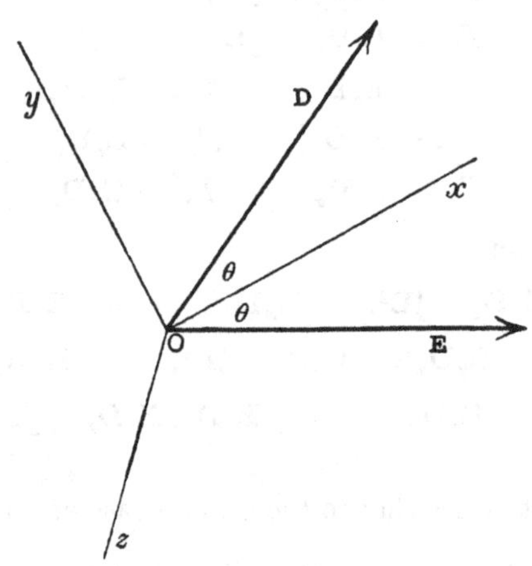

Fig. 6

and thus the matrix is now

$$\begin{vmatrix} ED\cos^2\theta - \frac{1}{2}E^2, & ED \sin\theta\cos\theta, & 0 \\ - ED \sin\theta\cos\theta, & - ED\sin^2\theta - \frac{1}{2}E^2, & 0 \\ 0, & 0, & -\frac{1}{2}E^2 \end{vmatrix}$$

which can be dissected into parts.

* Cf. Maxwell, *Treatise*, II. p. 280.

(i) The terms $-\frac{1}{2}E^2$ make a uniform hydrostatic pressure $\frac{1}{2}E^2$ throughout the medium.

(ii) The two terms in the diagonal give a torque per unit volume $ED \sin 2\theta$ and the remaining terms represent

(iii) a tension along the x-axis $ED \cos^2 \theta$,

(iv) a pressure (−) along the y-axis $-ED \sin^2 \theta$.

In a diagram they are

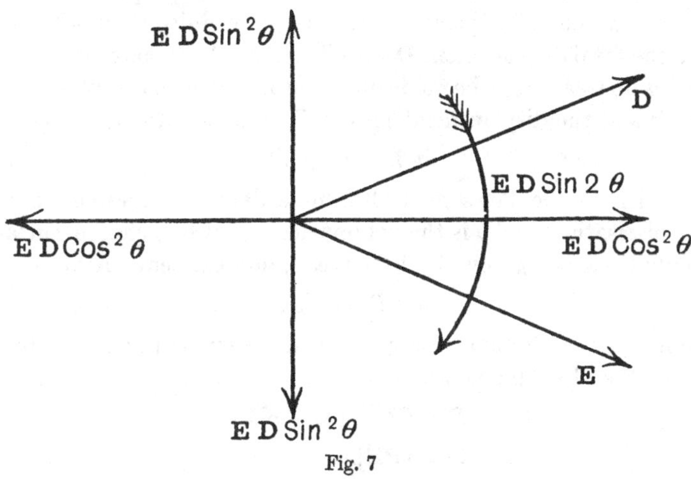

Fig. 7

This is the general result, but if we take the medium to be isotropic considerable simplification results. In this case $\theta = 0$ and the bodily torque vanishes. Moreover the other parts reduce to a pull or tension along the lines of force or induction equal to ED and a hydrostatic pressure all round equal to $\frac{1}{2}E^2$. This is identical with Maxwell's stress system, when the dielectric medium is free aether, which is the only case with which he deals*.

* The stress system here given is somewhat different from that adopted by recent authors, where the uniform pressure constituent occurs as $\frac{1}{2}$ (**ED**) in the simplest cases and as $\int(\mathbf{E}\,d\mathbf{D})$ in the more general case of a non-linear inductive law. This alternative form is usually derived from the general theory of electrostatic dielectric forces originated by Korteweg (*Wied. Ann.* 9 (1880)), formulated in general terms by von Helmholtz (*Wied. Ann.* 13 (1882); *Abhandl.* I. p. 798) and further developed by Lorberg, Kirchhoff (*Wied. Ann.* 24, 25 (1885); *Abhandl.*, Nachtrag, p. 91), Hertz (*Wied. Ann.* 41 (1890)) and others. This general theory, based on the method of energy, is usually made to lead to the aforementioned type of stress, but the

89. The reduction of the bodily forcive on dielectric media to a representation by means of an imposed stress system, which we have just discussed, is valid only in so far as the medium is without discontinuities. When there are in the electric field interfaces of transition between different media, there will also exist surface tractions on them which may be evaluated either as the result of the Maxwellian tractions towards the two sides of the interface or by considering an actual, somewhat abrupt, interface to be the limit of a rapid continuous variation of the properties of the medium which takes place across a layer of finite thickness. Let then the total displacement **D** with its circuital characteristic where there is no free charge be made up of the dielectric material polarisation **P** and the displacement proper **E**. We have then

$$\operatorname{div} \mathbf{E} = (\rho + \rho'),$$

wherein ρ' is the Poisson ideal volume density corresponding to the polarisation, and ρ is the volume density of free charge, surface distributions being now, by hypothesis, non-existent. Also

$$\operatorname{div} \mathbf{P} = -\rho'.$$

The mechanical forcive acting on the dielectric is per unit volume a force **F** and couple **G** where

$$\mathbf{F} = (\mathbf{P}\,.\,\nabla)\,\mathbf{E} + \rho\mathbf{E},$$

$$\mathbf{G} = [\mathbf{PE}].$$

The linear force acting on the whole transition layer is the value of $\int \mathbf{F}dv$ integrated through it. This integral is finite although the volume of integration is small, on account of the large values of the differential coefficients which occur in the expression of $\nabla\mathbf{E}$. To evaluate it we endeavour by integration by parts to reduce the magnitude of the quantity that remains under the sign of volume

analytical argument by which this is obtained has been criticised by Larmor (*Phil. Trans.* A, 190 (1897), p. 280). Cf. also Livens, *Phil. Mag.* 32 (1916), p. 162, who shows that, properly interpreted, it can only lead either to the type of stress given in the text or to an altogether impossible one. This criticism appears however to have been entirely overlooked and the method is tacitly accepted and reproduced by all recent writers on the subject. Cf. Cohn, *Das electromagnetische Feld*, p. 87 (Leipzig, 1900); Abraham, *Die Theorie der Elektrizität*, I. p. 434 (2nd ed. Leipzig, 1907); Jeans, *Electricity and Magnetism*, p. 172 (1st ed. Cambridge, 1908); also the articles by Lorentz, Pockels and Gans in the *Encyclopädie der mathematischen Wissenschaften*, Bd. v.

integration, so that in the limit we may be able to neglect that part: thus we obtain

$$\int \mathbf{F}\,dv = \int_f \mathbf{E}\,.\,\mathbf{P}_n\,df + \int (\rho + \rho')\,\mathbf{E}\,dv.$$

Now by the definition of the electric force **E** it is the force due to a volume distribution of density $\rho + \rho'$ and to extraneous causes; so that in the limit when the transitional layer is indefinitely thin, we have, by Coulomb's principle,

$$\int (\rho + \rho')\,\mathbf{E}\,dv = \tfrac{1}{2}\int (\sigma + \sigma')\,(\mathbf{E}_1 + \mathbf{E}_2)\,df$$

$$= \tfrac{1}{2}\int (\mathbf{E}_{1_n} - \mathbf{E}_{2_n})\,(\mathbf{E}_1 + \mathbf{E}_2)\,df,$$

$\mathbf{E}_1$, $\mathbf{E}_2$ being the vectors of electric force on the two sides of the layer and $\mathbf{E}_{1_n}$, $\mathbf{E}_{2_n}$ the normal components of these forces, all measured towards the side 1, while σ and σ' are the surface densities constituted in the limit by the aggregates of ρ and ρ' respectively taken throughout the layer. Hence in the limit

$$\int \mathbf{F}\,dv = \left|\int \mathbf{P}_n \mathbf{E}\,df\right|_1^2 + \tfrac{1}{2}\int (\mathbf{E}_{1_n} - \mathbf{E}_{2_n})\,(\mathbf{E}_1 + \mathbf{E}_2)\,df.$$

Thus the electric traction on the interface of transition may be represented by a pull towards each side, along the direction of the resultant force **E**; this pull is, on the side 2, of intensity

$$\mathbf{P}_{2_n}\mathbf{E}_2 + \tfrac{1}{2}(\mathbf{E}_{1_n} - \mathbf{E}_{2_n})\,\mathbf{E}_2,$$

or what is the same thing

$$\mathbf{P}_{2_n}\mathbf{E}_2 - \tfrac{1}{2}(\mathbf{P}_{2_n} - \mathbf{P}_{1_n} - \sigma)\,\mathbf{E}_2 = \tfrac{1}{2}(\sigma + \mathbf{P}_{1_n} + \mathbf{P}_{2_n})\,\mathbf{E}_2,$$

in the direction of $\mathbf{E}_2$; on the face 1 the pull is

$$\tfrac{1}{2}(\sigma - \mathbf{P}_{1_n} - \mathbf{P}_{2_n})\,\mathbf{E}_1,$$

now in the direction of $\mathbf{E}_1$. As the tangential component of the electric force **E** is under all circumstances continuous across the interface the total traction on both sides is along the normal and equivalent to

$$\tfrac{1}{2}(\mathbf{P}_{1_n} + \mathbf{P}_{2_n})\,(\mathbf{E}_{2_n} - \mathbf{E}_{1_n}),$$

together with tractions $\tfrac{1}{2}\mathbf{E}_2\sigma$, $\tfrac{1}{2}\mathbf{E}_1\sigma$ acting on the true charge σ, all the quantities being now measured positive in any the same direction. In the case where there is no surface charge this simply reduces to a normal pull towards the side 1 of amount

$$\tfrac{1}{2}\mathbf{P}_{2n}{}^2 - \tfrac{1}{2}\mathbf{P}_{1n}{}^2.$$

When the interface is between a dielectric 1 and a conductor 2, the traction is only towards the side 1 and is equal to $\frac{1}{2}(\sigma - \mathbf{P}_{1_n})\mathbf{E}_1$, or $\frac{1}{2}\mathbf{E}_1{}^2$, per unit area, along the normal, which is now the direction of the resultant force.

90. Thus under the most general circumstances as regards electric field, the forcive on the material due to its electric excitation consists of the interfacial tractions thus specified together with a force **F** and a torque **G** per unit volume given by the above formula, viz.

$$\mathbf{F} = (\mathbf{P}\nabla)\,\mathbf{E},$$

$$\mathbf{G} = [\mathbf{PE}].$$

In the case of a fluid medium, the bodily part of the forcive produces and is compensated by a fluid pressure

$$\int (\mathbf{P}d\mathbf{E}),$$

where **P**, being polarisation induced by the electric force **E**, is for a fluid in the same direction as **E** and a function of its magnitude. This pressure will be transmitted statically in the fluid medium to the interfaces (i.e. a reacting pressure $\int(\mathbf{P}d\mathbf{E})$ exerted by the interface will keep the medium in internal equilibrium); combining it there with the surface tractions proper, it appears that the material equilibrium of fluid media is secured as regards forces of electric origin, if extraneous force is provided to compensate a total normal traction towards each side of each interface, of intensity

$$-\tfrac{1}{2}\mathbf{P}_n{}^2 - \int (\mathbf{P}d\mathbf{E}),$$

or when the interface is between a charged conductor and dielectric the extraneous force necessary is

$$\tfrac{1}{2}(\sigma - \mathbf{P}_n)\,\mathbf{E} + \int (\mathbf{P}d\mathbf{E}) = \tfrac{1}{2}\mathbf{E}^2 + \int (\mathbf{P}d\mathbf{E}).$$

In the case usually treated, in which a linear law of induction is assumed, so that the relation between **P** and **E** is

$$\mathbf{P} = (\epsilon - 1)\,\mathbf{E},$$

the mechanical result of the electric excitation of the fluid medium is easily shown to be the same as if each interface were pulled towards each side by a Faraday-Maxwell stress made up of a pull

$\frac{\epsilon}{2}\mathbf{E}^2$ along the lines of force and an equal pressure in all directions at right angles.

This imposed geometrical self-equilibrating stress system would not however be an adequate representation of the mechanical forcive in a solid medium; for then the bodily forcive, instead of being wholly transmitted, is in part balanced on the spot by reactions depending on the solidity of the material. The forcive acting on isotropic material may however in every case, whether the induction follows a linear law or not, be expressible as an extraneous or imposed system, made up of a bodily hydrostatic pressure $\int(\mathbf{P}d\mathbf{E})$ (which in the case of a fluid only relieves the ordinary fluid pressure and so diminishes the compression) together with normal tractions on the interfaces between dielectric media, of intensity $-\frac{1}{2}\mathbf{P}_n{}^2 - \int(\mathbf{P}d\mathbf{E})$ acting towards each side, and tractions

$$\tfrac{1}{2}\mathbf{E}^2 + \int(\mathbf{P}d\mathbf{E})$$

on the surfaces of conductors acting towards the dielectric.

91. On electric displacement*. Since the vectors of the present theory satisfy exactly the same conditions as those of the original Maxwell-Faraday theory, they must ultimately represent the same quantities. But the theory just developed is based on elementary physical notions regarding the behaviour of the dielectric medium when introduced into an electric field, and by means of it therefore we should be able to obtain some insight into the physical nature of 'electric displacement.' This is best accomplished by considering a particular problem, viz. that of a parallel plate condenser with large surfaces with equal positive and negative charges; a plane slab of some dielectric substance (constant ϵ) is inserted parallel to the plates and the air spaces are regarded as vacua. If then the surface densities of charge on the plates of the condenser are $\pm\,\sigma$, the electric force is σ all across the air spaces and is $\frac{\sigma}{\epsilon}$ in the dielectric. (The lines of force go

* Cf. Larmor, *Aether and Matter*; Lorentz, *Versuch einer Theorie der electrischen und optischen Erscheinungen in bewegten Körpern.*

straight across by symmetry and at the surface of the dielectric $\epsilon\mathbf{E}_2 = \mathbf{E}_1$.)

Now on our present views the dielectric substance is polarised, the molecules have a positive and a negative pole, and owing to the presence of the field the axes have a convergence towards a definite direction, viz. straight across between the plates, so that their moments no longer cancel. The intensity of polarisation of the medium is thus at each point directed straight across between the plates.

If now we consider a small rectangular volume element of the substance parallel to the lines of force the little molecular moments in it can be summed up into a uniform polarisation, or the irregular molecular distribution can be smoothed out into a uniform average. We thus see that the polarisation of all the molecules in it is equivalent to a small polar distribution in the volume, which is just the same as if it had a positive charge of density $+\sigma_t$ on one end and a negative charge of density $-\sigma_t$ on the other. At least this is an effective representation of the matter. It does not mean that we assert that there is an actual charge on each end of the little element but that the aggregate of the polarisation in the element can be replaced by these charges when investigating its action at external points. The essential thing for this purpose is the electric moment of the element and any distribution giving the right moment is an effectively correct one.

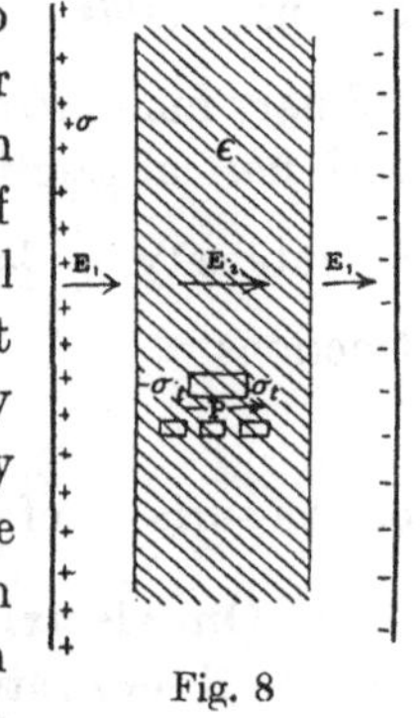
Fig. 8

Now by combining all these small rectangular elements so polarised into the finite piece of dielectric we see that there will be an uncompensated part* of the surface density (which is not necessarily the same for each element) where one rectangular block abuts on the next one, and at an end at the boundary of the dielectric itself there remains the complete surface polarity. This amounts to what we have called the ideal electric distribution of Poisson; the outstanding parts throughout the medium correspond to the volume density and the complete polarity remaining at the surface of the

* In the particular case examined this uncompensated charge is zero.

medium corresponds to the ideal surface density. Regarded in this way it is obvious that this theoretical distribution and the actual one will not give the same field in the immediate neighbourhood of an element of the substance. The ideal distribution has been smoothed out from the other and it is only at a distance that it is effectively equivalent to the actual polarity.

In the example under immediate consideration the field in the dielectric is uniform and so the intensity of polarisation will also be uniform throughout the medium: thus the charges on the ends of adjacent small elements will be the same and thus when put together there will be no uncompensated polarity: we shall merely have a surface density of ideal electric charge $-\sigma'$ on one face of the dielectric and $+\sigma'$ on the other. In the old-fashioned way of describing these things σ (the charge density on the plates of the condenser) might be called the *free* charge and σ' the *bound* charge (as it cannot be moved); σ' is only the end aspect of the polarisation in the medium which has a counterpart at the other side of the medium and they cancel across.

On our theory σ' is equal to the normal component of the polarisation at the surface and this is

$$(\epsilon - 1)\,\mathbf{E}_2;$$

but since $\mathbf{E}_2 = \dfrac{\mathbf{E}_1}{\epsilon} = \dfrac{\sigma}{\epsilon}$ we have

$$\sigma' = \left(1 - \frac{1}{\epsilon}\right)\sigma,$$

so that

$$\sigma - \sigma' = \frac{\sigma}{\epsilon}.$$

92. Let us now examine another point. The polarisation of the element can be expressed by saying that an electric displacement in the element from one end to the other has taken place. Initially the positive and negative charges effectively coincide and cancel, but on the application of an electric field they are separated and the electric moment can be considered to arise from the electric displacement of one charge relative to the other. We can at least theoretically imagine it to be like this. There is thus an actual movement of electric charge. Essentially the movement consists in the molecules being really strained round a bit, but when we aggregate these up for the small rectangular volume element as before, the effect is the same as if the positive charge were moved from one face of the volume element to the opposite one. If this is the case, how can we measure the displacement? The proper measure is the product of each charge element by the distance through which it is moved and the total sum of the quantities so obtained, because if we moved the same charge in each case through half the length it ought to give half the measure of the displacement. Thus the total electric displacement in our small rectangular volume element of end area $\delta\epsilon$ and length δl is

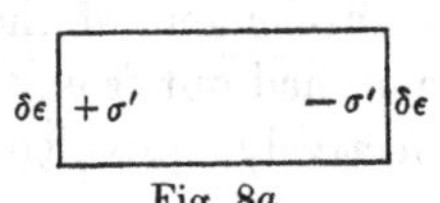

Fig. 8a.

$$\sigma'\delta\epsilon \times \delta l = \sigma'\delta\tau,$$

where $\delta\tau$ is the volume of the element; but this is the moment produced in the element. Thus an effective measure of the displacement in the volume element is the intensity of polarisation multiplied into the volume.

Thus for the slab of dielectric in the example considered the result of the total electric displacements in the medium is merely to displace a surface charge σ' from one side of the slab to the other straight across. The flux of electricity measured in this way is a true electric displacement.

But on the Faraday-Maxwell theory of electric action the electric displacement in any small volume dv is taken to be $\epsilon\mathbf{E}$ which is equal to

$$(1 + \epsilon')\mathbf{E}dv = \mathbf{E}dv + \mathbf{P}dv,$$

so that in addition to the true electric displacement represented by the term $\mathbf{P}dv$ as in the present theory there is something else which still exists in empty space when there are no dielectrics or conductors present so that it cannot possibly be ascribed to an electric displacement. To distinguish this fictitious part of the total flux vector of the theory we call it *aethereal* displacement. It has the same properties as the true electric part.

Thus in order to retain the analogy between the simple displacement theory of Maxwell and the polarisation theory just developed we must introduce this new type of displacement, and the total electric displacement of Maxwell includes then the true electric displacement of the present theory and the aethereal displacement.

93. The real significance of the matter is however best exhibited in another manner. Consider again the condenser with the dielectric slab and the process involved in charging it. As far as we are at present concerned the condenser may be charged by transferring a positive charge $+Q$ round a wire connecting the two plates thereby leaving, in defect, a charge $-Q$ on the one plate and creating an excess of charge $+Q$ on the other plate. While this is being accomplished a displacement is taking place in the dielectric (the polarisation is being gradually set up) and a charge Q' is displaced across the medium from one side to the other. This is all the electric motion that takes place; an actual charge Q moves round the wire and a change of polarisation in the dielectric corresponds to a motion of a charge Q' across the dielectric.

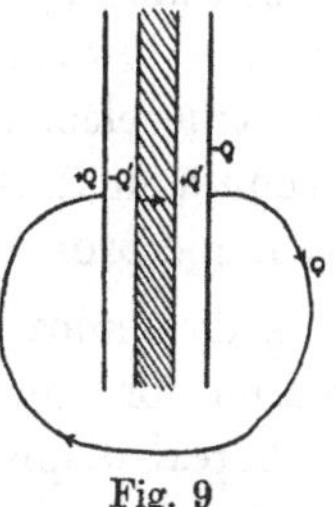

Fig. 9

But since $Q = A\sigma$ and $Q' = A\sigma'$, where A is the equal area of the parallel faces of condenser plates and dielectric (the charge is assumed uniformly distributed), we have that

$$Q' = \left(1 - \frac{1}{\epsilon}\right) Q,$$

so that the displacement of charge across the dielectric is not equal to that round the wire: and there is nothing at all in the free spaces between.

To avoid the complexity of the circumstances in this and other cases Maxwell assumed that the displacement which takes place during any electric change is always circuital; that is it always takes place in complete circuits.

In the example above, he would therefore postulate a hypothetical total displacement equal to Q in the air and $\frac{Q}{\epsilon}$ in the dielectric; this being all that is required to complete the flow of the quantity Q all round. Estimated per unit volume this would mean adding a displacement $\mathbf{E}$ at each point of space between the condenser plates. (It is assumed that $\mathbf{E} = 0$ everywhere except between the plates.)

94. This is easily seen to be the general result. If $\mathbf{E}$ is the force intensity at any point of an electric field, Maxwell's theory adds a displacement equal to **E** at that point to any true electric displacement that may occur there. If we do this then the flux of displacement is always in closed cycles. This additional displacement is not true electric displacement at all, as it exists at points in a vacuum; it is an aethereal displacement possessing all the electrodynamic properties of true electric flux.

A dynamical theory of electromagnetic actions should give a reason for this action in the aether, for the existence of this aethereal displacement which has the same properties as a flow of electricity but is not itself a flow of electricity. The hypothesis is however experimentally correct and it simplifies the theory immensely, and there we shall leave it for the present.

On this view of the matter the aether is to be regarded as the seat of part of the energy associated with any electrostatic field, viz. that part associated with the production of the aethereal displacement. On a previous analogy this part may be taken as distributed throughout the field with a density at any point equal to

$$\tfrac{1}{2}(\mathbf{E}\,.\,\mathbf{E}) = \tfrac{1}{2}\mathbf{E}^2,$$

a result which is verified by the fact that all the energy in the field is located in the aether if no dielectric medium is present.

95. The relation of inductive capacity to density. One of the most successful ways of testing a constitutional theory of the present type is to formulate on the same basis the connection between the constitutional relations involved in it and the physical or chemical constitution of the medium. In the present case the whole constitutive character of the theory is involved in the one constant introduced in it, viz. the specific inductive capacity ϵ. If therefore we can formulate a connection between this constant and the constitution of the medium we shall have a definite means of testing the general validity of our theory. It is quite easy to obtain a relation between the constant ϵ and the density of the medium in certain simple cases and we shall find that it agrees very well with our experimental knowledge on the same question.

Let the dielectric medium contain n molecules per unit volume, these molecules being presumed to be merely concentrated when a change of density of the medium occurs. Each of the molecules becomes polarised to a moment $\mathbf{p}$ by the field of the electric force;

this field is made up of the extraneous exciting field and that of the polarised molecules themselves; the latter again consists of a part arising from the polarised medium as a whole and a part involving only the immediate surroundings of the point considered; to obtain an estimate of these various parts let us consider again the method of their separation.

96. The total electric force acting on a single molecule is derived from the aggregate potential

$$\phi = \frac{1}{4\pi} \Sigma \, (\mathbf{p}\nabla) \frac{1}{r}.$$

This potential, when the point considered is inside the polarised medium, involves the actual distribution of the surrounding molecules; and thus the force derived from it changes rapidly, at any instant of time, in the interstices between the molecules. But when the point considered is outside the polarised medium, or inside a cavity formed in it (whose dimensions are large compared with molecular distances) the summation in the expression for ϕ may be replaced by continuous integration; so that $\mathbf{P}$ denoting the intensity of polarisation in the molecules of the dielectric medium

$$\phi = \frac{1}{4\pi} \int (\mathbf{P}\nabla) \frac{dv}{r},$$

and the force thus derived is perfectly regular and continuous. This expression may be integrated by parts since, the origin being now outside the region of the integral, no infinities of the function to be integrated occur in that region. Thus

$$4\pi\phi = \int \frac{\mathbf{P}_n}{r} \, df - \int \frac{\operatorname{div} \mathbf{P}}{r} \, dv,$$

that is, the potential at points in free aether is due to Poisson's ideal volume density $\rho = -\operatorname{div} \mathbf{P}$ and a surface density $\sigma = \mathbf{P}_n$. When the point considered is in an interior cavity, this surface density is extended over the surface of the cavity as well as over the outer boundary. Now when it is borne in mind that, at any rate in a fluid, the polar molecules are in rapid movement and not in fixed positions which would imply a sort of crystalline structure, it follows that the electric force on a molecule in the interior of the material medium, with which we are concerned, is an average force involving the average distribution of these molecules, and is therefore properly due to an ideal continuous density like Poisson's, even as regards the elements of volume which are very close up to the point considered. To compute the average force which causes the polarisation of a given molecule we have thus to consider that molecule as situated in the centre of a spherical cavity whose radius is of the order of molecular dimensions; and we have to take account of the effect of a Poisson averaged continuous local polarisation surrounding the molecule, whose intensity increases from nothing at a certain distance from the centre up to the full amount **P** at the limit of the molecular range, this intensity being practically uniform in direction and a function of the distance only.

We therefore assume spherical stratification in the distribution of the Poisson ideal volume density near the point under investigation. To estimate the effect of an elementary shell in this stratification, the charge in it can be reckoned as a surface density on it of intensity

$$\delta\mathbf{P}\cos\theta,$$

$\delta\mathbf{P}$ denoting the small increment of **P** as we pass through the shell, and θ the polar angle between the direction of **P** and the normal at the point of the shell. This shell thus contributes a force at the centre in the direction of **P** equal to

$$\frac{1}{4\pi}\iint\frac{r^2\delta\mathbf{P}}{r^2}\cos\theta\sin\theta\,d\theta\,d\phi$$

$$=\tfrac{1}{3}\delta\mathbf{P}.$$

Thus on the whole the local part of the forcive is

$$\tfrac{1}{3}\int_0^P\delta\mathbf{P}=-\tfrac{1}{3}\mathbf{P}.$$

The force polarising the molecules is therefore

$$\mathbf{E}+\tfrac{1}{3}\mathbf{P},$$

E denoting the total electric force. Now if the polarisation produced be presumed to be proportional to the polarising force

$$\mathbf{p}=\epsilon'(\mathbf{E}+\tfrac{1}{3}\mathbf{P}),$$

and thus since $\mathbf{P}=\Sigma\mathbf{p}=n\mathbf{p},$

we have $\mathbf{P}=n\epsilon'(\mathbf{E}+\tfrac{1}{3}\mathbf{P}),$

and by the definition of ϵ we have

$$\mathbf{P}=(\epsilon-1)\,\mathbf{E}.$$

Thus $$\epsilon - 1 = n\epsilon'\left(1 + \frac{\epsilon - 1}{3}\right) = n\epsilon'\left(\frac{\epsilon + 2}{3}\right).$$

Thus $$\frac{\epsilon - 1}{\epsilon + 2} = \frac{n\epsilon'}{3},$$

from which we see that the function

$$\frac{\epsilon - 1}{\epsilon + 2}$$

must be proportional to the density of the medium. This is the usual Lorentz formula* which has been satisfactorily verified in numerous cases.

97. Pyro- and piezo-electricity†. We have generally assumed in the preceding discussions that the elements of all dielectric media are always permanently neutral as regards their electrical effect on external systems so long as they are not under the influence of an external polarising field. This would imply that anything in the nature of permanent polarity, which is such an important feature in the correlative subject of magnetism, is non-existent, or at least negligible, in the electrical case. Whether any such presumption is really justifiable it is difficult to say, but there are certain phenomena which seem to suggest at least the possibility that it is not valid in every case.

Several substances like quartz and tourmaline which crystallise in asymmetric forms always appear to be polarised immediately after their temperature is changed, and in opposite directions according as it is raised or lowered. The polarisation exhibits itself mainly as an apparent separation of charge on the outer surface of the piece of the substance under investigation, one part of which appears positively charged and the opposite negatively charged. This is the phenomenon of *pyro-electricity*. If the substance is maintained for any period at the new temperature, the polarisation gradually disappears and soon ceases to be observable at all.

* This formula was determined for the optical case by L. Lorenz, *Ann. Phys. Chem.* (3), 11 (1880), p. 77; 20 (1883), p. 19; and independently by H. A. Lorentz, *Ann. Phys. Chem.* (3), 9 (1880), p. 642. The mode of deduction here given is due to Larmor, *Phil. Trans.* A, 190 (1897), p. 232.

† A complete account of these phenomena with all the associated experimental and theoretical details can be found in Voigt, *Lehrbuch der Kristallphysik* (Leipzig, 1910). Cf. also the same author's *Kompendium der theoretischen Physik*, Bd. II, Teil 4, §§ 11–15, 20 (Leipzig, 1896).

98. Lord Kelvin explains* this phenomenon by assuming that the elements of the crystal substance are permanently polarised to an extent, however, depending on the temperature, and that they are arranged with their electric axes in regular order in the crystalline media with which the phenomenon is associated. If the material of the crystal and in particular its outer surface are not perfectly non-conducting, the polarisation will ultimately give rise to a surface distribution of charge which neutralises the effect of its electric field at all external parts of the field. If the polarisation of the medium is altered by changing its temperature and the establishment of the neutralising charge takes place slowly, it should be possible to detect the polarisation before its external field is again neutralised.

This explanation of the phenomenon appears to be perfectly consistent with all the characteristic properties of the effect and, in addition, with the results of numerous experiments—based mainly on the independent variability of the polarisation and its neutralising charge—which have been performed with a view to testing it. It would thus appear to be highly probable that the underlying assumption of permanent molecular polarity is largely justified.

99. There is an inverse effect associated with the phenomenon of pyro-electricity, the existence of which was predicted by Lord Kelvin† but which was not observed until quite recently. If there is a relation of dependence between the polarisation of a medium and its temperature there must be a path of transformation open between the kinetic energy of thermal agitation of the molecules and the organised electric energy of their polarisations, and if the transformation can be carried out in either direction (i.e. if the effect is a reversible one) an alteration of the electric energy should produce a corresponding change in the thermal energy. The electrical energy of the polarisations may be altered by moving the body about in an electric field and thus we conclude that any such movement will give rise to a temperature variation in the substance. This electrocaloric effect has been observed by Straubel ‡

* *Nicol's Cyclopedia of Physical Science*, 1860; *Math. and Phys. Papers*, I. p. 315.

† *Math. and Phys. Papers*, I. p. 316, 1877.

‡ *Göttinger Nachr.* (1902), Heft 2.

and Lange*, who find that the quantitative relation established for the phenomenon by Lord Kelvin by thermodynamic reasoning is satisfactorily verified.

Another effect of an analogous nature and of even more wide-spread character than the purely thermal effects just described has been observed in a large number of substances†. In these cases the observed polarisation of the substance is produced not by changing the temperature but by the application of pressure on opposite sides of the substance. This pressure gives rise to an additional strain in the material the main effect of which is that the constituents of the permanent polar elements take up new positions in the substance and the old neutralising surface charge is no longer effective in balancing their field at external points.

Associated with this so-called *piezo-electric effect* there is an inverse phenomenon ‡ in which an alteration in the pure elastic strain in the medium is effected simply by changing the energy of the electric polarisations by moving the substance about in an electric field of variable intensity.

100. These ideas have been put into mathematical form by Voigt § with the aid of empirical equations, and the form of these equations has been supported by the recent work of Born on the elastic properties of crystal lattices, at least as far as primary effects are concerned.

Voigt expresses the electric moment or polarisation produced by given stresses or strains in the form

$$\mathbf{P}_x = \epsilon_{11}\mathbf{E}_x + \epsilon_{12}\mathbf{E}_y + \epsilon_{13}\mathbf{E}_z + \eta_{11}e_{11} + \eta_{12}e_{22} + \eta_{13}e_{33} + \eta_{14}e_{12} + \eta_{15}e_{23} + \eta_{16}e_{31}$$

and two similar equations for $\mathbf{P}_y$, $\mathbf{P}_z$; in these equations the co-efficients η_{rs} are the *piezo-electric constants*, the coefficients ϵ_{rs} the polarisation constants analogous to those giving the polarisation in terms of the electric field when there is no strain, and the six quantities e_{rs} are the usual components of strain.

* Dissertation, Jena, 1905.

† J. and P. Curie, *Paris C. R.* 91 (1880), pp. 294, 383.

‡ Lippmann, *Ann. chim. phys.* (5), 24 (1881), p. 164; *Journ. de phys.* (1), 10 (1881), p. 391. Cf. also Riecke, *Gött. Nachr.* (1893), pp. 3–13; Voigt, *Gött. Nachr.* (1894), Heft 4.

§ *Ann. d. Physik*, Bd. 48 (1915), p. 443. Cf. also *Lehrbuch der Kristallphysik*.

Voigt also expresses the general stress components in the form
$$X_x = -c_{11}e_{11} - c_{12}e_{22} - c_{13}e_{33} - \ldots - c_{16}e_{31} - \beta_{11}\mathbf{E}_x - \beta_{12}\mathbf{E}_y - \beta_{13}\mathbf{E}_z,$$
with five similar equations wherein the coefficients c_{rs} are the elastic constants. Substituting the expressions for the strains derived from these equations in the preceding equations, the polarisation may be expressed directly in terms of the generalised stresses and the electric force, the coefficients of the stress being the *piezo-electric moduli*.

As in the case of magnetism, these relations cannot be regarded as entirely satisfactory because the phenomena are complicated by a type of hysteresis*. When there is no external electric field the electric field arising from the polarisation produced by an applied stress will itself produce a secondary polarisation, so we have to distinguish between primary and secondary effects; but the secondary effects are usually small; they have, however, been studied by Voigt in some special problems.

The pyro-electricity of a crystal may also be supposed to arise from the strains introduced by a change of temperature; but if a true pyro-electricity is found to exist it may be necessary to add terms proportional to the temperature in each of the equations.

The calculations of Voigt† indicate that in the case of a crystal without a centre of symmetry it is not possible by means of observation to determine the magnitude of the permanent electric moments of the crystal elements, because in any deformation the effects of the geometrical and physical changes are added together in such a manner that they cannot be separated. In the case of a crystal with a centre of symmetry the theory requires modification, and a separation of the geometrical and physical effects seems possible.

101. Although these phenomena of pyro- and piezo-electricity seem to require for their explanation the assumption of permanent polar elements in the substance, it is possible that the occurrence of such elements may be a result merely of the mutual interaction of the molecules, for it is only observed in crystalline substances in which the molecular structure is perfectly regular, and in which

* Cf. J. Valasek, *Phys. Review*, 17 (1921), p. 475; 19 (1922), p. 478.

† *Phys. Zeitschr.* Bd. 17 (1916), pp. 287, 307; Bd. 18 (1917), p. 59.

therefore the local interaction between any molecule and its neighbours would always be related in a definite manner to the crystal structure. Such a view is supported by the fact that the polarity, which in such cases is essentially a phenomenon of molecular grouping, depends on the physical conditions as to temperature and strain in the medium, which are just the conditions which are circumscribed by the mutual interaction of the molecules.

Associated with these reversible phenomena of pyro- and piezo-electricity, which depend essentially on the presence of permanent polar elements in the medium, there are irreversible phenomena arising from the induced polarity*, when the extent of the induction in a given field is a function of the temperature and strain conditions of the medium. These two new effects can of course only be exhibited in their inverse aspects and appear as a temperature and strain condition variation resulting from the polarisation of the medium induced by an external field. The former of these effects has never yet been detected and the latter is usually inseparably mixed up with the strain produced by the mechanical forces proper on the medium resulting from its polarisation (the effect of *electro-striction*), although arrangements can be devised by which it can be observed†.

* The thermal one was predicted by Lippmann, *Ann. chim. phys.* (5), 24 (1881), p. 171, and the mechanical one by Larmor, *Phil. Trans.* 190 A (1897), § 83. The magnetic aspect of these phenomena is more important. Cf. below, ch. IV.

† Bidwell, *Phil. Trans.* A (1888), p. 228; *Proc. R. S.* (1894).

CHAPTER III

ELECTRIC CURRENTS

102. Introduction. When a conductor is introduced into an electric field a separation of electricity takes place until the field in its interior is compensated. Until this state is attained there is a flow of electricity, or a current, as we say. To illustrate the matter more fully let us consider a conductor somewhat in the form of that shown in the figure. The end a is charged with a positive charge $+q$ and the end b with a negative charge $-q$. There is then an electric field partly inside and partly outside the conductor, the

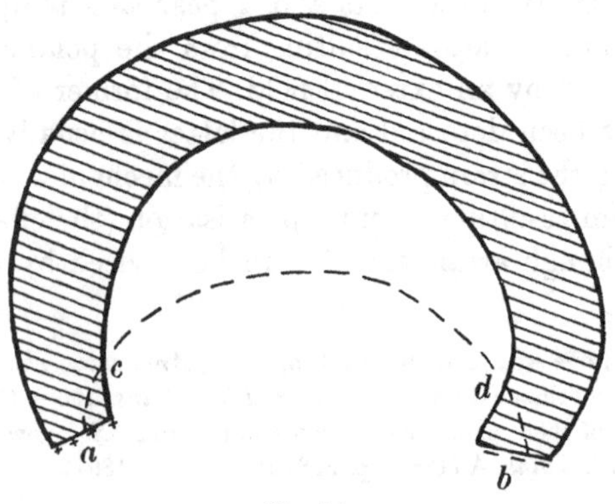

Fig. 10

lines of force in which may cross the surface of the conductor. Such a state of affairs if initially established is however not a possible equilibrium one so that there follows immediately a separation of the charges at each point of the conductor the total result of which is the final annulling of the charges at a and b. We shall now suppose that we can continually renew the charges at a and b in such a way as will maintain a constant potential difference between the two ends of the conductor. The manner in which this is accomplished will be hereinafter discussed.

The force driving the charge is the electric force of the field and so the initial charge flux must follow the lines of force. There will thus be initially a displacement of electricity along a line of force such as that shown in the figure by the dotted line *acdb*. But this displacement can only proceed as far as *c* where the surface of the conductor is reached. There will thus be an initial accumulation of positive charge on the surface of the conductor and this excites a field in the interior of the conductor, whose normal component at the surface is opposed to the normal component of the original force which was directed along the line *acdb*. This charge at *c* accumulates until this normal component is actually compensated, i.e. until the original line of force is altered into one running from *a* to *b* inside, and nearly parallel to the surface of the conductor.

Thus the first part of the electric flow is concerned merely with charging the surface of the conductor so that all the lines of force starting inside it from *a* remain inside it till *b* is reached.

In external space the normal component of the force originally along the line (*a*, *b*, *c*, *d*) is not compensated by the field of the surface charge, since they are both in the same direction. There is in fact in the external space a complicated electrostatic field composed of the original field superposed on that due to the surface charge.

It must however be said that the charges and force intensity in the field thus brought into existence are both extremely small. The electric elements in a conductor are so extremely mobile that it requires only a very small electrostatic force to produce an appreciable current.

The current can now flow undisturbed from *a* to *b* in the interior of the conductor along the new line of force, and if the charges at *a* and *b* are continually supplied so as to maintain the constant potential difference a condition of stationary streaming is attained. The field in the whole space then remains constant. Moreover the amount of electricity crossing any section of the conductor per unit time must be the same, as otherwise there would be an accumulation of charge in the conductor and a slight accumulation would create a back electromotive force which would tend to stop the current. A very slight accumulation would produce a sufficient back electromotive force to stop the current.

The process of starting the current thus requires a very slight accumulation of charge on the conductor which is just enough to make the flow steady or the current uniform and stationary. In this steady state the stream lines of the electric flow are also the lines of force of the electric field inside the conductor.

This idea of a current as a flow of electricity did not exist even for a considerable time after the discovery of batteries. It was Ohm (1827) who started the notion*.

103. Definition of an electric current. We must now define the electric flow in such a manner as to render it susceptible of calculation. If we adopt the general method we should specify the flux of electricity in any direction at a point in the conductor by the amount Cdf which crosses per unit time a small surface df placed perpendicular to the direction with its mean centre at the point. We can easily show that this defines a vector quantity if we choose rectangular axes with their origin at the point under consideration and consider the flow in and out of the small tetrahedral volume $OABC$ at the origin of coordinates with edges $\delta x \delta y \delta z$ along the axes. The area ABC has projections on the axial planes equal to

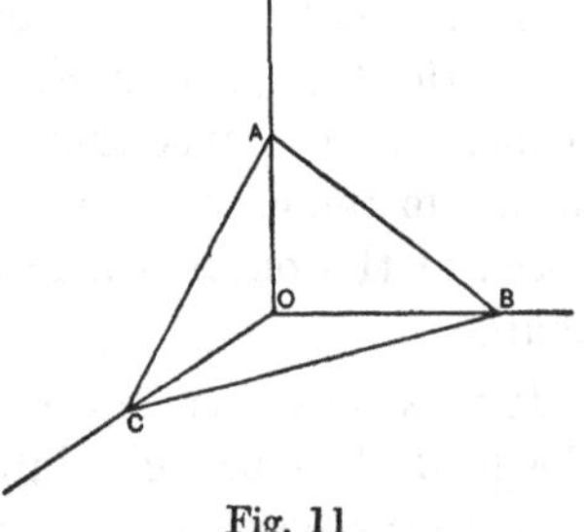

Fig. 11

$$\tfrac{1}{2}(\delta y \delta z, \delta z \delta x, \delta x \delta y)$$

or $$(\mathbf{n}_{1_x}, \mathbf{n}_{1_y}, \mathbf{n}_{1_z})\,\delta f,$$

where δf is the area ABC and $\mathbf{n}_1$ its direction vector.

The equation of continuity of flow expresses that the aggregate flux out of this volume is equal to *minus* the rate at which the total charge inside is increasing. If ρ is the density of charge inside

$$\delta v \frac{\partial \rho}{\partial t} = -(\mathbf{C}_x \mathbf{n}_{1_x} + \mathbf{C}_y \mathbf{n}_{1_y} + \mathbf{C}_z \mathbf{n}_{1_z})\,\delta f + \mathbf{C}_n \delta f,$$

where $\mathbf{C}_n$ is the flux component normal to δs and $\mathbf{C}_x$, $\mathbf{C}_y$, $\mathbf{C}_z$ those normal to the axial planes. The volume $\delta v\,(ABCO)$ is infinitely small of the third order and the surface δf is infinitely small of the second order and thus ultimately when the volume is very minute the left-hand side of this equation is zero and thus

$$\mathbf{C}_n = \mathbf{C}_x \mathbf{n}_{1_x} + \mathbf{C}_y \mathbf{n}_{1_y} + \mathbf{C}_z \mathbf{n}_{1_z},$$

which proves that $\mathbf{C}$ is a vector with components $(\mathbf{C}_x, \mathbf{C}_y, \mathbf{C}_z)$.

* *Die galvanische Kette, mathematisch bearbeitet* (Berlin, 1827). Translated in Taylor's *Scientific Memoirs.*

104. This is the general method of definition. In order however to obtain a closer insight into the true nature of the flux involved we must proceed in a slightly different manner. Consider again the small surface δf and construct on it a small cylinder of length δl with its axis parallel to the resultant motional velocity v_1 of the electric charge at the point (the direction of the lines of force at the point), this direction making an angle α with the normal at δf. If the density of the positive electricity at the point is ρ_1 the quantity of positive electricity in this cylinder is $\rho_1 \delta l \delta f \cos \alpha$ and during a time $\delta t = \frac{\delta l}{v_1}$ all of this electricity flows out across δf. Thus for this surface

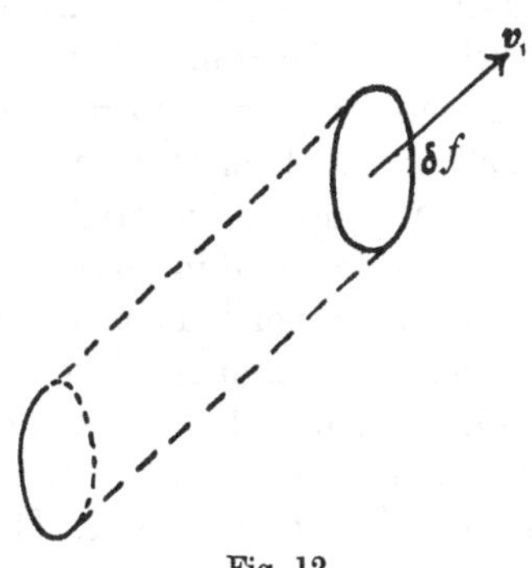

Fig. 12

$$\mathbf{C}_1 \delta f = \frac{\rho_1 \delta l \delta f \cos \alpha}{\delta t} = v_1 \rho_1 \cos \alpha \delta f,$$

or

$$\mathbf{C}_1 = \rho_1 v_1 \cos \alpha$$

measures the current of positive electricity in the direction normal to δf at the point.

If there is at the same time a flux of negative electricity of density $-\rho_2$ we know that it takes place in the opposite direction to that of the positive although perhaps with a different velocity v_2. The current across δf normally in the same direction due to the negative charge is therefore

$$\mathbf{C}_2 = -\rho_2 (-v_2) \cos \alpha = \rho_2 v_2 \cos \alpha$$

and the total current in this direction is

$$\mathbf{C} = \mathbf{C}_1 + \mathbf{C}_2 = (\rho_1 v_1 + \rho_2 v_2) \cos \alpha,$$

from which again we see that $\mathbf{C}$ is the component of a vector $(\rho_1 v_1 + \rho_2 v_2)$; which we call the *current density* of the electric flow at the point.

105. Ohm's Law*. The force driving the charge and imparting to it the motional velocity is the electric force: the positive elements of charge are moving in the positive direction of the lines of force

* This law was anticipated by Cavendish in 1781. Cf. his *Electrical Researches.*

and the negative ones in the opposite direction, and the two motions having the same effect constitute the current. This being the case there must be some relation between the electric force and the current. Ohm tried to reason the relation out by considering the phenomena of currents as analogous to the conduction of heat down a temperature gradient. There are two kinds of streaming motion recognised in physics: that of steady diffusion in which the *velocity* is proportional to the driving force, and that of free motion in which the change of velocity is proportional to the force. In diffusion the motion is so modified by impeding frictional forces that a state of steady motion is attained in which the velocity is proportional to the force. The conduction of heat is the typical example of a process of steady diffusion.

If now we assume with Ohm that the electric flow in the conductor is a process of steady diffusion under the action of the electric force we must also assume with him that both v_1 and v_2 are *proportional* to the driving electric force $\mathbf{E}$. Thus if we now use $\mathbf{C}$ for the resultant electric current we have in the vector sense

$$\mathbf{C} = \kappa\mathbf{E},$$

where κ is a physical constant, which is usually called the *conductivity* of the substance at the point.

At least this is the simplest case when the medium is completely isotropic. When this is not the case all that Ohm's law asserts is that the components of $\mathbf{C}$ are linear functions of the components of the driving force $\mathbf{E}$, or in symbols

$$\begin{aligned}\mathbf{C}_x &= \kappa_{11}\mathbf{E}_x + \kappa_{12}\mathbf{E}_y + \kappa_{13}\mathbf{E}_z,\\ \mathbf{C}_y &= \kappa_{21}\mathbf{E}_x + \kappa_{22}\mathbf{E}_y + \kappa_{23}\mathbf{E}_z,\\ \mathbf{C}_z &= \kappa_{31}\mathbf{E}_x + \kappa_{32}\mathbf{E}_y + \kappa_{33}\mathbf{E}_z.\end{aligned}$$

Maxwell called the coefficients $\kappa_{11}, \kappa_{22}, \kappa_{33}$ the coefficients of *longitudinal conductivity* whilst the cross coefficients κ_{rs} are called the coefficients of *transverse conductivity*. These latter indicate current produced in one direction by an electromotive intensity in a perpendicular direction, a phenomenon that is only apparent in media of crystalline structure. We have reason to believe that these latter coefficients satisfy the condition that

$$\kappa_{rs} = \kappa_{sr},$$

but for the present we shall disregard the complications of this general case and consider only isotropic media.

106. Now consider the case of a steady current flowing in a wire of finite thickness. The surfaces of the wire form a tube of force for the internal field and also a tube of flow for the current. The

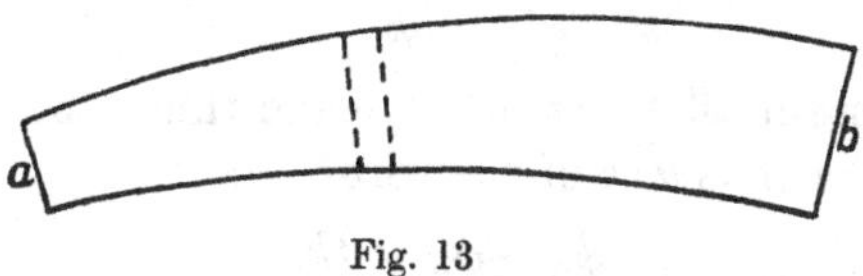

Fig. 13

ends a, b of the wire are presumed to cut the lines of force of the internal field everywhere normally and the same assumption tacitly underlies the choice of any other cross-section subsequently made. These sections will then be equi-potentials of the internal field.

Now consider any cross-section of the wire of total area f and suppose it resolved into small elements of area df. If $\mathbf{C}$ is the current density at a point in the wire, the total quantity of electricity crossing the section per unit time is

$$J = \int_f \mathbf{C}_n \, df;$$

J is called the strength of the current in the wire, or simply the current.

Again since the normal to df is in the direction of the line of force in the field at the place

$$\mathbf{C}_n = \mathbf{C} = \kappa \mathbf{E},$$

so that

$$J = \int_f \kappa \mathbf{E} \, df.$$

Now at any infinitely small distance from the section f draw another equi-potential section f' and let δs be the distance between corresponding points of the sections, then

$$J = \int_f \frac{k\mathbf{E}\,\delta s}{\delta s} \, df.$$

But $\mathbf{E}\,\delta s = \delta\phi$ is a constant over the whole surface f, viz. the constant difference of potential between the two surfaces f and f'. Thus

$$J = \delta\phi \int \frac{\kappa df}{\delta s}.$$

The quantity
$$\delta k = \frac{1}{\int_f \frac{\kappa df}{\delta s}}$$
is called the 'resistance' of the portion of the conductor between the cross-sections f and f': thus at this point of the wire
$$J = \frac{\delta\phi}{\delta k}.$$
But J is a constant all along the wire and thus if ϕ_a and ϕ_b are the potentials at the ends a, b of the wire
$$\phi_a - \phi_b = Jk,$$
where $k = \int_a^b \delta k$ is the resistance of the wire between the two ends.

This is Ohm's law in its original form. The procedure adopted by Ohm was however rather different from that sketched above. He tried to extend the mathematics just previously developed by Fourier for the conduction of heat down a temperature gradient. In doing this he had of course to assume something analogous to temperature and it did not require much to convince him that the potential was the required quantity. The current in a wire is proportional to the fall in potential from one end to another
$$J = \frac{\phi_1 - \phi_2}{k} *.$$
This idea that currents go by diffusion was at first merely an hypothesis, but on the modern theory of electrons it appears as the actual state of affairs.

107. There is an important hydrostatic analogy which enables us to picture the process more clearly. If liquid is forced through a tube blocked by a number of small obstacles so that no eddies can be formed and if the motion is a steady pushing through with the hydrostatic pressure as the driving force, the amount of the flow is
$$= \frac{\text{difference of pressures}}{\text{resistance of channel}}.$$
This is the more direct analogy with the electrical case. The term electric resistance is coined on this basis.

* These considerations are of course confined to a steady system: it is only when a steady state of flow has been attained that a potential exists (see chapter IV).

The analogy goes even further and enables us to talk of the driving force in the electrical case as an electrostatic pressure. The modern theory of the flow of electricity basing the current on the flow of electrons is in fact a direct application of this analogy. A current consists largely of free electrons being pushed through among the obstacles presented by the molecules of the matter.

The notion that electric pressure is the same as potential dates back to Volta's time. He knew that it was electrostatic pressure that pushed the current and he made a condensing electroscope sufficiently sensitive enough to show this. The distinction between free motion and diffusion was however due to Ohm.

108. The Volta potential difference. We have so far assumed that we are able to maintain a constant potential difference between two points on the surface of a conductor. We can now discuss how this is attained. The foundation of the method is Volta's discovery that when two conductors or generally any two different substances are in contact, there is a definite potential difference between them*.

If we place two different conductors in contact, then an adjustment of charge will take place so that the one conductor will acquire a definite negative charge and the other a definite positive charge. The quantities of electricity involved in the rearrangement depend on the different conditions such as the form, size and relative position of the bodies, but with the same two substances the potential difference thus set up between them has a definite value so long as we always work at the same temperature.

If the substances are conductors, as is usually the case, then in the equilibrium condition the potential of the electrostatic field which arises has a constant value at all points inside either conductor except very near the surface where it is in contact with the other. The change from the potential of the one conductor to that of the other thus takes place in the infinitely thin contact surface so that we can speak of a sudden jump of the potential. It therefore also follows that the origin of the action is situated in the immediate neighbourhood of the surface of contact. Between the two substances at the adjacent faces there is a certain stress of chemical affinity which results in an inequality in the forces exerted by each

* *Ann. de Chim.* 40, p. 225. Cf. also *Gilbert's Annalen,* 9 (1801), p. 380; 10 (1802), p. 425; 12 (1803), p. 498.

metal on the elements of charge in the other. We could for example explain the potential difference between zinc and copper by imagining that there is a greater attraction on the positive charges exerted by the zinc than the copper, and on the negative charges exerted by the copper than the zinc. This electrochemical stress results in an electric displacement either by polarisation of the molecules (in non-conductors) or by actual finite separation of the charge between them. The resultant of the electric displacement is that one of the two contact surfaces appears positively charged and the other negatively, the double sheet thus created accounting for the jump of potential.

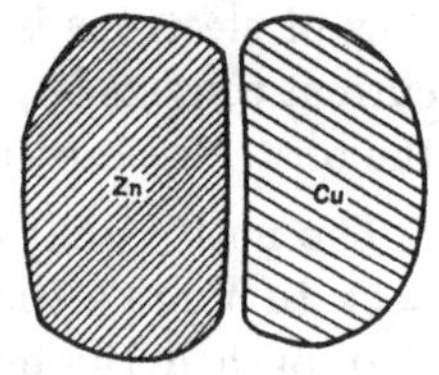

Fig. 14

109. When the substances are conductors the charges on each will distribute themselves over the surfaces of the separate conductors. The charges on the conductors must be equal and opposite (their total must be zero) and so will be practically all concentrated on the adjacent surfaces at the surface of contact. The potentials of the conductors being ϕ_1 and ϕ_2, the difference $\phi_1 - \phi_2$ is always the same for the same conductors under the given conditions; and is usually called their *volta difference* of potential.

When equilibrium has been established, which usually requires only an extremely short time, there is an electrostatic field surrounding the conductors which obeys all the laws of electrostatics. In the interior of the surface of contact between the adjacent surface charges the electric field is however compensated by the contact forces of chemical affinity. Thus for example in the Zn-Cu case mentioned the zinc attracts the positive charge from the copper and the copper the negative charge from the zinc, the result being that the zinc becomes positively charged and the copper negatively. But each addition to the positive charge on the zinc repels the remaining positive charge on the copper and so lessens the total attraction of the zinc on it. The separation thus goes on till the attraction of the zinc for any further positive charge becomes balanced by the repulsion of that charge by the positive electricity on the zinc.

The electrostatic field of the conductors is practically that due to the double sheet on the surface of separation, the small remaining charges on the further parts of the surface having no effect.

110. Now consider several such conductors joined in a ring. There would be a potential difference at each junction of two different conducting surfaces caused by the creation of a double sheet as indicated above. There would thus be a potential gradient at each junction in the circuit and thus the system is ready for an electric current to flow. But no current can flow or else the principle of the conservation of energy would be violated. The non-existence of the current may be explained by the fact that the currents arising from the single impressed electromotive forces at the separate junctions so flow as to cancel one another out.

The energy principle is no longer violated and a current can result if only we could supply energy from some external source at one of the junctions. We shall presume the possibility of this supply, postponing the discussion of the exact method in which it is applied. We should then have a current in the circuit, its density at any point being determined by $\mathbf{C} = \kappa \mathbf{E}$,

$\mathbf{E}$ being the electric force intensity of the field in the interior of the conductor; but $\mathbf{E}$ is composed of parts $\mathbf{E}_1$, $\mathbf{E}_2$, ... so that

$$\mathbf{E} = \mathbf{E}_1 + \mathbf{E}_2 + \dots,$$

where $\mathbf{E}_1$, $\mathbf{E}_2$, ... are the components in the direction of the resultant $\mathbf{E}$ of the several force intensities in the separate fields arising from the double sheet distributions at the various junctions.

111. Adopting the notation of the previous paragraph and considering for the present the current of strength dJ flowing through an elementary tube of flow (or tube of force) in the circuit of cross-section df at any place, we have

$$J = \int_f \frac{dJ}{df}\, df,$$

where $$dJ = \mathbf{C} df = \frac{\kappa(\mathbf{E}_{1_s}\delta s + \mathbf{E}_{2_s}\delta s + \dots)\, df}{\delta s},$$

or by integration round the whole circuit, i.e. with respect to s,

$$dJ \int \frac{ds}{\kappa df} = \int_s (\mathbf{E}_1\, ds + \mathbf{E}_{2_s} ds + \dots).$$

But
$$\int_s \mathbf{E}_{1_s} ds = \phi_1 - \phi_2 = \phi_{12}$$
is the volta potential difference of the two metals at the first junction of the circuit reckoned in a definite sense round the circuit. In this integral the element of the circuit between the double sheet concerned is of course not included, although the current crosses through this sheet; the argument for this may be stated as follows. If there were equilibrium with no current flowing the electric force intensity **E** in the contact sheet is exactly balanced by the contact forces (of chemical affinity), say **S**, so that
$$\mathbf{S} - \mathbf{E} = 0.$$
If however the charges separated by the contact forces can break away and flow off as an electric current, there is no longer an exact balance. But even in this case the outstanding difference between **S** and **E** is small compared with either of them, because sufficient charge accumulates in any case to establish the volta potential difference. Thus since **S** and **E** are both large of the first order the difference **S** − **E** can be at most finite, say **E**′, so that
$$\mathbf{S} - \mathbf{E} = \mathbf{E}'.$$
Since now the value of κ in the contact surfaces is certainly not large (it is at most finite) the values of $\mathbf{E}'d$ and $\kappa\mathbf{E}'d$ may be neglected in comparison with finite quantities, d being the thickness of the sheet. Thus practically the only driving force for the current is that in the interior of the metals. We have thus
$$dJ \int \frac{ds}{\kappa df} = \phi_{12} + \phi_{23} + \dots,$$
or
$$J = \frac{1}{k}(\phi_{12} + \phi_{23} + \dots),$$
where
$$k = \int_f \frac{df}{\int_s \frac{ds}{\kappa df}}$$
represents the total resistance of the circuit. This is of course Ohm's law for the circuit.

112. If no energy is supplied at any part of the circuit then $J = 0$ and thus
$$\phi_{12} + \phi_{23} + \dots + \phi_{n1} = 0,$$
which is *Volta's law*. This sum is however no longer zero when a

current is flowing or when energy is being added from outside at one of the junctions.

A direct consequence of this result is that the volta difference of potential of two metals in contact is exactly the same as the potential difference of the same two metals connected through a whole series of other metals. No direct electrometer reading will therefore ever detect the potential difference here described, because the instrument merely records the difference between the pieces of metal forming its quadrants (which are of course of the same metal*).

113. If we integrate in the positive direction round only part of the circuit, say that between the two sections α and β, we get

$$\int_\alpha^\beta E ds = \phi_\alpha - \phi_\beta + \Sigma\phi_{r,r+1},$$

where ϕ_α, ϕ_β are the potentials at the sections α, β and $\Sigma\phi_{r,r+1}$ refers to the volta potential difference for all the junctions occurring in the section of the circuit between α and β. Consequently we have also

$$J = \frac{\phi_\alpha - \phi_\beta + \Sigma\phi_{r,r+1}}{k_{\alpha\beta}},$$

where $k_{\alpha\beta}$ is the resistance of the part of the circuit concerned.

If the sections α, β are very near but on opposite sides of a given junction, the resistance $k_{\alpha\beta}$ is very small and so

$$\phi_\alpha - \phi_\beta + \phi_{r,r+1} = 0,$$

or

$$\phi_\alpha - \phi_\beta = -\phi_{r,r+1},$$

so that the potential in the circuit jumps at each junction by the corresponding volta difference of potential for that junction. If we put the sections α, β close together but so that practically the whole circuit is included between them and then remove the small portion of the circuit not included $J = 0$ and thus

$$\phi_\alpha - \phi_\beta = -\Sigma\phi_{r,r+1},$$

the potential difference between the ends of an open circuit is equal to the electromotive force operative in the same circuit when closed.

* By constructing electrometers with opposite quadrants of different metals it is possible to determine the difference of potential between the metals (Kelvin).

114. The general practical method of obtaining an effective electromotive force of this kind in the circuit is obtained by the insertion in it of a voltaic element or cell. The general principle involved may be illustrated by the following particular example. If we put into a vessel which contains dilute sulphuric acid a plate of copper and a plate of zinc at a small distance apart, it will be found by connecting the ends of the metals projecting from the acid to the quadrants of an electrometer that the pair of quadrants connected with the copper is at a higher potential than the other. In this experiment we have a series of conductors; brass, copper, acid, zinc and brass. The potential in the interior of each conductor has the same value at each point but it jumps at each surface of contact and the observed potential difference is the algebraic sum of the jumps. If ϕ_B, ϕ_C, ϕ_S, ϕ_Z, ϕ_B denote the potentials of the metals in order, and if also we use ϕ_{BC} for $\phi_B - \phi_C$, then we have the observed difference equal to

$$\phi_{BC} + \phi_{CS} + \phi_{SZ} + \phi_{ZB}.$$

One or more of these terms may be negative and if the acid were replaced by a metal the sum would be zero. It is only because we have an acid (or fluid conductor) in the series that the expression has a definite positive value different from zero.

The apparatus here described, and many others of a like nature which are composed of rigid and fluid conductors and which possess the property of creating a potential difference between two pieces of the same metal, is called a galvanic element or cell. They were first invented and used by Volta and are called after him*.

With such an apparatus it is possible to produce a permanent steady current by connecting the metal ends projecting from the liquid through a simple metallic circuit. The chemical affinities of the elements of metal and those of the acid produce an electric separation in the manner previously described and the double sheet so produced makes the sudden jump of potential in crossing from the metal to the liquid. Chemical attractions prevent the separate induced charges combining across the liquid and so they have to go round the metals closing the circuit. The double sheet thus dissipated by these charges going round is then continually renewed by chemical action and this makes a permanent current. The energy

* Cf. the letter in *Phil. Trans.* (1800), p. 402. Also *Gilbert's Annalen*, 6, p. 340.

supplied at the junctions here is the chemical energy of combination of the one substance with the other. The precise nature of the action involved will be discussed later.

115. On impressed forces in mechanics*. The mechanical distinction between the two cases here discussed, when the current can exist and when it cannot, is easily recognised. The closed circuit of physically and chemically homogeneous conductors in itself is a self-contained mechanical system so that no current can flow in it unless there are external forces of some kind acting. Thus if a current flows in such a circuit there must on the whole be an impressed electromotive force equal to Jk (as this is the electromotive force in the circuit) which arises from actions of a purely external nature.

This conception of electromotive force corresponds exactly to the conception of external or impressed forces in the mechanics of ponderable bodies. The idea is that of a force, not determined by the conditions which govern the system under consideration, but which nevertheless acts on it in an arbitrary manner and by means which are in no way in essential connection with the system. For example in a system of elastic bodies, the elastic stresses and internal forces are in perfect accord with one another and with the internal deformations and motions; but external forces may act on the system in any arbitrary manner. Of course, these forces influence the distribution of the internal stresses but only in the sense that they alter the conditions under which the system exists.

116. There are two reasons why we introduce the idea of external forces into ordinary mechanics. The first is that we can thereby limit our discussions to the consideration of a particular system by itself. If in the above example of elastic bodies the external forces are produced by weights, their action is determined by mechanical laws; such forces would of course be internal forces if gravity were also included in the mechanical system. Another important example will be given later, it involves the electromotive forces which can be produced by the oscillations of a magnetic field in the neighbourhood of a current circuit or to the motion of that circuit through a magnetic field. Such forces are external as long as we prefer to leave the magnetic field out of the

* Cf. Abraham, *Theorie der Elektrizität*, I. p. 199.

calculation. They become internal forces, whose action is determined by the ordinary laws of electrodynamics, as soon as the magnetic field is considered part of the system.

The second reason for introducing external forces into mechanics is that they often represent actions, which cannot be sufficiently well explained on a purely mechanical basis. Examples are provided in connection with the motions of magnets or electrically charged bodies. If mechanics attempted to explain all the motions of natural bodies, it would tacitly ignore all those motions which cannot be explained on a purely dynamical basis. In the cases mentioned, for example, a complete dynamical description of the motions is not possible until we have a mechanical explanation of the magnetic and electric actions. Mechanics treats only of one side of natural phenomena, it is useless when we are dealing with phenomena of a non-mechanical nature. A modified usefulness is however attained by admitting ignorance of their fundamental basis but representing the actions of these non-mechanical processes by means of impressed forces. The present type of impressed electromotive force in a circuit is of this nature. Such forces could not be treated from the standpoint of pure electric theory, because we are in reality involved in them in chemical and thermal phenomena, the laws in which are known only in a few special cases. Thus if we limit ourselves to the description of electrodynamic phenomena, we must in such cases resort to the idea of the impressed electromotive force to explain the action of such processes on those under direct review.

117. On the energy relations of an electric current*. We must now enquire into the amount of work expended in driving the current. Consider for this purpose any conductor in which a current J is flowing. Let ϕ_1 and ϕ_2 be the electrostatic potentials of the internal electrical field at two sections of this conductor between which the resistance is k_{12}. In a time δt an amount of electricity equal to $J\delta t$ is transferred from the one section to the other through the conductor (or at least this is the effective result of the electrical flow during this time). This means that an amount of electrical energy

$$(\phi_1 - \phi_2)\, J\delta t$$

* Kelvin, *Phil. Mag.* Dec. 1851.

has been lost in this part of the conductor during the time δt. Per unit time this is

$$J(\phi_1 - \phi_2).$$

What has become of this energy? The driving of the current is an affair of diffusion, the electric force is pushing the electric charges along among a large number of obstacles. The electric atoms get up a velocity, but impart it by collision to the molecules of the matter, so that their own motion becomes irregular. This is the essence of frictional resistance; the motion of the electric elements becomes irregular through collision with the obstacles. The wasted energy thus appears again as irregular motion of the electric charges and also partly of irregular motion of the molecules of the matter, that is it appears as heat in the conductor. Thus the heat developed per unit time in the portion of the conductor considered is

$$J(\phi_1 - \phi_2).$$

If we introduce Ohm's principle that

$$J = \frac{\phi_1 - \phi_2}{k_{12}},$$

the heat developed appears as of amount

$$J^2 k_{12}$$

per unit time. Thus in the whole circuit the total heat developed is

$$J^2 k,$$

where k denotes the resistance of the circuit. This expression however not only gives the total amount of heat but also its location.

118. If a voltaic cell is supplying the current the energy which appears as heat in the circuit must also come from the cell. Moreover this energy is available energy, for if we had conductors of small resistance we could turn it into work. (If the conductors are of big resistance the work is entirely wasted in them.) This work is introduced into the circuit in the battery at the places where the substances are decomposed (i.e. at the surfaces of metals in liquid). The supply of available energy comes in from the liquid and may be used to drive a machine somewhere if it does not waste. The location of the energy supply is different from that of its emergence. Thus an electric current is a means of transmitting power.

If a voltaic cell is the source of the current the total heat developed in the circuit appears as the work required to raise the quantity of electricity supplied by the current through any cross-section of the circuit through the various potential jumps at the contact surfaces in the circuit, because $(\phi_1 - \phi_2)$ for a whole circuit is simply

$$\Sigma\phi_{r, r+1}$$

for that circuit. The work per unit time is thus

$$J \,.\, (\Sigma\phi_{r, r+1}),$$

or since $\Sigma\phi_{r, r+1}$ is the quantity measured as the electromotive force of the cell, it is measured by the product of the current by the electromotive force of the cell.

It is of course assumed that all the work done on the electric charges in driving them forward is spent in increasing their velocity and is thus dissipated by collision and appears as heat. This only applies when none of the energy is turned into mechanical work, as is usually the case in circuits containing dynamos.

119. The above results, based on the idea of diffusion, were experimentally tested and verified by Joule 85 years ago*. From his results he formulated his law expressing that the total amount of heat developed as expressed above is correct and also that the distribution given is correct.

He was also able to formulate the principle for electric flow which in its generalised form states that in all cases of diffusion the flow distributes itself so as to give the least possible heat for a given current. In other words if a given current is introduced into a network of conductors it distributes itself among the conductors so that the energy wasted is least.

The generalised principle in the form that in any steady dynamical motion of a given material system, when the forces are only frictional, the motion is such as to make the waste of energy the least possible, was given and proved by Lord Kelvin. The particular case here quoted is however usually called *Joule's law of minimum dissipation.*

120. The proof of this law in the electrical case is easy. Let there be n electrodes joined by $\frac{n(n-1)}{2}$ wires and let there be given electromotive forces in the wires and given conditions of supply and withdrawal of current at the electrodes so that the currents in the wires are steady.

* *Phil. Mag.* 19 (1841), p. 260.

Let $\phi_1, \phi_2, \dots \phi_n$ be the potentials of the n points; $Q_1, Q_2, \dots Q_n$ be the amounts of electricity supplied per unit time at these points so that in a steady state

$$Q_1 + Q_2 + \dots + Q_n = 0,$$

and let E_{rs} be a possible internal electromotive force in the wire joining the rth and sth points so that

$$E_{rs} = - E_{sr};$$

and also let K_{rs} denote the reciprocal of the resistance in this same wire so that $K_{rs} = K_{sr}$; we also use other symbols $K_{11}, K_{22}, \dots$ having no physical significance but which are such that

$$K_{11} + K_{12} + \dots + K_{1n} = 0,$$
$$K_{21} + K_{22} + \dots + K_{2n} = 0,$$
etc.

In applying Ohm's law to each conductor and examining the flow at each point we have

$$Q_1 = K_{12}(E_{12} + \phi_1 - \phi_2) + K_{13}(E_{13} + \phi_1 - \phi_3) + \dots\dots$$
$$+ K_{1n}(E_{1n} + \phi_1 - \phi_n),$$
$$Q_2 = K_{21}(E_{21} + \phi_2 - \phi_1) + \dots\dots + K_{2n}(E_{2n} + \phi_2 - \phi_n),$$
....

These equations, usually ascribed to Kirchhoff*, can be written in the form

$$K_{11}\phi_1 + K_{12}\phi_2 + \dots + K_{1n}\phi_n = - Q_1 + K_{12}E_{12} + \dots + K_{1n}E_{1n},$$
....

The sum of the left- and right-hand sides of these equations is zero and they therefore reduce to $(n - 1)$ independent ones.

These are the conditions of flow, if the currents obey Ohm's law, and we have now to show that the heat developed in this case is the least possible.

121. We multiply the above equations in order by $\phi_1, \phi_2, \dots \phi_n$ and then add the n equations together; this gives us

$$- K_{rs}(\phi_r - \phi_s)^2 = - \sum_{r=1}^{n} Q_r\phi_r + \Sigma K_{rs}(\phi_r - \phi_s) E_{rs},$$

whence also $\Sigma Q_r\phi_r = \Sigma K_{rs}(\phi_r - \phi_s)(\phi_r - \phi_s + E_{rs}),$

* Kirchhoff, *Ann. Phys. Chem.* 64 (1845), p. 512; 72 (1847), p. 497; Wheatstone, *Phil. Trans.* 2 (1843), p. 323; *Ann. Phys. Chem.* 62 (1844), p. 535. Cf. also Maxwell, *Treatise*, vol. I.

but if J_{rs} is the current in the conductor joining the rth and sth points we have

$$K_{rs}(\phi_r - \phi_s + E_{rs}) = J_{rs},$$

and thus

$$\Sigma Q_r\phi_r = \Sigma J_{rs}(\phi_r - \phi_s),$$

or also if we use $k_{rs} = 1/K_{rs}$

$$\Sigma Q_r\phi_r = \Sigma k_{rs}J_{rs}^2 - \Sigma J_{rs}E_{rs},$$

either of which expresses the energy equation. The total heat lost in the wires is equal to the total energy supplied at the junctions plus that drawn from the cells in the circuits.

In these equations J_{rs} is the actual current in the typical conductor determined by Ohm's law. Suppose now that $J_{rs} + x_{rs}$ be a modification of this current which is compatible with the conditions of supply and output; so that

$$x_{12} + x_{13} + \dots + x_{1n} = 0,$$
$$x_{21} + x_{23} + \dots + x_{2n} = 0,$$
$$\dots\dots\dots\dots\dots\dots\dots\dots$$

and then we have

$$\Sigma k_{rs}(J_{rs} + x_{rs})^2 - \Sigma k_{rs}J_{rs}^2 = \Sigma k_{rs}x_{rs}^2 + 2\Sigma x_{rs}k_{rs}J_{rs}$$
$$= \Sigma k_{rs}x_{rs}^2 + 2\Sigma x_{rs}(\phi_r - \phi_s),$$

if there are no internal electromotive forces in the circuits to complicate matters. But then the last term on the right is

$$2\Sigma(x_{rs}\phi_r + x_{sr}\phi_s),$$

and is zero in virtue of the linear relations among the x's, and thus since

$$\Sigma k_{rs}(J_{rs} + x_{rs})^2 - \Sigma k_{rs}J_{rs}^2$$

is essentially positive in this case the heat developed in the actual state, with no internal electromotive forces, is less than that in any other state. The more general theorem when internal electromotive forces are included will be discussed on a future occasion.

122. The thermal relations of an electric current. We have mentioned so far the voltaic cell as the only means of producing by chemical action the necessary energy to drive a steady current in a linear conducting circuit. Experience however has shown that heat is also very effective as a generating agent for electric flow. In fact a current is at once observed in a circuit consisting entirely of pieces of metal if two junctions between different metals or even two points of the same metal are maintained at different temperatures*.

The simplest case is that in which the circuit consists of two pieces of different metals A and B joined at P_1 and P_2 into a complete circuit.

Any difference of temperature between P_1 and P_2 then causes a current to flow in the circuit and the direction of the current is reversed by inverting the temperature difference. If the junctions are at the same temperature there is absolutely no sign of any electric motion whatever.

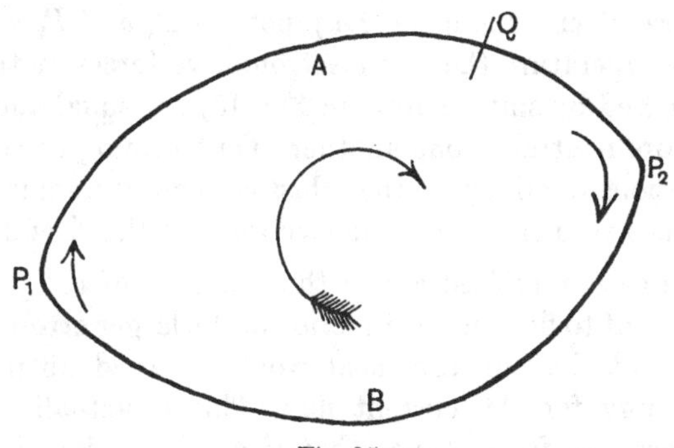

Fig. 15

It is possible to arrange all the metals in a series so that in a circuit of the type described the current flows across the warmer junction in the direction from the metal higher in the series to that which is lower down. A few of the metals arranged in this order are: bismuth, platinum, lead, copper, gold, silver, zinc, antimony.

123. If the circuit just described is broken at a point Q and if the junctions P_1 and P_2 are maintained at the different temperatures θ_1 and θ_2 the same cause which makes the current to flow in the closed circuit will create a potential difference between the ends of the broken circuit even if these ends themselves are at the same temperatures. This potential difference, which may be directly measured, will then serve as a measure of the electromotive force in the circuit. It appears that for small differences of temperature the electromotive force is proportional to $(\theta_1 - \theta_2)$ but for larger

* Seebeck, *Gilbert's Ann.* 73 (1823), pp. 115 and 430; *Pogg. Ann.* 6 (1823), pp. 1, 133, 253.

differences the simple proportionality is not even approximately verified.

These thermoelectric currents of course obey Ohm's law as regards the relation between the electromotive force and current and they also obey the series laws relating to compound circuits similar to those formulated for the volta potential difference.

124. In order to explain these phenomena we can suppose that the heat motion in the junction between the metals A and B drives the electricity towards the former metal (which we may take to be antimony, B is bismuth), with a force which is a function of the temperature. Then as long as the junctions P_1 and P_2 are kept at the same temperature the two electromotive forces at these junctions (indicated by small arrows in Fig. 15) are equal and opposite and therefore neutralise one another. On warming or cooling one of the junctions relatively to the other this equilibrium is disturbed and there is a resulting electromotive force in the circuit.

If in the case exhibited above the warming of the junction P_1 causes a current to flow in the direction of the larger arrow we should expect that at this junction heat would be used up in order to provide energy for the current flux. This is actually the case: Peltier* observed for instance that if a current is driven by an applied electromotive force across a junction between Bismuth and Antimony in the direction B—A the junction is always cooled: whereas if the current is driven in the opposite direction the junction is warmed. This local development or absorption of heat is of course a function of the temperature and if the complete circuit contains two such junctions at the same temperature the development of heat at one junction exactly balances the absorption at the other, whereas if the junctions are at different temperatures so that more heat is absorbed than developed the heat which disappears will appear elsewhere in the circuit either as Joule's heat or as heat of chemical transformations or it may be used up in other more effective ways as purely electrical energy.

125. Next let us take a simple circuit again consisting now of a piece of iron and a piece of copper and steadily heat the one junction up. The electromotive force in the circuit gradually increases, at

* *Ann. de Chim. et de Phys.* 56 (1834), p. 371; *Pogg. Ann.* 43, p. 324.

first proportional to the temperature, but subsequently more slowly until the hot junction reaches the temperature 280°, beyond which point an increase of temperature reduces the electromotive force, finally changing its sign after passing through the zero value*. We might again explain this in terms of our local electromotive forces at the junctions which we may now call ϕ_1 and ϕ_2 at the respective junctions. The electromotive force in the circuit is

$$\phi_{21} = \phi_2 - \phi_1.$$

Now suppose that at ordinary temperatures with $\theta_1 > \theta_2$ then $\phi_1 < \phi_2$ so that ϕ_{21} is positive, and that ϕ_1 decreases as the temperature θ_1 is increased, attaining however a minimum value zero at the temperature $\theta_1 = 280°$ after which it gradually increases again, then ϕ_{21} would behave exactly as described, so that so far the explanation is effective.

The temperature $\theta = 280°$ C.

is called the neutral temperature of the copper-iron combination of metals. Similar neutral temperatures exist for all combinations of metals. Moreover it is found that at the neutral temperature there is no Peltier effect, so that if an electric current is passed across the junction at this temperature then no development or absorption of heat takes place. This agrees generally with our explanation, for it is only when the local electromotive force at the junction is zero ($\phi_1 = 0$) that there is no development or absorption of heat.

126. Suppose now we have a circuit of the kind described in which the first junction has a temperature equal to the neutral temperature, i.e. $\phi_1 = 0$. The total electromotive force in the circuit is then

$$+\phi_2.$$

The direction of the current will then be that in which the Peltier effect at the upper temperature (θ_1) exists as a heat absorption and at the lower temperature as a heat development. But in the present instance no heat absorption does take place at the temperature θ_1. In spite of this however electrical energy will be transformed into heat energy at the lower junction and developed there, and in addition there will be the usual development of Joule's heat. We are therefore driven to the conclusion that at some position in the

* Cumming, *Annals of Philosophy*, 6 (1823), p. 427.

circuit other than at the junctions heat must be absorbed and transformed into electrical energy. This implies that even in a single wire whose ends are unequally heated an electromotive force may arise as the result of an absorption of heat, and if this phenomenon is reversible there must be an additional development of heat in the circuit when a current passes along an unequally heated conductor. This phenomenon was theoretically predicted by Kelvin* and he was soon enabled experimentally to justify the prediction: it is therefore usually known as the Kelvin or Thomson thermoelectric effect.

127. The effectiveness of the explanation here suggested for these phenomena is supported and further substantiated by the application to the transformations of energy of which they are the expression of the general laws of thermodynamics governing all such transformations. There are however difficulties of a fundamental nature involved in any such application in the present case: it may in fact be argued that the whole thermodynamic procedure may be invalid because it is applied to a case in which degradation is continually going on, in the form of conduction of heat, along the same circuit which conducts the current, and of amount depending on the first power of the temperature differences: and it does not appear that this fundamental objection to the procedure can be safely ignored, considering that conductivity for heat is closely connected with conductivity for electricity. It would of course be removed if the heat conduction proceeds in entire independence of the electric current, except as regards the transfer of the electric elements, the influence of which is reversible and is taken into account in the Kelvin effect. The electric cycle can moreover be completed in so short a time that the thermal transfer by ordinary conduction may possibly be neglected. Waiving these difficulties however the argument of Lord Kelvin† may be put in the following form‡.

Suppose that in the transfer of the amount $\delta\theta$ of electricity from a place where the temperature is θ to a place where it is $\theta + \delta\theta$ the amount $\sigma\,\delta\theta\,\delta Q$ of heat is absorbed by the current and converted

* *Phil. Mag.* [4], 11 (1856), pp. 214 and 281.

† The thermodynamic reasoning was first developed by Clausius (*Pogg. Ann.* 90 (1853), p. 513), but in an incomplete form.

‡ Cf. Larmor, *Aether and Matter*, p. 306.

into electrical energy of motion of the electricity; σ is called the 'specific heat of electricity' for the conductor and it may be either positive or negative.

128. Let us then—ignoring the finite degradation by heat conduction, but realising that the electric flow may be made so slow that the electric degradation, proportional to the square of the current, is negligible and that therefore the operations are certainly electrically reversible in Carnot's sense—apply the principle of energy and Carnot's principle to a circuit, formed of two metals and including as a part of itself the dielectric of a condenser having these metals for its coatings, the temperature θ varying from point to point along the circuit. When the plates of the condenser are moved closer together without alteration of temperature its charge increases, as the difference of potential ϕ between the plates remains constant: so that there is an electric flow round the circuit and there is at the same time a gain of mechanical work and of available energy each equal to $\frac{1}{2}\phi\,\delta Q$, or in all ϕ per unit total flow. Thus the plates of the condenser (Fig. 16) being at the same temperature θ_2, we have, by the energy principle and Carnot's principle, considering unit electric flow round the circuit

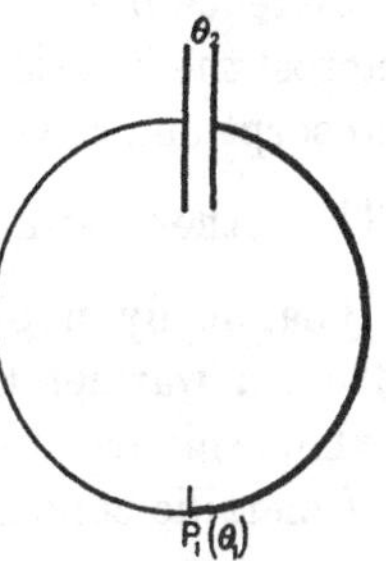

Fig. 16

$$\phi = \Pi_1 + \int_{\theta_1}^{\theta_2} (\sigma - \sigma')\, d\theta,$$

$$0 = \frac{\Pi_1}{\theta_1} + \int_{\theta_1}^{\theta_2} \left(\frac{\sigma}{\theta} - \frac{\sigma'}{\theta}\right) d\theta,$$

where Π_1 is the Peltier effect at the temperature θ_1 of the junction of the two metals, that is the amount of heat absorbed or set free on the passage of unit current across the junction at this temperature; and σ, σ' are the specific heats of electricity in them. Thus

$$\sigma_1 - \sigma_1' = \theta_1 \frac{d}{d\theta_1}\left(\frac{\Pi}{\theta_1}\right),$$

and
$$\phi = \Pi - \int^{\theta_1} \theta \frac{d}{d\theta}\left(\frac{\Pi}{\theta}\right) d\theta = \int^{\theta_1} \frac{\Pi}{\theta}\, d\theta.$$

Hence for a temperature θ of the junction, everything can be expressed in terms of the curve connecting the electromotive force ϕ of the circuit with θ, by the simple relations

$$\frac{\Pi}{\theta} = \frac{d\phi}{d\theta}, \quad \frac{\sigma - \sigma'}{\theta} = \frac{d^2\phi}{d\theta^2}.$$

The Peltier effect appears in the expression for ϕ, in the form of an electromotive force at the junction, as surmised in the simple explanation offered above. The chemical mutual attractions of the molecules across the interface produce in fact a polar electric orientation of these molecules which gives rise to an abrupt potential difference of contact equal to Π, and each unit of charge passing across the junction thus introduces an energy effect Π which involves absorption or evolution of heat at that place in the Peltier manner. The other term in the potential, viz. $\int(\sigma - \sigma')\,d\theta$, is thermodynamically involved in a convection of heat by the current passing from a warmer to a colder part of the wire: the exact mode in which this arises appears better in the discussions on the mechanism of metallic conduction in Appendix II.

129. Electrolysis. So far we have been dealing with the relations of electric flow without troubling much about the actual generation of that flow. We have, it is true, been led to certain conditions which are essential to the flow, but the exact way in which they are satisfied did not appear. We shall now attempt an exposition of this other side of the subject and explain how a current is actually generated. As a preliminary we must first explain in detail some important facts connected with the flow of currents through liquid conductors*.

Chemical compounds which can exist either as salts or acids, or have the same general characteristics as these bodies, can conduct an electric current when in the liquid form or in solution. In order to observe this it is merely necessary to insert two conductors as electrodes into the liquid or solution and connect them with the poles of a voltaic cell. A current will be found to flow round the circuit and if this experiment is carefully examined it will be found that:

* A complete account of the theory of conduction and the correlated phenomena in liquids is given with references by Whetham, *The Theory of Solutions* (Cambridge, 1902).

(i) No substance can conduct an electric current without being resolved into two constituents of which the one appears at the positive electrode and the other at the negative. This resolution is called *electrolysis*: the substances which admit of such, *electrolytes*, and the two constituents, the *ions*.

(ii) The metal or the stuff which takes its place in the compounds (Hydrogen in the acids) is always one of these constituents and it always appears at the electrode where the current leaves the electrolyte (the negative electrode or cathode). We call it the *cation*, the other part the *anion*.

(iii) Secondary actions may follow the deposition of the ions on the electrodes, but they are not connected with the current.

If for example copper sulphate ($CuSO_4$) is in solution between copper electrodes; when the current flows one electrode is covered with copper but from the other a portion of the copper combines with the free SO_4 ion again. The total quantity of dissolved copper sulphate thus remains the same but the one electrode increases in weight and the other decreases by the same amount.

130. We can obtain a very simple picture of this phenomenon in the following way. Suppose AB and CD are the electrodes and suppose the current flowing from left to right. We can now imagine that one constituent P of the substance remains at rest but that the other moves with the current. There will thus appear a definite quantity of this latter constituent in the free state at the electrode CD and an equivalent amount of P left free at AB; whilst each volume element in the middle of the liquid contains just as at first equal amounts of both constituents.

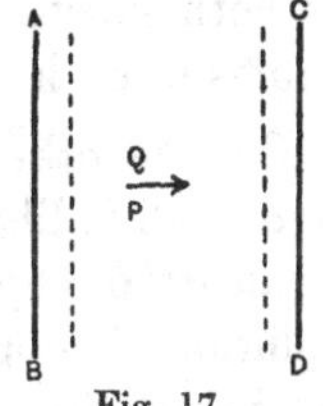

Fig. 17

As far as the resolution of the substance is concerned we might have imagined that the constituent Q was at rest and the part P moved to the left. The phenomenon really only depends on the relative motion of the two constituents and both P and Q might move, which will in general be the case.

Of course the actual motions in the electrolytes are very complicated. Before the current is sent through, each molecule has its irregular heat motion and the constituents may have relative motion inside the molecule. Single molecules may even be already dissociated into atoms or atomic groups and capable of existence as such for a short time. But in all of these phenomena no direction can be chosen for which the motion has a preference. All this however alters as soon as the current crosses the liquid. The atoms of the constituent P will then move to the left in large numbers or with larger velocities than to the right. Similarly the atoms of Q, although moving in reality in every direction, will have a slight preference for going to the right.

131. In a careful examination of all the circumstances governing this phenomenon of electrolysis Faraday found the following laws to be always true*:

(i) The quantity of a substance which is resolved by an electric current, and thus also the quantity of each constituent which is liberated, is proportional to the current strength.

(ii) If different substances are traversed by equally strong currents, then the quantities resolved in each case are chemically equivalent: the same is true also of the different constituents.

If for example solutions of zinc sulphate and copper sulphate are included in the same circuit, then the same number of molecules of both substances will be deposited and also the same number of atoms of each metal.

On the other hand from dilute sulphuric acid a quantity of hydrogen is liberated which contains twice as many atoms as the quantity of copper which the same current would separate from a solution of copper sulphate. But then one atom of divalent copper is chemically equivalent to two atoms of univalent hydrogen.

132. All of these facts tend to show that in electrolytes the motion of the electricity is invariably connected with the motion of the matter, in as far as that motion is conceived as a displacement of the constituent ions relative to one another. There is only one explanation of this connection; we must assume that of the two parts into which a molecule can be separated (the ions) the metal or corresponding part has a positive charge and the other an equal negative charge. It is thus clear how under the influence of a potential difference in the liquid the metal is driven to the cathode

* *Experimental Researches*, Ser. 7 (1834). § 662 *et seq.*

and the other constituent to the anode. We must in addition assume that the ions give up their charges as soon as they reach the electrodes, i.e. they are uncharged as soon as they appear in a free state.

This hypothesis means that the motion of electricity in electrolytes is a convection but with the characteristic that the small particles which convey the current need not receive any charge to convey, they merely give up what they already have. If this convection is the only kind of electrical motion involved, then the quantity of electricity which goes over from the cathode through a connecting wire is equal to the sum of the charges of all the metal particles which reach the cathode. The first Faraday law is thus explained by assuming that in a definite electrolyte each metal-atom is combined with a definite invariable charge.

As far as the second law is concerned it leads to a consequence which is more than a connection between the chemical and electrical phenomena. For example when two voltameters, the one with copper-sulphate and the other with zinc-sulphate, are connected in the same circuit, then equal quantities of electricity go to the cathode in each voltameter. Since this is accomplished by the deposition of the same number of atoms we see that the copper atom in the one salt and the zinc in the other must have equal charges. The negative charges which belong to the SO_4 ion in each molecule are also equal and equal to the positive charge on a copper or zinc atom. Consequently the same group in sulphuric acid has exactly the same charge, so we see that an atom of hydrogen in this acid has a charge equal to half that of a copper atom. Thus the atom of a divalent element carries a charge twice as large as that of a univalent element.

It is these Faraday laws which suggest that electricity like matter has an atomic structure and that the ions are combinations of chemical and electrical atoms; and they were first publicly interpreted in this light by von Helmholtz*. The suggestion however was not well received at first chiefly owing to the apparent lack of the necessity for it, and it was left for the modern theory of electrons definitely to adopt the conception into electrical theory. The remarkable success which has attended the developments on

* *Faraday Lecture*, 1881.

both the theoretical and the experimental side of this theory now hardly leaves any doubt as to the fundamental basis on which it is constructed.

133. The explanation of the action of a voltaic element. We can now attempt an explanation of the action previously ascribed to the voltaic element, viz. that of being able to supply a current. We know that in any such element as soon as a current is flowing there is a chemical action and we must therefore regard these actions or rather the forces which are in play with them as the origin of the electrical motion. We are strengthened in this view by the law of conservation of energy. This connects the heat developed in the current circuit with the decrease in the energy of the cell. Since new chemical combinations are formed and thereby energy lost, there is no doubt that we must seek for the source of the heat developed in the circuit in this direction and experiment proves that the one perfectly accounts for the other. We thus conclude that the chemical attraction between the atoms, which combines them together, is the origin of the electromotive force which drives the electricity in the element to the electrodes.

How can these chemical attractions create the potential difference in the circuit? How also is it that the chemical actions depend on the presence of a wire outside the cell connecting the two electrodes? Or how is it that a heat development can be found at a place different from that where the process giving rise to it is operative?

Without being able to give a perfect answer to this question we can make a fairly good representation of how the phenomena work. We shall assume that in the electrolytes the ions are electrically charged and that in consequence the motion of the ions and the motion of the electricity are invariably connected with each other. When decomposed these ions can be driven by the electric force; they may however also be moved by forces of another kind: in any case then there is a motion of electricity.

When we dip a copper plate and a zinc plate in sulphuric acid solution without completing the circuit the zinc attracts the SO_4 ions; some molecules of SO_4 will give way to the attraction and combine with the metal. This cannot however proceed very far, for the molecules as they combine give up their negative charge to the zinc and so there gradually arises a repelling force between the zinc and the next arriving SO_4 molecules. A condition is in fact very soon attained in which the chemical attraction is in equilibrium with the electric repulsion. A similar process goes on at the copper plate, although the SO_4 ions are attracted much more by the zinc than the copper, so that the copper plate will have a much less negative charge than the zinc when equilibrium is attained.

It is then clear that between the zinc and the copper there will be a potential difference such as is actually observed.

134. We can now if we like restart the chemical action which ceases when equilibrium is attained as above. We have only to put a positive charge on the zinc to neutralise its negative charge. This is accomplished by connecting it by a wire to the copper plate. If this is done the equilibrium is broken. The negative charge on the zinc reduces and thus the attraction between the zinc and the SO_4 ion is greater than the electric repulsion. Thus the SO_4 ion moves towards the zinc and communicates its negative charge to it and this is continually being neutralised by positive charge supplied from the copper through the wire.

It would thus appear that the resultant difference in the attraction of the zinc for the SO_4 and the copper for the SO_4 is the moving force in the circuit.

This idea also provides us with an explanation of the heat development phenomena. When a particle of SO_4 moves towards the zinc, only under the influence of the chemical attraction, it acquires a kinetic energy which we would notice as heat. But in the cell the attraction is practically balanced by the electric repulsion so that the molecule of SO_4 reaches the zinc without a large velocity, i.e. the combination takes place with only a small heat production.

Similar considerations are also true for cells of other types. A chemical attraction between one electrode and the negatively charged element of the surrounding electrolyte always tends to create a current in a definite direction, whilst an opposed current can arise through an attraction between the other electrode and the positive ion. If the electrode attracts both ions it is only a question as to which is the preponderating action. In the general case we should also have to take into account the forces exerted by the second electrode on the ions, so that the phenomena can be, and are in reality, more complicated than our above description. We have however been able to illustrate the underlying essential physical principles and our object is thus accomplished.

135. The effectiveness of the explanation of the action of the voltaic cell is further substantiated by results of the application of the general laws of thermodynamics to the thermal transformations which take place in them*. There are several of the galvanic cell elements in which the chemical transformations consequent on the passage of a current are perfectly reversible, i.e. elements in which the reverse current produces exactly the reverse chemical changes; there are others where secondary actions of a purely chemical character accompany the passage of a current through the element so that the actions are not properly reversible.

We shall confine our present considerations to the former type of element, for which alone the general thermodynamical procedure is directly applicable.

If a reversible element at a temperature θ has the electromotive force ϕ we may extract current from it until the total quantity δQ of electricity has passed round the circuit. The mechanical work gained in this process from the chemical reactions is $\phi\,\delta Q$. In this case a definite quantity of metal from one electrode has dissolved and an equivalent quantity of the other metallic electrode has been deposited. The amount of heat corresponding to these transformations we shall suppose to be $H\delta Q$. If then $\phi < H$ the whole of the energy obtained from the chemical processes is not all transformed into available electrical energy; part of it has been dissipated as heat in the circuit. Conversely if $\phi > H$ heat must have been absorbed from the circuit to account for all the electrical energy.

Now increase the temperature of the cell to $\theta + \delta\theta$: its electromotive force will in general be different, say $\phi + \delta\phi$. Then at this higher temperature join the element to a cell of another type and pass in the opposite direction the quantity δQ of electricity through it. In doing this we use up electrical energy from the second element of amount $\delta Q\,(\phi + \delta\phi)$, and an amount $\delta Q\,(H + \delta H)$ of

* Cf. W. Thomson, *Phil. Mag.* [4], 2 (1881), pp. 429, 551. The theory is originally due to Helmholtz and W. Gibbs. Cf. Whetham, *Theory of Solutions*, p. 235.

energy will be absorbed on account of the reverse chemical reactions taking place in the cell; finally the amount of energy

$$\delta Q\,\{(H + \delta H) - (\phi + \delta\phi)\}$$

has been developed in the cell as purely thermal energy.

Next cool the cell to the original temperature θ; the initial conditions in it are then completely re-established*. We have thus performed a completely reversible Carnot cycle with the cell, and since electrical and mechanical energy are completely transformable and can therefore be regarded as equivalent, we may apply the ordinary principle of Carnot to the process. But in the cycle we have absorbed the quantity of heat $\delta Q\,(\phi + \delta\phi - H - \delta H)$ at the higher temperature and given up the quantity $\delta Q\,(\phi - H)$ at the lower temperature, and the excess of electrical energy gained is $\delta Q\,.\,\delta\phi$. Since this process is reversible we must have

$$\frac{\delta\phi\,.\,\delta Q}{\delta Q\,(\phi - H)} = \frac{d\theta}{\theta},$$

or

$$\theta\frac{d\phi}{d\theta} = \phi - H.$$

Thus when the electromotive force of an element increases with the temperature the element must be cooled when a current is passed through it. Conversely when heat is developed in the cell by the passage of a current the electromotive force decreases with the temperature. These conclusions have been completely verified by experiment.

136. The conduction of electricity through gases. We must finally consider the question of the conduction of electricity through gases. The electrical conductivity of a gas in its normal state and at atmospheric pressure is extremely small, and it is only within the last few years that it has been established irrefutably that gases conduct at all; it can however be increased by subjecting the gas to certain influences: for instance in the neighbourhood of a red-hot body a gas conducts quite well†, as also do the gaseous products of combustion proceeding from a red-hot flame‡: again

* This of course assumes that the thermal capacity of the cell is so large as not to be appreciably affected by the slight changes taking place as described.

† Becquerel, *Ann. de Chimie et de Physique* [3], 39 (1853), p. 355.

‡ Cf. Wiedemann, *Die Lehre der Elektrizität*, IV. B; Giese, *Wied. Ann.* 17 (1882), p. 519.

when the so-called Roentgen rays pass through a gas they render it a good conductor of electricity, or to speak more accurately they increase its conductivity; if the rays cease to act the conductivity dies away rapidly and after a few seconds the gas is no more conducting than before exposure to the rays: the conductivity imparted by these rays can also be removed by passing the gas through a plug of cotton-wool.

The relation between the potential difference between the boundaries of the gas and the current through it, when it is rendered and maintained conducting by the uniform action of some agent as above described, is deserving of attention: it is shown by the diagram in Fig. 18. As the potential difference is increased from zero the current gradually increases, but it does not increase so fast as the potential difference (as it would in a conductor obeying Ohm's law): the increase of current for a given increase of potential difference decreases as the potential difference is increased until finally no increase in the current is produced by an increase in the potential difference.

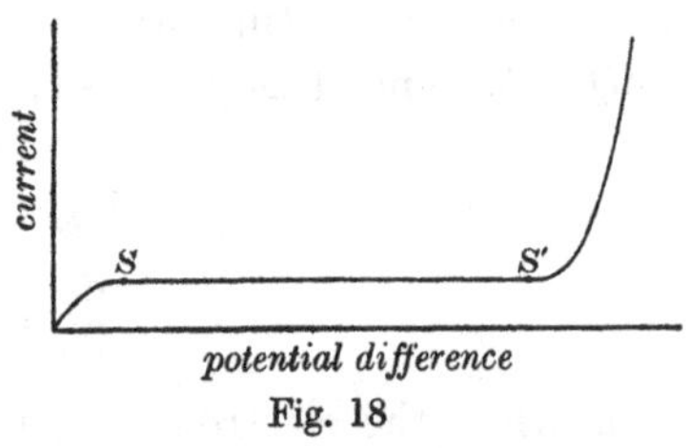

Fig. 18

This limiting current, the strength of which is independent of the electric intensity and depends only on the volume and nature of the gas and on the rays acting, is termed the 'saturation' current. It is represented by the part of the curve SS'. If the electric field be increased still further a stage will be reached at which a spark will pass through the gas and the current will increase once more. These facts were described and an explanation of them offered by J. J. Thomson and Rutherford in 1896*. They supposed that when a gas is rendered conducting there are introduced into it, by some means or other, a number of particles charged, some with positive and some with negative electricity. When the gas is subjected to an electric field the positive particles move to the negative boundary and the negative to the positive: this motion of the charge constitutes the current in the gas. But, if the gas is left to itself under the action of no external field, the oppositely charged particles

* *Phil. Mag.* [5], 42 (1896), p. 392.

attract each other and coalesce in pairs, or *recombine*. When once the particles have all recombined and their charges are neutralised, they move no longer under the action of a field and the gas loses its conductivity. However if the gas is kept under the action of the rays, fresh charged particles are produced continuously: the number of particles in the gas at any time is such that the number recombining in one second is equal to the number produced in the same time: the gas is then in a steady state, and the number of particles in the gas does not change with the time. In this condition the gas shows no sign of possessing a charge as a whole, and hence the total charge on all the positive particles must be equal to the total charge on all the negative.

When the gas is passed through a plug of cotton-wool, the charged particles, which differ from the molecules of the gas, are retained, while the molecules pass through. It is not necessary to conclude that the particles are larger than the molecules, for the fact that they are charged, while the molecules are neutral, might account for their retention.

137. Consider now the variation of the current with the potential difference in such a mixture of molecules and charged particles as has been imagined. The greater the strength of the field, the greater is the velocity with which the particles move and the shorter the time that is required for them to get from any part of the field to the boundary. But while any particle is moving to the boundary, it is liable to encounter a particle of the opposite sign and coalesce with it, losing thereby its conducting power. The shorter the time of passage, the less likely will such an encounter be, and the greater the number of particles which start from any part of the field and arrive, still charged, at the boundary: hence the current, which is measured by the number of charged particles arriving at the boundary, will increase with the potential difference. But it is clear that the current cannot increase indefinitely: it must reach a limit when the time occupied by the particles in reaching the boundary is so short that none recombine and the number arriving at the boundary per second is the number produced in the gas in the same time. This number multiplied by the charge on each particle gives the *saturation current* through the gas. The further increase in the current observed with still stronger fields can only be attributed to an increase in the rate of production of the particles.

According to this explanation, the process of conduction is essentially the same, whether it takes place in a metal or in an electrolyte or in a gas. In each case the current consists of a stream of charged particles, and the great difference between the laws governing the conduction in different states of matter is due only to a difference in the nature of the charged particles, in their mode of production and in their relation to the medium that surrounds them. In view of the similarity between gaseous and electrolytic conduction, the nomenclature of the latter has been applied to the former. The moving charged particles are called *ions* and a gas, when it is rendered conducting, is said to be *ionised*: the process of ionising a gas is named *ionisation*, a word that is also used quantitatively to express the concentration of the ions in the gas or the number per unit volume. The negative electrode is called the kathode and the positive the anode.

138. The charge and velocity of the ions. The charge on each ion in a gas can be ascertained if (1) the total charge on all the ions present in a gas at any time, and (2) the number of such ions, can be measured. The first quantity presents no difficulty. Let the gas be subjected to an ionising influence, such as Roentgen rays, until a state of equilibrium is reached between the number of ions produced per second by the rays and the number disappearing in the same time through combination. Let the action of the rays be stopped suddenly and the gas immediately exposed to the action of a strong electric field. All the ions of either sign present in the gas will be driven to the boundary of the field, and, if the field be sufficiently strong, the time occupied in reaching the boundary will be so short that no appreciable recombination will have taken place in the meanwhile: the number reaching the boundary is equal to the number present in the gas before the field was put on. If the charge received by the boundary is measured by connecting it to an electrometer, the total charge on all the ions of one sign present in the gas can be ascertained. It may be remarked that if the ionising agent consists of Roentgen rays it is found that the charge on all the ions of either sign is the same.

139. The measurement of the number of ions present is a much more difficult matter, but has been effected by a most ingenious method devised by J. J. Thomson and based on a discovery due to C. T. R. Wilson*.

It is known that the quantity of water which can exist in the state of vapour in a given volume of gas depends greatly on the temperature of the gas, and that, if a volume of gas saturated with water vapour is cooled, the excess of the water, or the amount representing the difference between that which the gas could hold at the higher temperature and that which it can hold at the lower, is deposited in the form of rain or mist. It was found by Aitken that this deposition of the superfluous water depends on the presence in the gas of solid particles, or dust, which acts as nuclei for the formation of liquid drops: if a gas is rendered perfectly free from dust by filtering it through cotton-wool, it can be cooled to a temperature very much lower than that at which a cloud would form in the presence of dust without any consequent condensation. The necessity for the presence of nuclei in the formation of drops can be shown readily to be a consequence of a surface tension in liquids, and it appears that the larger the particle of dust the more efficient is it as a nucleus and the smaller the supersaturation of water vapour that it is possible to produce in its presence.

C. T. R. Wilson found that ions possess peculiar properties with respect to the formation of clouds in gases supersaturated with moisture. In his investigations he cooled the gases by expanding them adiabatically, thus producing a fall in temperature which could be calculated from the known values of the initial and final volume and the ratio of the specific heats of the gas. He found that it requires less supersaturation to produce a cloud in air, when it is ionised, than when it is not ionised: the ions act as nuclei for condensation. This does not imply necessarily that the ions are larger than the molecules which are always present and available as nuclei, for it can be shown on theoretical grounds that a charged body should act as a more efficient nucleus than an uncharged body of the same size. In the absence of ionisation eightfold supersaturation† is required to produce condensation, whereas fourfold

* *Phil. Trans.* A, 189 (1897), p. 265; 192 (1899), p. 403.

† By 'supersaturation' is meant the ratio of the mass of water actually present in the gas to that which it could hold without condensation in the presence of large nuclei.

supersaturation is sufficient for the same purpose in the presence of negatively charged ions, and sixfold supersaturation in the presence of positively charged ions. These values are independent of the degree of ionisation and also, with few exceptions, of the means by which it is produced. Further, no increase in the amount of condensation is caused by increasing the supersaturation between fourfold and sixfold.

These experiments show that there is a difference in the properties of positive and negative ions; that the properties of the ions are not determined solely by the method by which they are produced, and that the properties of all the ions of the same sign are the same.

140. Wilson's discovery was used by Thomson to determine the number of ions in a gas. The ionised gas, in which the total charge on all the ions had been determined, was expanded adiabatically and cooled to such an extent that a supersaturation between sixfold and eightfold was produced: a cloud formed round all the ions of the gas, but not on the molecules. As soon as the cloud formed it began to fall, because the drops were heavier than the air surrounding them: owing to viscosity of the air they soon attained a small and constant velocity which could be measured by observing the rate at which the upper boundary of the cloud moved down the vessel. Now Stokes has calculated the steady velocity with which a body of known dimensions moves through a medium of known viscosity under the action of a constant force, such as its own weight. He finds that if F is the force, a the radius of the body (supposed to be spherical), ρ its density, μ the coefficient of viscosity of the medium, v the steady velocity is given by

$$v = \frac{F}{6\pi a\mu},$$

or if F is the weight of the body

$$v = \frac{2}{9}\frac{ga^2\rho}{\mu}.$$

By applying this formula to the case under consideration, where the velocity, the force and the viscosity are known, a, the radius of the drop, can be determined. The total mass of all the drops is the mass of the water set free by the cooling, for the mass of the ion contained in each drop is so small compared with the whole mass of the drop, that it may be neglected. The calculation of this total mass is somewhat complicated, but it depends on well-known principles, and for further details the reader may be referred to Thomson's account. If, then, the total mass of all the drops is M, the number of the drops n, which is the number of ions, can be calculated from the relations

$$n = \frac{M}{m}, \quad m = \tfrac{4}{3}\pi\rho a^3, \quad a = \sqrt{\frac{9\mu v}{2g\rho}}.$$

The charge on each ion, e, is the total charge on all the ions divided by the number of the ions.

As a final result of this research the charge on each ion in a gas was found to be $e = 3{\cdot}4 \,.\, 10^{-10}$ electrostatic units. Experiments made in air and hydrogen gave results identical within the limits of experimental error. More recent and improved methods have however increased this estimate to $4{\cdot}77 \,.\, 10^{-10}$ units.

141. H. A. Wilson* has used a slight modification of the method, which obviates the necessity for the cumbrous calculation of the total quantity of water set free. The total charge on the ions is measured as before. The gas is then expanded and cooled so as to give a supersaturation between four and six and to cause condensation on the negative and not on the positive ions. The velocity v with which the cloud falls is observed. The experiment is then repeated, with the difference that the falling cloud is exposed to the influence of a vertical electric field E, which exerts on the charged drops a force Ee, accelerating or retarding the fall. The new velocity v' is measured.

Since the velocity of the drop is proportional to the force on it, we have

$$\frac{v'}{v} = \frac{Xe + mg}{mg} \quad \text{where } m = \tfrac{4}{3}\pi\rho a^3;$$

but from above

$$v = \frac{2}{9}\frac{ga^2\rho}{\mu},$$

hence

$$e = \frac{9\pi\sqrt{2}}{E}\sqrt{\frac{\mu^3 v^3}{g\rho}}\,\frac{v' - v}{v}.$$

* *Phil. Mag.* [6], 5 (1903), p. 429.

By this method Wilson found $e = 3{\cdot}1 \,.\, 10^{-10}$. A few drops were found to bear a double and triple charge but their numbers were too small to have exerted any influence in Thomson's experiments.

Since Thomson measured the average charge on both positive and negative ions and Wilson that on the negative ions alone, the coincidence between their results shows that the charge on an ion is independent of its sign.

142. The velocity of the ions can be measured in various ways, of which the most convenient depends on the following principles. Suppose that two plates A and B, parallel to each other, are maintained at a constant difference of potential, and that ions are produced by some means close to the plate A. Under the action of the electric field the ions of one sign will give up their charges to A while the others will move towards B. Just as they are about to settle on B let the direction of the field be reversed, so that the plate which was formerly positive is now negative: these ions will then be driven back towards A while those of opposite sign move towards B: as these are about to touch B let the direction of the field be again reversed. If then the direction of the field is reversed at constant intervals and these intervals are so short that they do not permit the ions to travel from one plate to the other before the field is reversed, the plate B will receive no charge: but if the interval is longer, so that the ions have time to get across before the reversal of the field, B will receive a charge. By measuring the time of reversal for which B just begins to receive a charge, the time required for the ions to pass from one plate to the other is known, and hence the velocity of the ions ascertained. In this way it is found that the velocity of an ion is proportional to the strength of the field in which it moves, i.e. to the force acting on it: the essential condition for a diffusive motion is thus satisfied. The velocity of the negative ions is slightly greater in all cases than that of the positive ions.

The evidence thus far points to the fact that the carriers of the current in gases are more like those in liquids than the electrons in metals. But a study of the discharge of electricity through gases at low pressures has thrown a new light on this question.

143. Suppose that a long cylindrical glass vessel is taken and filled with a gas at atmospheric pressure, and suppose the potential difference between the kathode (K) and the anode (A) is gradually increased. Only an extremely small current will pass through the gas, but at a certain high limit a 'spark' passes through the gas and a very considerable current begins to flow. If the pressure of the gas is lower the limit at which the spark passes is also lower and the spark itself is broader and more diffuse than before. As the pressure is reduced the diffuseness of the spark increases until finally the luminosity of the discharge fills the whole volume of the tube between the electrodes; but it is seen on careful inspection that it is not continuous. Covering the kathode is a thin luminous layer, the 'kathode layer': next comes a dark space, the 'Crookes dark space': a luminous layer follows, the 'negative glow,' then another dark space, the 'Faraday dark space,' and lastly, a region of light, which is sometimes divided into striae, extending up to the anode and known as the positive column. As the pressure is still further reduced, the Crookes dark space grows at the expense of all the other regions, which diminish in extent, except the kathode layer, which remains practically the same. When the dark space has become so large that its boundaries touch the glass of the walls a curious green phosphorescence is excited in the glass. At first this phosphorescence appears on a few isolated patches near the kathode, but by decreasing the pressure sufficiently the whole of the tube may be filled by the Crookes dark space, with the exception of thin layers on the kathode and anode, and then all the glass walls of the tube with the exception of the parts behind the kathode and anode glow with the green phosphorescence: it is in this condition that the tube is emitting Roentgen rays plentifully.

If the pressure be still further reduced, the current diminishes and finally vanishes: at sufficiently high vacua no current can be made to pass.

144. In 1859 Pluecker* discovered that if a magnet is brought near to the tube in which the luminous discharge is observed all the luminous portions are distorted in some measure from their original positions. But the most marked distortion was produced in the patches of phosphorescence near the kathode. This discovery at once directed the attention of physicists to these patches. Soon afterwards Hittorf discovered that if a solid body is placed in the

* *Pogg. Ann.* 103 (1859), p. 88.

tube, near the cathode, the phosphorescence ceases at all points which were shielded from the kathode by the solid body: a shadow of the body is thrown on the walls of the tube. It thus appears that the influence which caused the phosphorescence on the walls must proceed in a straight line from the surface of the kathode to the walls of the tube, and it was therefore described as due to certain kathode rays. It may also be noted that it was found that the nature of the shadow was such as to prove that the rays do not proceed in all directions from the kathode but are emitted at right angles to its surface.

It was surmised by Goldstein* that these kathode rays were a type of aethereal radiation, like light; but the view of Sir William Crookes† that they were streams of particles electrically charged has been proved to be the more correct of the two. Before it can be definitely asserted that the rays consist of charged particles some further information is required: the term 'particle' connotes a finite mass and a finite number of carrying agents, and to settle the point it is necessary to determine these two quantities. The number is known if we can measure the charge carried by each particle: hence we want to determine the mass of each particle and the charge carried by it. Unfortunately no direct method has been devised for dealing with individual particles in the kathode rays and neither of these quantities has been determined directly, but the ratio of the mass to the charge of a kathode particle can be ascertained by a method devised by Thomson‡.

145. Suppose that a particle of mass m carrying a charge e is moving along parallel to the x-axis of a conveniently chosen rectangular coordinate system with a velocity v under the action of a magnetic field of intensity H parallel to the axis of y. Then as we shall see later the particle will be subject to a force $\frac{Hev}{c}$ perpendicular to both H and v, i.e. along the z-axis. The acceleration of the body parallel to the axis of z will be $f = \frac{Hev}{cm}$, and if it moves

* *Wied. Ann.* 1880–1884.

† *Phil. Trans.* 1879–1885.

‡ *Phil. Mag.* [5], 44 (1897), p. 293. Cf. also Wiechert, *Sitzungsber. der phys. kon. Gesellsch. zu Königsberg*, 38 (1897), p. 1; *Wied. Ann. Beiblätter*, 21 (1897), p. 443.

for a time t it will have travelled a distance $\frac{1}{2}ft^2 = \frac{Hevt^2}{2mc}$ in that direction. In the same time it will have travelled a distance vt parallel to the x-axis. Hence the particle while travelling a distance l parallel to the x-axis is deflected a distance δ parallel to the z-axis, where

$$\delta = \frac{l^2}{2c} \cdot H \cdot \frac{e}{m} \cdot \frac{1}{v};$$

if we know the dimensions of the discharge tube we can thus find e/m.

We are not able to measure directly the charge on the kathode ray particle; in fact it is only when the carriers of the current are atoms or molecules of matter that it is possible for us to determine the charge on the separate carrier; but since it is found that the value of this charge in all cases where it can be determined is always identical, it is natural to assume that the charges on the kathode particles have the same constant value, viz.

$$4{\cdot}774 \,.\, 10^{-10},$$

so that the mass is $9{\cdot}042 \,.\, 10^{-28}$ gms.,

this is much smaller than any known molecule of matter: a molecule of hydrogen has for instance a mass of about

$$3 \,.\, 10^{-24} \text{ gms.}$$

The kathode particle whose existence is thus presumed is the negative electron: it is identical with the negative particle thrown off by radio-active substances and these facts combined with certain other evidence to be subsequently discussed confirm us in the view that these electrons are common constituents of all matter.

146. While it thus appears that in the low pressure discharge through gases the most important part is played by negative electricity there is no doubt that positively charged particles also play some part in the mechanism of the phenomena. The complicated appearance of the discharge and the difference which it shows in different gases suggests that some agent other than the universal negative electron must have an influence on the phenomena; but for some time after the discovery of kathode rays no such agent had been directly detected. However in 1886 Goldstein working with a discharge tube in which the kathode was a plate perforated by several holes, observed faintly luminous streaks stretching out from the holes in the kathode into the space remote from the anode. Where these streaks meet the walls of the tube, they excite a slight phosphorescence, usually of a mauve colour, but always totally different from the green phosphorescence excited by the kathode. He imagined that these streaks represented the path of rays, similar in some respects to the kathode rays, to which he gave the name Kanalstrahlen. Since these rays are travelling in a direction from the anode to the kathode, it was natural to suppose that they are positively charged, but at first no experiment could detect any sign of charge. However in 1898 Wien* showed that, if fields of sufficient magnitude be employed, a deflection of the rays could be obtained, and its direction being in the direction opposite to that of the electrons indicates that the charge is positive. The results of the experiments were however otherwise somewhat indefinite and it remained for Thomson† to attack the problem with an improved apparatus. By using very low pressures Thomson was able to determine distinctly the types of particles involved and he found that they were in almost every case molecules of the gas in the tubes carrying the monovalent charge equal to that of a single electron. They are therefore merely the atoms or molecules of the substance from which an electron has been extracted.

These experiments unfortunately still leave us in the dark concerning the nature of the positive electricity in the atoms and so far no definite evidence is forthcoming on this point from any other branch of the work. Crowther sums up the state of affairs in his excellent little book‡ by the statement that 'the term positive electrification' remains a not too humiliating method of confessing ignorance. And this is where we must leave it for the present.

* *Wied. Ann.* 65 (1898), p. 440. Cf. also *Ann. der Phys.* [4], 8 (1902), p. 244.

† *Phil. Mag.* 13 (1907), p. 561; 16 (1908), p. 657; 18 (1909), p. 821.

‡ *Molecular Physics* (London, 1914).

CHAPTER IV

THE MAGNETIC FIELD

147. Introduction and definitions. An electric flow manifests itself in ways other than those discussed in the preceding chapter, for the surrounding space is always a field of action for so-called *magnetic* bodies. Similar action had long before been known to be associated with certain bodies called *magnets*, and a statical theory somewhat analogous to electrostatics had been built up* to coordinate the different phenomena associated with it.

Magnets when suspended so as to turn freely about a vertical axis at any point of the earth's surface, except the magnetic poles, will in general set themselves in a certain azimuth, and if disturbed from this position they will oscillate about it. On closer investigation it is found that the force which acts on the body tends to cause a certain line in the body to become parallel to a certain direction in space. This line we call the axis of the magnet.

Let us now suppose the axes of several magnets have been determined and that the end of each which points north is marked. Then if one of these magnets be freely suspended and another brought near to it, it is found that two marked ends repel each other as do also two unmarked ends, but a marked and an unmarked end attract. If the magnets are in the form of long rods or wires uniformly magnetised along their length, it is found that the greatest manifestations of force occur when the end of one magnet is held near the other, and that the phenomena can be accounted for by supposing that like ends of the magnets repel each other, that unlike ends attract each other, and the intermediate parts of the magnets have no sensible effect.

The ends of a long thin magnet are commonly called poles. In the case of an indefinitely thin magnet uniformly magnetised in its length the extremities act as centres of force and the rest of the magnet appears devoid of magnetic action. In all actual magnets

* Poisson, *Mém. de l'Acad.* 5 (1826), pp. 247, 488; 6 (1827), p. 441. Cf. also Maxwell, *Treatise*, II. § 385.

however the magnetisation deviates from uniformity so that no single points can be taken as poles. Coulomb however by using long thin rods magnetised with care succeeded in establishing the law of force between two poles.

The force between two magnetic poles is in the straight line joining them and numerically proportional to the product of the strength of the poles divided by the square of the distance between them*.

This law of course assumes that the strength of each pole is measured in terms of a certain unit, the magnitude of which may be deduced from the terms of the law. The absolute unit pole is such that when placed at unit distance in air from a similar pole, it repels it with a unit force, but for theoretical purposes it will be more convenient again to use a somewhat larger unit increased from this absolute one by the factor $\sqrt{4\pi}$.

The quantity called the strength of a pole may also be called a quantity of magnetism (positive or negative), provided we attribute no properties to it other than those observed in magnets. The quantity of magnetism at one pole of a magnet is always equal and opposite to that at the other, so that the total quantity in any magnet is zero.

Since the expression of the law of force between given quantities of magnetism has exactly the same mathematical form as the law of force in electrostatic theory much of the mathematical treatment of magnetism must be similar to that of electricity. We shall in fact transfer the results of our previous analysis directly to the subject now before us.

148. If the middle of a long thin magnet be examined, it is found to possess no magnetic properties, but if the magnet be broken at that point, each of the pieces is found to have a magnetic pole at the place of fracture, and this new pole is exactly equal and opposite to the other pole belonging to the piece. It is impossible to procure, by any means, a magnet whose poles are unequal.

If we break the long thin magnet into a number of short pieces we shall obtain a series of short magnets, each of which has poles of nearly the same strength as those of the original long magnet.

* Coulomb, *Mém. de l'Acad.* 1785, p. 603. Cf. also Biot, *Traité de physique*, t. III.

Let us now put all the pieces of the magnet together as at first. At each point of junction there will be two poles exactly equal and of opposite kinds, placed in contact, so that their action on any other pole will be null. The magnet, thus rebuilt, has the same form as at first.

Since in this case we know the long magnet to be made up of little short magnets, and since the phenomena are the same as in the unbroken magnet, we may regard the magnet, even before being broken, as made up of small particles, each of which has two equal and opposite poles. The same idea is also found to be generally true for magnets of any shape. We are thus induced to transfer to this case the analysis and methods of Chapter II on polarised media. We regard each particle of the body (molecule or molecular group) as a little magnet possessing two magnetic poles of equal strengths at a small distance apart. The magnetic field of force of such a particle is deduced in a manner exactly analogous to that already employed. The potential for instance at any point distant r from its centre in a direction making an angle θ with its axis is

$$\phi = \frac{m \cos\theta}{4\pi r^2},$$

m being the moment of the doublet. This may also be written in the form

$$\phi = \frac{1}{4\pi} (\mathbf{m}\nabla) \frac{1}{r}$$

when we take cognisance of the vector sense of the moment m.

The magnetic force intensity is then deduced as before by differentiation of the potential.

149. The field of a finite magnet is now obtained by regarding it as composed of a large number of small magnetic particles of the above kind.

The condition of magnetisation at any point of the finite magnet is completely specified by a vector $\mathbf{I}$, called the intensity of magnetisation, which is such that if δv is any small element of volume of the substance at any point then $\mathbf{I}\delta v$ is the resultant effective moment of all the little bi-polar elements (magnets) in it. If the axes of these elementary magnets are distributed anyhow in all different directions then $\mathbf{I} = 0$, but if there is any degree of convergence to a definite direction $\mathbf{I}$ has a finite value.

The discussion of the potential and force in the field of this finite magnet then follows exactly the same lines as in Chapter II.

The potential of the magnet at points external to the distribution of polarisation is obtained by the addition of the potentials of each of its constituent elements and is therefore

$$\psi = \frac{1}{4\pi}\int (\mathbf{I}\nabla)\frac{dv_1}{r_1},$$

and the force is obtained from this potential in the ordinary way, i.e. as its negative gradient. The application of these expressions at internal points however fails firstly on account of the uncertainty as to the law of action of a doublet very close up to it and secondly, as regards the expression for the force, on account of the non-absolute convergence of the integral. We are then led to the introduction of Poisson's ideal *magnetic matter** consisting of a volume density

$$\rho = -\operatorname{div}\mathbf{I}$$

at any part of the solid magnet together with a surface distribution of density

$$\sigma = \mathbf{I}_n$$

on the surface of the magnet. This distribution effectively replaces the distribution of bi-poles as far as the determination of the magnetic field outside the magnetic matter is concerned, a determination which is valid up to within a physically small differential distance from the matter. If however the point at which we wish to investigate the field is inside the magnet we must as usual put a physically small cavity round it and define the field there as the field inside this cavity due to the distribution beyond it which is effectively represented as above by the distribution of ideal magnetic matter and which therefore includes a part due to the distribution $\sigma = \mathbf{I}_n$ of magnetic matter on the surface of the cavity: together with the field due to the matter inside the cavity. This latter part of the field with the distribution on the walls of the cavity are the parts, and the only parts, of the field at the internal point which depend appreciably on the local molecular configuration, and which therefore in any real case are quite unknown.

* *Mém. de l'Acad.* 5 (1826), pp. 247, 488; 6 (1827), p. 441. Cf. also Maxwell, *Treatise*, II. § 385.

150. We might now follow the usual course and ignore the local part of the field altogether as being ineffective in a proper physical theory; but unfortunately this course, which has up to the present been followed, leads to a certain amount of confusion as between the mechanical and molecular applications of the subject. We shall find in fact that it is necessary in a consistent theory to define the complete magnetic force **B** at any point in the field by the relation

$$\mathbf{B} = \mathbf{H} + \mathbf{I},$$

wherein **H** is that part of the complete force which is obtained when the effects of the local part are ignored and which is therefore completely defined as due to the distribution of ideal magnetic matter throughout the volume of the magnet and on its surface. The force **B** has to be used in all molecular applications of the theory, but when the average mechanical effect is alone under review the local part of this complete force can be omitted and the force **H** used. It is this force **H** which in statical cases is derived as the gradient of the potential in the field, and to distinguish it from the complete force we shall henceforth call it the *magnetic induction*, but it must be remembered that it is the induction of *mechanically effective force* that is implied.

151. Special distributions of magnetisation. There are certain limiting distributions with which we have frequently to deal in practice and for which the defining field functions take certain simple forms. The first of these is the magnetic filament or thin wire magnetised at each point along its length. In this case the strength of the filament at any point is the product of the transverse section of the filament and the mean intensity of the magnetisation across it.

A magnetic filament of this type so magnetised that its strength is the same at every section is often called a *solenoid*.

If m is the strength of the solenoid at any point distant s from the negative end of the filament, r the mean distance of the element ds from a given point in the field, then the potential at this field-point of the element ds is from the general formula

$$-\frac{m}{4\pi r^2}\frac{dr}{ds}ds.$$

The potential of the whole filament, obtained by integration, is therefore

$$\phi = -\frac{1}{4\pi}\int m\frac{dr}{r^2}$$

and for a solenoid m is constant so that

$$\phi = \frac{m}{4\pi}\left(\frac{1}{r_1} - \frac{1}{r_2}\right),$$

r_1, r_2 being respectively the distances to the field-point from the positive and negative ends.

The potential and consequently all the magnetic effects of such a solenoid depend only on the position of its ends and not at all on its form, whether straight or curved, between those ends.

If a solenoid forms a closed curve the potential due to it at every point is zero, so that such a solenoid exhibits no magnetic effects.

If the magnetisation m of a filament is not uniform all along its length we can still express its potential in the form

$$\phi = -\frac{1}{4\pi}\int \frac{m}{r^2}\frac{dr}{ds}ds,$$

but now m is a function of s. By an integration by parts this may be transformed to

$$\phi = \frac{m_1}{4\pi r_1} - \frac{m_2}{4\pi r_2} - \frac{1}{4\pi}\int \frac{1}{r}\frac{dm}{ds}ds,$$

so that in addition to the actions at the two ends, which may in this case be different, there is also an action due to the departure from uniformity in the strength at each point of the filament. Interpreted in terms of ideal magnetic matter we may say that in addition to the magnetic action of the free unequal poles at the ends of the filament there is also in this case an action due to a distribution of magnetic matter all along the wire. In the solenoid proper there is no free magnetism except at the ends.

152. The other type of distribution of importance is the magnetic shell or surface magnetised at each point normally to itself. In this case the elementary magnets of the distribution are arranged side by side normal to the surface, in contrast with the end-on arrangement of the previous case. The strength of the shell at any point is the product of its normal thickness and the mean intensity of magnetisation (normal) across it.

If the strength of the shell at any point is m then the small magnets standing on the area df have a resultant moment equal to $m\,df$ directed along the normal to df. The potential of this element at an external point is therefore

$$d\phi = \frac{m}{4\pi}\,df\,\frac{\partial}{\partial n}\left(\frac{1}{r}\right),$$

the element dn of the normal being positive in the direction from the negative to the positive side of the sheet. For the whole shell therefore

$$\phi = \frac{1}{4\pi}\int_f m\,\frac{\partial}{\partial n}\left(\frac{1}{r}\right) df.$$

If the shell is uniform, and this is the most important theoretical case

$$\begin{aligned}\phi &= \frac{m}{4\pi}\int_f \frac{\partial}{\partial n}\left(\frac{1}{r}\right) df \\ &= \frac{m}{4\pi}\int_f \frac{\cos \widehat{rn}}{r^2}\,df \\ &= -\frac{m\omega}{4\pi},\end{aligned}$$

where ω is the solid angle subtended by the surface at the point where the potential is calculated, the solid angle being taken so that the positive direction across the shell corresponds generally to the direction of increasing ω, that is, the positive direction for the magnetic force.

The important aspect of this solution is that the potential and therefore also the entire action of the shell depends only on the shape of the curve bounding it, and not at all on the way the surface abuts on this curve. Any two uniform shells of the same strength and with the same boundary curve would give exactly the same action at all points external to both.

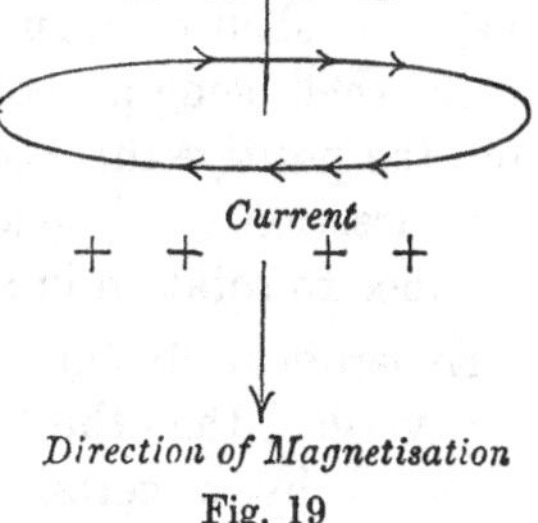

Fig. 19

153. The two typical distributions here analysed have given rise to a discrimination and designation of two other special types of the more general distribution of magnetisation.

(i) If any magnet or magnetised body can be divided into solenoids, all of which either form closed curves or have their ends in the outer surface of the magnet, the distribution of the magnetisation is said to be *solenoidal*, and since the action of the magnet now depends entirely upon the ends of the solenoids, the distribution of imaginary magnetic matter will be entirely superficial. The condition that the magnetisation is solenoidal is therefore

$$\operatorname{div} \mathbf{I} = \frac{\partial \mathbf{I}_x}{\partial x} + \frac{\partial \mathbf{I}_y}{\partial y} + \frac{\partial \mathbf{I}_z}{\partial z} = 0.$$

(ii) If a magnet can be divided into simple magnetic shells either closed or having their edges on the surface of the magnet, the distribution of magnetisation is called *lamellar*. In such a case if M is the sum of the strengths of all the shells traversed by a point passing from a given point to a point (x, y, z) by a line drawn within the magnet, then clearly

$$\mathbf{I} = \operatorname{grad} M,$$

$$\mathbf{I}_x = \frac{\partial M}{\partial x}, \quad \mathbf{I}_y = \frac{\partial M}{\partial y}, \quad \mathbf{I}_z = \frac{\partial M}{\partial z}.$$

The analytical condition satisfied by $\mathbf{I}$ in this case is

$$\operatorname{curl} \mathbf{I} = 0.$$

154. The field of a linear current. The space around any flow of electricity is always the seat of a magnetic field resulting from that flow. The simplest case and the practical one is that of a current flowing in a very thin wire circuit. If the magnetic field of such a circuit be mapped out in the usual way by lines and tubes of force it will be found that it is identical with the field of a uniform magnetic shell on a surface bounded by the circuit, the strength of the shell being proportional to the current flowing in the circuit and the positive direction of magnetisation in the shell bearing to the direction of circulation of the current the same relation as advance to rotation in a right-handed screw.

Experimentally this result is obtained in another manner. It is first verified that the field of a small plane closed circuit can be annulled by a certain small magnetic needle stuck normally through its centre, the moment of the magnet being proportional to the strength of the current and the area of the plane enclosed by the small circuit. This effectively proves the rule for small circuits and enables us to say that the magnetic field of such a small circuit is equivalent to that of a magnetic shell bounded by it. The extension to finite circuits is then accomplished in the manner indicated by Ampère*.

155. Conceive any surface f bounded by the circuit and not actually passing through the point P at which the field is examined. On this surface draw two series of lines crossing each other so as to divide it into elementary portions, the dimensions of which are small compared with their distance from P and also with the radii of curvature of the surface.

Round the boundary of each of these elements conceive a current of strength equal to that of the original current to flow, the direction of circulation being the same in all the elements as it is in the original circuit.

Along every line forming the division between two contiguous elements two equal currents flow in opposite directions. But the effect of two equal and opposite currents in the same place is absolutely zero, in whatever aspect we consider them. Hence their magnetic effect is zero. The only portions which are not neutralised in this way are those which coincide with the original circuit. The total effect of the elementary circuits is therefore equivalent to that of the original circuit.

Now since each of the elementary circuits may be considered as a small plane circuit whose distance from P is great compared with its dimensions we may substitute for it an elementary magnetic shell of strength proportional to that of the current whose bounding edge coincides with the elementary circuit. But the whole of these elementary shells constitute a magnetic shell of equal strength coinciding with the surface f and bounded by the original circuit and thus the magnetic action of the whole shell at P is equivalent to that of the circuit.

The magnetic force due to the circuit carrying a current J at any point is therefore identical with that due to a uniform magnetic shell bounded by the circuit and not passing through the point, the strength of the shell being numerically equal to $\frac{1}{c}J$. The constant $\frac{1}{c}$ thereby

* "Mémoire sur la théorie mathématique des phénomènes électrodynamiques," *Mém. de l'Institut*, 6 (1820). Cf. also *Ann. de Chim.* 15 (1820).

introduced is an absolute constant whose magnitude will depend on the units employed, and which we can if necessary make equal to unity by a suitable choice of such units.

The restricting phrase of the previous paragraph that the shell must not pass through the field-point emphasises the one limitation tc the general equivalence of the two fields. At points inside the magnetic shell the field is very large. This however does not vitiate the general usefulness of the relation because we can deform the shell as we please without altering its potential, so long as its strength remains the same and it retains the circuit as a boundary.

The potential of a uniform magnetic shell of strength $\frac{1}{c}J$ is

$$\psi = -\frac{1}{4\pi c} J\omega,$$

where ω is the solid angle subtended by the point in the field, the solid angle being taken so that the positive direction across the shell corresponds generally to the direction of increasing ω.

This then is also the potential in the magnetic field of a closed current of strength J, ω now being interpreted as the solid angle subtended by the circuit at the field-point.

156. Suppose now we follow a point along a path encircling the current circuit starting from any point P where the potential is ψ_P and eventually returning to this same point P; it will be found that the value of the potential on arriving back at P is not ψ_P but $\psi_P \pm \frac{1}{c}J$, the $\pm$ sign depending on the direction in which the path is traversed. If we traverse the same or a similar path again in the same direction, the potential becomes $\psi_P \pm \frac{2}{c}J$ and so on; on the other hand traversing any path which does not link with the current does not result in any such change on arriving back at P.

The magnetic potential of a closed current is thus a *cyclic potential* and there is really no unique value at any point in the circuit. Mathematically the cyclic property arises from the fact that the current forms a line singularity in the magnetic field close up to which the magnetic force is very large.

This cyclic property of the potential distinguishes it from the similar potential of the magnetic shell. The potential function of a magnetic double sheet, being obtained from a sum

$$\Sigma \frac{m}{r},$$

is essentially a single-valued function. It is true that as we go round a circuit as shown by the line from P to P' (near the sheet one on each side on the same normal) we find that there is a drop of potential equal to $\frac{J}{c}$ but this drop is recovered in going through the sheet from P' to P. The potential changes very rapidly in going through the sheet. In the case of the current however there is no place where the change in potential produced in going round a circuit as from P to P' (which are ultimately very close together) can be recovered, so that the cyclic change $\pm \frac{1}{c'} J$ persists on arrival at P from P'.

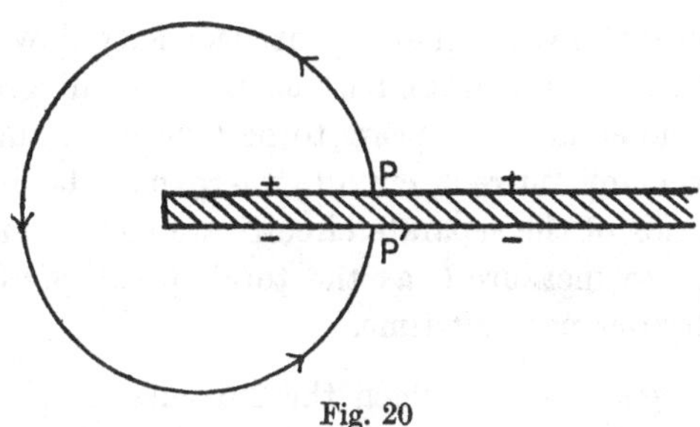

Fig. 20

157. Ampère's circuital relation. The cyclic property of the potential function of the magnetic field of a linear electric current was extended by Ampère to cover the most general case of electric flow. First the relation in the simplest case is expressed in an integral form. If $\mathbf{B}$ is the magnetic force vector at any point in the field, then we have

$$\begin{aligned} \mathbf{B} &= \mathbf{H} + \mathbf{I} \\ &= -\nabla\psi + \mathbf{I}, \end{aligned}$$

where we include the most general case with magnetic media in the field.

Thus if $d\mathbf{s}$ is the vector element of arc in any path encircling the current in the positive direction, that is, the direction of increasing

ω or decreasing ψ, then the total decrease in ψ, the magnetic potential, is

$$\int_s (\mathbf{B} - \mathbf{I}, d\mathbf{s}) = \int_s (\mathbf{H}\, d\mathbf{s}) = \frac{1}{c} J.$$

If the closed path does not encircle the current we have

$$\int_s (\mathbf{H}\, d\mathbf{s}) = 0.$$

These two results can now be expressed in another form. Having chosen a definite path let us imagine it closed over by a simple barrier, then the condition that the path encircles the current or not is simply the condition that the current cuts through this barrier surface or not. But wherever the current cuts through this surface a quantity J of electricity will flow across the surface per unit of time; if the path does not cut across the surface no electricity flows across this surface on its account. Thus the above two equations may be expressed in the form

$$\int_s (\mathbf{H}\, d\mathbf{s}) = \frac{1}{c} C,$$

where C denotes the total quantity of electricity flowing per unit time through the circuit around which the first integral is taken. In this form the equation appears to be true in the most general case however many current circuits there may be in the field, because the fields of the separate circuits are additive in any case. We have only to measure C as the total quantity of electricity crossing the barrier per unit time.

158. In the general case when the currents are distributed in space with a volume density $\mathbf{C}$ at each point we have

$$C = \int_f \mathbf{C}_n df,$$

so that

$$\int_s (\mathbf{H}\, d\mathbf{s}) = \frac{1}{c} \int_f \mathbf{C}_n\, df,$$

where f is any barrier surface bounded by the curve s. This is the general integral form of *Ampère's circuital equation.*

We can convert this relation to a simple differential form, for the integral on the left can be converted by Stokes' rule into the surface integral of $\text{curl}_n\, \mathbf{H}$ also over the surface f and we then get

$$\int_f \left(\text{curl}\, \mathbf{H} - \frac{1}{c}\mathbf{C}\right)_n df = 0.$$

But since the surface f and the boundary curve s were arbitrarily chosen we must have

$$\operatorname{curl} \mathbf{H} = \frac{1}{c}\mathbf{C},$$

which is equivalent to three vectorial equations.

159. Another point of some interest follows from these equations: they may be written in the form

$$\operatorname{curl} \mathbf{B} = \frac{1}{c}(\mathbf{C} + c \operatorname{curl} \mathbf{I}).$$

Now Ampère himself made the suggestion that the term $c \operatorname{curl} \mathbf{I}$ added to the current in this last equation and arising from the magnetic polarity is in fact due to a true distribution of electric flux; his view was that if such currents did exist their magnetic field would in fact be identical with that observed; and since the magnetic field is after all the true essence of the affair there is really no need to complicate the theory by postulating any other independent constitution for the magnetism. Of course the currents so imagined must circulate within the molecule or molecular aggregate which is the ultimate element of the average magnetic polarity; they must be minute current whirls not involving continuous flow in any direction.

The equation in its last form is

$$\int_s (\mathbf{B}\,d\mathbf{s}) = \frac{4\pi}{c}\int (\mathbf{C} + c \operatorname{curl} \mathbf{I})_n\, df,$$

and this points the distinction between the force vector **B** and the induction vector **H**. The magnetic induction **H** being, in the most general case with linear currents only, derivable from a potential is cyclic with respect to the finite linear currents only, but the complete magnetic force is cyclic with respect not only to these finite currents but also with respect to the indefinitely great number of molecular circuits.

160. The mathematical relations of the field. We have now a consistent scheme of definitions to which to apply our analysis and the procedure is almost identical with that of Chapter II. The magnetic induction **H** (mechanically effective force) and the potential ψ are connected by the relation

$$\mathbf{H} = -\operatorname{grad} \psi = -\nabla\psi,$$

whilst the magnetic potential ψ itself at each point of the field satisfies the equation

$$\nabla^2\psi = -\rho$$
$$= \operatorname{div} \mathbf{I}.$$

Thus we have
$$-\operatorname{div} \mathbf{H} = \nabla^2\psi = \operatorname{div} \mathbf{I},$$
or
$$\operatorname{div} (\mathbf{H} + \mathbf{I}) = 0,$$
so that the vector of complete magnetic force, viz.
$$\mathbf{B} = \mathbf{H} + \mathbf{I} = 0,$$
is everywhere solenoidal. This means simply that the surface integral of the normal force over any closed surface whatever in the field is zero. We now see why it is that the force $\mathbf{B}$ is chosen as the complete magnetic force, for on analogy with the electrostatic case, the surface integral of such force must necessarily determine the amount of magnetism inside that surface, which is always zero. The di-polar constitution of magnetism of course necessarily involves the existence of separate poles, but it is a matter of actual experience that it is not possible ever to separate these poles. The alternative view which explains magnetism as the result of molecular current whirls necessarily involves this result as it gives the magnetism an extension not linearly along its axis, but superficially in a plane perpendicular to that axis.

161. When there are surfaces of discontinuity in the field the magnetic induction $\mathbf{H}$ satisfies the usual conditions
$$\mathbf{H}_{1_n} - \mathbf{H}_{2_n} = -(\mathbf{I}_{1_n} - \mathbf{I}_{2_n}),$$
and even the existence of these conditions does not vitiate the solenoidal character of the complete force $\mathbf{B}$, for if we consider the integral $\int \operatorname{div} \mathbf{B}\, dv$ taken throughout any region bounded by the surface f, it must vanish, for at each point $\operatorname{div} \mathbf{B} = 0$; but by Green's theorem it consists of
$$\int_f \mathbf{B}_n df$$
together with the surface integrals arsing from the discontinuities when we pass across the surfaces of different magnetic matter; these are the integrals of
$$\mathbf{B}_{1_n} - \mathbf{B}_{2_n}$$

over the surfaces concerned, or of

$$(\mathbf{H}_{1_n} - \mathbf{H}_{2_n}) + (\mathbf{I}_{1_n} - \mathbf{I}_{2_n})$$

which is zero.

If then we plot the magnetic field in terms of the lines of magnetic force and form them into tubes, we see that the tubes of force are always closed, there being no magnetic matter for them to start or end on. Moreover by the usual argument it is easily seen that the product of the intensity of the force and the cross-section of the tube is constant along each such tube, and it is the product which is presumed to measure the number of the so-called 'unit' tubes enclosed by any one such tube.

162. When there are currents present in the field we have at each point of space

$$\text{curl}\,\mathbf{H} = \frac{1}{c}\,\mathbf{C},$$

$\mathbf{C}$ denoting the current density at the point. Thus at points of the field where there is no current flowing we have

$$\text{curl}\,\mathbf{H} = 0,$$

so that $\mathbf{H}$ is still derived from a magnetic potential ψ at such points, or

$$\mathbf{H} = -\nabla\psi,$$

whilst we still have

$$\nabla^2\psi = \text{div}\,\mathbf{I}$$

with the equivalent boundary conditions at the surfaces of discontinuity in the magnetic media. But now the potential ψ has to satisfy the additional condition of being cyclic with the proper period with respect to the current flow in the field; actually it will consist of two parts; an acyclic potential due to the magnetism alone and a cyclic potential due to the current flow which satisfies in all space outside this flow the characteristic equation

$$\nabla^2\psi = 0,$$

and these two parts are superposed to form the complete potential.

When currents are present in the field it is however generally not convenient to use this potential method at all; the potential function ψ only exists outside the space occupied by the current flow: inside these spaces we must use the vectors $\mathbf{H}$ and $\mathbf{B}$ to define the field, there being no potential from which either is derived. This irregularity of definition for the different parts of the field is at least very confusing and for these general cases we therefore proceed on entirely different lines. When there are no currents in the field we can of course use the present method with great advantage.

163. The vector potential. Since the magnetic force is always solenoidal the value of the integral of its normal component over any part of a surface depends solely on the boundary curve of that surface and must therefore be expressible in terms of that boundary alone. This is accomplished by means of Stokes' conversion of a surface integral into a line integral and requires us to find a vector which is such that its components $(\mathbf{A}_x, \mathbf{A}_y, \mathbf{A}_z)$ satisfy

$$\mathbf{B}_x = \frac{\partial \mathbf{A}_z}{\partial y} - \frac{\partial \mathbf{A}_y}{\partial z}$$

and two similar equations: this is expressed in the form that

$$\mathbf{B} = \operatorname{curl} \mathbf{A},$$

and then for any surface f and the boundary curve s

$$\int_f \mathbf{B}_n df = \int (\mathbf{A}\, d\mathbf{s}),$$

the former integral being taken over f and the latter round its boundary curve.

The vector $\mathbf{A}$ is called the *magnetic vector potential** and its significance will subsequently appear.

164. If a magnetic particle is placed along the axis of z at the origin the potential is

$$\psi = \frac{mz}{4\pi r^3},$$

so that by definition of the vector potential $\mathbf{A}$

$$\frac{\partial \mathbf{A}_z}{\partial y} - \frac{\partial \mathbf{A}_y}{\partial z} = \mathbf{B}_x = -\frac{\partial \psi}{\partial x} = \frac{3mxz}{4\pi r^5},$$

$$\frac{\partial \mathbf{A}_x}{\partial z} - \frac{\partial \mathbf{A}_z}{\partial x} = \mathbf{B}_y = \qquad = \frac{3myz}{4\pi r^5},$$

$$\frac{\partial \mathbf{A}_y}{\partial x} - \frac{\partial \mathbf{A}_x}{\partial y} = \mathbf{B}_z = \qquad = -\frac{m}{4\pi r^3} + \frac{3mz^2}{4\pi r^5}.$$

These equations are solved by

$$\mathbf{A}_x = -\frac{my}{4\pi r^3}, \quad \mathbf{A}_y = \frac{mx}{4\pi r^3}, \quad \mathbf{A}_z = 0.$$

* Maxwell, *Treatise*, II. § 405.

The result is that the vector potential in the case of a magnetic particle is at right angles to the axis of the magnet and to the radius to the point and is of magnitude $\frac{m \sin\theta}{4\pi r^2}$, where θ is the angle between the axis of the magnet and the radius to the point and its sense is that of positive rotation round the axis of the magnet. The expression $\frac{m \sin\theta}{r^2} = \frac{mr \sin\theta}{r^3}$ shows that we may break it up into components, for $mr \sin\theta$ may be interpreted as twice the area of the triangle formed by m and r and the vector is at right angles to the area and its components are the projections. If we now use generally $\mathbf{r}$ to denote the radius vector from the centre of the magnet to any external point in the field, $\mathbf{m}$ the vectorial moment of the particle and $\mathbf{r}_1$ the unit vector along $\mathbf{r}$, then we see that

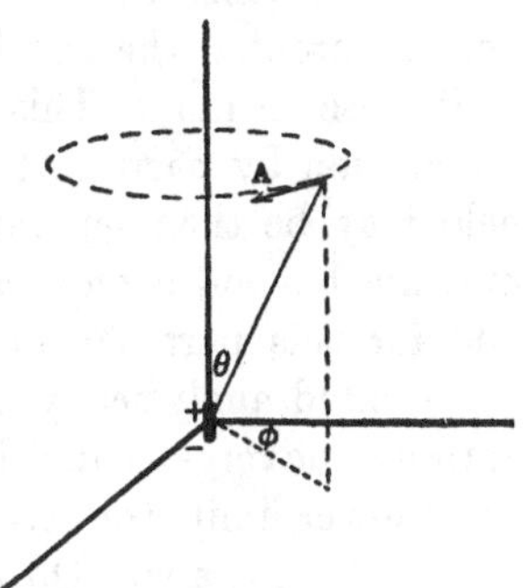

Fig. 21

$$\mathbf{A} = \frac{[\mathbf{m}\mathbf{r}_1]}{4\pi r^2} = \frac{1}{4\pi}[\mathbf{m}\nabla]\frac{1}{r}$$

is the general expression for the vector potential at the point in the field.

165. The above formulae for the vector potential of a magnetic particle may now be used to enable us to write down a vector potential for any finite magnet. We have merely to replace $\mathbf{m}$ by $\mathbf{I}dv$ and then integrate over the whole body. We get at once

$$\mathbf{A} = \frac{1}{4\pi}\int [\mathbf{I}\nabla]\frac{dv}{r}\,*,$$

a formula which certainly applies at points outside the magnetism. At points inside the magnetism r can vanish in an element of the integral but the integral is nevertheless quite convergent and $\mathbf{A}$ is thus representable quantitatively, in a physical theory which neglects purely local actions, by this integral at every point of the field. But as regards its differential coefficients, by means of which $\mathbf{B}$ is derived from it, it is dependent on the unknown distribution of the local polarity. This is of course quite in keeping with the fact

* Maxwell, *Treatise*, II. § 405.

that the magnetic field inside the magnet, considered as due to the aggregate of the molecular magnetic elements, by means of which the vector potential is defined through the relation curl $\mathbf{A} = \mathbf{B}$, itself involves this local contribution. But in the previous theory we were able to discard the local part of the magnetic force, depending on the molecular character of the distribution at the point, from which alone indefiniteness arises. It may be surmised that we should in like manner discard from the vector potential the purely local contribution which is the source of its discontinuity. This may be effected as usual by the aid of integration by parts. At a point inside the material medium the field may be then separated into two parts; the first due to the medium beyond a physically small closed surface surrounding it, and for this part the function $\mathbf{A}$ and its first gradients can be represented analytically by integrals of the above type which are entirely convergent and determinate since r cannot be less than a finite lower limit; the second part is due to the elements inside the surface thus drawn. On integration by parts the contribution of the former to the expression for $\mathbf{A}$ may be put in the form

$$\mathbf{A} = \frac{1}{4\pi}\int \frac{\text{curl}\,\mathbf{I}}{r}\,dv + \frac{1}{4\pi}\left|\int_f [\mathbf{n}_1\mathbf{I}]\frac{df}{r}\right|_1^2,$$

and is thus expressible as a volume integral together with an integral over interfaces of transition of the magnetism, and also an integral over the surface of the cavity: the volume integral is convergent and does not depend on the form of the cavity, while the integral over the surface of the cavity is finite and with the part due to the distribution inside the small surface drawn is the sole representative of the influence of the local molecular configuration; in our present procedure it depends on the form of the cavity; in actual practice it depends on the local molecular configuration. By the general principle, the mechanically effective functions are the analytical integrals obtained by excluding this undetermined local part. This leads to an expression for the total vector potential of the medium treated as continuous

$$\mathbf{A} = \frac{1}{4\pi}\int \frac{\text{curl}\,\mathbf{I}}{r}\,dv + \frac{1}{4\pi}\int \frac{[\mathbf{n}_1\mathbf{I}]}{r}\,df^*,$$

the latter surface integral being taken over the transition boundary of the magnetism.

* Larmor, *Aether and Matter*, p. 260.

166. The values of the magnetic potential in two special cases may be noted. In the case of the magnetic filament we may apply the first formula of paragraph 164 and we find then that

$$4\pi\mathbf{A} = \int_s [m\,d\mathbf{s}, \nabla]\frac{1}{r}.$$

If the filament lies in the axis Oz for example we have

$$4\pi\mathbf{A}_x = -\frac{\partial}{\partial y}\int\frac{m\,ds}{r}, \quad 4\pi\mathbf{A}_y = \frac{\partial}{\partial x}\int\frac{m\,ds}{r}, \quad \mathbf{A}_z = 0,$$

(x, y, z) being the coordinates of the point at which the potentials are calculated. If m is uniform these give

$$\mathbf{A}_x = -m\frac{\partial\phi}{\partial y}, \quad \mathbf{A}_y = m\frac{\partial\phi}{\partial x}, \quad \mathbf{A}_z = 0,$$

where
$$4\pi\phi = \int_s\frac{ds}{r}.$$

A is therefore equal to the gravitational force at the same point due to the same filament loaded with mass m per unit length, but it is directed perpendicular to the plane of this force and the wire.

The uniform magnetic shell case is still simpler. This is obtained from the form for the particle by replacing **m** by $\mathbf{n}_1\tau\,df$ and integrating over the surface of the shell: this gives

$$4\pi\mathbf{A} = \tau\int_f [\mathbf{n}_1\nabla]\frac{df}{r},$$

which on reconversion by Stokes' theorem gives

$$4\pi\mathbf{A} = \tau\int\frac{d\mathbf{s}}{r},$$

these integrals being taken round the boundary of the shell. The form of this result emphasises the fact that the effect of the shell depends only on its strength and the shape of its boundary curve.

167. Since a closed linear current of strength J is equivalent as regards its magnetic field to a magnetic shell of strength $\frac{1}{c}J$ it follows that the vector potential in the field of such a current is simply

$$\mathbf{A} = \frac{1}{4\pi c}J\int\frac{d\mathbf{s}}{r},$$

the integrals being extended round the circuit.

We can even generalise this result to give us the magnetic potential of any distribution of currents in space; in fact regarding such distribution as being made up of current filaments laid side by side we get by simple addition of their fields

$$\mathbf{A} = \frac{1}{4\pi c}\int \mathbf{C}\frac{dv}{r},$$

the integral now being extended over the whole of space.

168. Induced magnetism. We have so far treated the magnetism of a body without reference to its origin, it being always supposed that the distribution of magnetisation in the field was given explicitly among the data of the problem. We made no assumption as to whether this magnetisation is permanent or temporary, except in those parts of our reasoning in which we have supposed the magnet broken up into small portions, or small portions removed from the magnet in such a way as not to alter the magnetisation of any part.

We must now consider, as in the analogous dielectric problem, the magnetisation of bodies with respect to the mode in which it may be produced and changed. A bar of iron held parallel to the direction of the earth's magnetic field is found to become magnetic, with its poles turned the opposite way from those of the earth, or the same way as those of a compass needle in stable equilibrium.

Any piece of soft iron placed in a magnetic field is found to exhibit magnetic properties. If the iron is removed from the field, its magnetic properties are greatly weakened or disappear altogether. On the other hand a piece of hard iron or steel retains its magnetic properties acquired when placed in a magnetic field.

If a magnet could be constructed so that the distribution of its magnetisation is not altered by any magnetic force brought to act on it, it might be called a permanently or rigidly magnetised body. There is no known magnetic substance which perfectly fulfils this condition, but it is nevertheless convenient for scientific purposes to make a distinction between the permanent and temporary magnetisation, defining the permanent magnetisation as that which exists independently of the magnetic forces, and the temporary magnetisation as that which depends on those forces. This distinction is however not founded on a knowledge of the intimate nature of the magnetisable substances: it is only the expression of a convenient hypothesis*.

169. The temporary magnetisation in a particle is always dependent on the magnetic force acting through it, and it is therefore called the *induced magnetisation* and the body is said to be magnetised by *induction*. In such a case the polarisation intensity must be a vector function of the magnetic force. Unless this relation is a linear one, that is, unless

$$\mathbf{I}_x = \mu_{11}'\mathbf{B}_x + \mu_{12}'\mathbf{B}_y + \mu_{13}'\mathbf{B}_z + \mathbf{I}_{0_x},$$
$$\mathbf{I}_y = \mu_{21}'\mathbf{B}_x + \mu_{22}'\mathbf{B}_y + \mu_{23}'\mathbf{B}_z + \mathbf{I}_{0_y},$$
$$\mathbf{I}_z = \mu_{31}'\mathbf{B}_x + \mu_{32}'\mathbf{B}_y + \mu_{33}'\mathbf{B}_z + \mathbf{I}_{0_z},$$

the mathematical theory is quite intractable, so we shall frequently assume these to be true. Fortunately for a large number of substances such linear relations appear to represent the facts exactly, but there are certain practically important cases where they do not even approximately represent the relation between **I** and **B**.

If $\mathbf{I}_0 = 0$ there is no permanent magnetism and if the medium is isotropic the nine coefficients μ_{rs}' reduce to one and in that case

$$\mathbf{I} = \mu'\mathbf{B} + \mathbf{I}_0.$$

The simple linear law will be approximately correct if the inducing field is small.

170. Let us now examine the magnetic field equations on the assumption that some such linear relation holds. The formulation involves three vectors: (i) **I** the intensity of magnetisation, (ii) **B**, the magnetic force which is always a stream vector, and (iii) **H**, the magnetic flux or induction vector which is the gradient of a potential in all statical cases. When

$$\mathbf{I} = \mathbf{I}_0 + \mu'\mathbf{B}$$

we have

$$\mathbf{H} = \mathbf{B}(1 - \mu') - \mathbf{I}_0$$
$$= \mu\mathbf{B} - \mathbf{I}_0.$$

* Various forms of the mathematical theory of magnetism have recently been constructed on the assumption of the existence of a distribution of permanent magnetic matter. In this case the magnetic induction vector is no longer solenoidal, its divergence determining the density of the magnetic matter. This procedure is adopted in order to secure a closer analogy with the electric case and to remove certain discrepancies supposed to exist in the more usual form of the theory. Cf. E. Cohn, *Das electromagnetische Feld*, p. 510; R. Gans, *Ann. der Physik*, 13 (1904), p. 634 and *Encyclopädie der math. Wissensch.* Bd. v. Art. 15.

The coefficient $\mu = 1 - \mu'$ is called generally the *coefficient of induction* of the medium; it is not now the same as Kelvin's permeability, for we have interchanged the significance of our vectors. If there is no permanent magnetism about μ is exactly the reciprocal of Kelvin's coefficient, but now it expresses the permeability to mechanically effective force. If $\mu = 1$ as is the case in space free from magnetisable media, the whole of the magnetic force is mechanically effective; but when, as is for example the case in iron, μ is small the material itself practically destroys the mechanical effectiveness of the complete field.

In the mathematical formulation of the scheme it is better to express everything in terms of the potential ψ; in the most general case

$$\operatorname{div} \mathbf{B} = 0$$

and

$$\mathbf{H} = -\operatorname{grad} \psi = -\nabla\psi,$$

so that

$$\operatorname{div}\left(\frac{\mathbf{H} + \mathbf{I}_0}{\mu}\right) = 0,$$

or

$$\operatorname{div}\left(\frac{1}{\mu} \operatorname{grad} \psi\right) = \operatorname{div}\left(\frac{1}{\mu}\mathbf{I}_0\right),$$

which if the medium is homogeneous reduces to

$$\nabla^2\psi = \operatorname{div} \mathbf{I}_0.$$

These are the characteristic equations of the potential in the theory. Any problem as to the effect of the presence of magnetisable media in a magnetic field now simply reduces to a determination of the appropriate solutions in the different regions of the field, these being chosen so as to agree with one another at the adjoining boundaries.

171. The boundary conditions are easily obtained in the usual way. In the first place it is clear that the potential function itself must be continuous over the whole of space, and this carries with it the equality of the tangential components of the induction on either side of a boundary surface; further the normal component of the force is always continuous across any surface or

$$\mathbf{B}_{1_n} = \mathbf{B}_{2_n},$$

the suffices 1 and 2 denoting the different media on the two sides of the surface. But in each media

$$\mathbf{B} = \frac{1}{\mu}(\mathbf{H} + \mathbf{I}_0),$$

so that
$$\frac{1}{\mu_1}(\mathbf{H}_1 + \mathbf{I}_{0_1})_n = \frac{1}{\mu_2}(\mathbf{H}_2 + \mathbf{I}_{0_2})_n,$$
or in terms of the potential
$$\frac{1}{\mu_1}\frac{\partial\psi_1}{\partial n} - \frac{1}{\mu_2}\frac{\partial\psi_2}{\partial n} = \left(\frac{1}{\mu_1}\mathbf{I}_{0_{1n}} - \frac{1}{\mu_2}\mathbf{I}_{0_{2n}}\right).$$
If, as is often the case, one of the media is air we can put $\mu_2 = 1$ and $\mathbf{I}_{0_2} = 0$ so that this condition reduces to
$$\frac{1}{\mu_1}\frac{\partial\psi_1}{\partial n} - \frac{\partial\psi_2}{\partial n} = \frac{1}{\mu_1}\mathbf{I}_{0_{1n}},$$
or in the complete absence of permanent polarisation
$$\frac{1}{\mu_1}\frac{\partial\psi_1}{\partial n} = \frac{\partial\psi_2}{\partial n}.$$

172. This potential method again becomes ineffective in the most general case when the magnetic field is in part due to a distribution of currents throughout it or any part of it. In these cases it is more convenient to define the field by its vector potential **A** which is defined by the volume integral
$$\mathbf{A} = \frac{1}{4\pi c}\int(\mathbf{C} + c\,\mathrm{curl}\,\mathbf{I})\frac{dv}{r},$$
in the absence of abrupt transition layers. From the form of this definition it is obvious, on analogy with the electrostatic potential case, that each component of **A** satisfies the characteristic equation
$$\nabla^2\mathbf{A} = \frac{1}{c}(\mathbf{C} + c\,\mathrm{curl}\,\mathbf{I}),$$
at each point of the field. Where there is no current or magnetism we have
$$\nabla^2\mathbf{A} = 0,$$
and thus we have similar characteristic equations to define the function **A** in the case of continuous distributions of currents and magnetism. In practice discontinuities only occur on certain closed surfaces in the field and we then solve the above equations separately for the different volumes into which these surfaces divide the space and then fit up the different solutions on the boundaries in the usual way.

If the magnetisation $\mathbf{I}$ is in part induced and in part permanent then in the case when a linear law of induction is valid and the medium is homogeneous and isotropic we have

$$\mathbf{I} = \mu'\mathbf{B} + \mathbf{I}_0,$$

and thus

$$\begin{aligned} \operatorname{curl} \mathbf{I} &= \mu' \operatorname{curl} \mathbf{B} + \operatorname{curl} \mathbf{I}_0 \\ &= \mu' \operatorname{curl} \operatorname{curl} \mathbf{A} + \operatorname{curl} \mathbf{I}_0 \\ &= \mu' (\nabla^2\mathbf{A} - \nabla \operatorname{div} \mathbf{A}) + \operatorname{curl} \mathbf{I}_0. \end{aligned}$$

Now the vector potential $\mathbf{A}$ is not completely defined by the relation

$$\mathbf{B} = \operatorname{curl} \mathbf{A},$$

and we may add the further condition

$$\operatorname{div} \mathbf{A} = 0,$$

which is in fact satisfied by the special values of this function given above, without restricting the generality of our analysis. If we do this we have

$$\operatorname{curl} \mathbf{I} = \mu'\nabla^2\mathbf{A} + \operatorname{curl} \mathbf{I}_0,$$

and thus the characteristic equation for $\mathbf{A}$ becomes

$$\nabla^2\mathbf{A} = \frac{1}{c}(\mathbf{C} + \mu' c\nabla^2\mathbf{A} + c \operatorname{curl} \mathbf{I}_0)$$

or

$$\mu\nabla^2\mathbf{A} = \frac{1}{c}(\mathbf{C} + c \operatorname{curl} \mathbf{I}_0),$$

wherein $\mu = 1 - \mu'$ is the permeability coefficient of the medium.

The boundary conditions at a surface of continuity can also be interpreted in terms of the values of the vector potential components. In fact again on analogy with the electrostatic case we see that the discontinuity in $\mathbf{A}$ at a surface where the magnetisation changes abruptly is the same as the discontinuity of the vector $[\mathbf{n}_1\mathbf{I}]$ at the surface, $\mathbf{n}_1$ being the direction vector for the normal to the surface at the point. If we care to take account of the induced nature of the magnetism and again assume a linear law we can say that the discontinuity in $\mathbf{A}$ is the same also as that of $\frac{1}{\mu}[\mathbf{n}_1\mathbf{I}_0]$, $\mathbf{I}_0$ being the part of the magnetisation which is not so induced.

At the only other type of discontinuity, a surface current sheet, the components of $\mathbf{A}$, like potentials in statics, are continuous (this ensures the continuity of the normal induction—a necessary condition), but the gradients of these components are discontinuous in such a way that the discontinuities in the tangential components of **B** are proportional respectively to the perpendicular components of the surface current flow.

173. The coefficients of mutual and self-induction. Before leaving these general relations we must return for a moment to the idea of the flux of magnetic force for a current circuit in any field. This is defined as the total number of tubes of magnetic force which thread through the circuit and analytically it is given by the integral (cf. page 168)

$$N = \int (\mathbf{A} d\mathbf{s})$$

taken round the circuit. If N is due in part to a series of linear conductors carrying currents $J_1, J_2, \ldots$ we know that the corresponding part of the vector potential $\mathbf{A}$ is given by

$$\frac{1}{4\pi c} \Sigma J_n \int_{s_n} \frac{d\mathbf{s}_n}{r_n},$$

so that the term in N due to these currents is $\Sigma a_n J_n$ where

$$a_n = \frac{1}{4\pi c} \int_s \int_{s_n} \frac{(d\mathbf{s}\, d\mathbf{s}_n)}{r}.$$

This constant a_n is the coefficient of mutual induction for the given circuit and the nth circuit giving rise to the field. It is obviously a true mutual coefficient depending on the form and relative configuration of the two circuits.

A single current circuit has of course a magnetic field of its own and therefore also it will have a coefficient of induction on itself, or as we say, a coefficient of *self-induction*, but a more refined analysis is needed in this case for its evaluation.

CHAPTER V

THE DYNAMICS OF THE MAGNETIC FIELD

174. The energy relations of the magnetic field. The analysis of the general energy and mechanical relations of the magnetic field with induced magnetism is to a certain extent analogous to the similar problem connected with polarised dielectric media discussed in Chapter II. In so far as this is the case we can therefore content ourselves by quoting results, interpreting them however in terms of the vectors of the present theory.

The work done by the magnetic forces in separating the poles of a small doublet $\mathbf{m}$ at a point in the field where the force is $\mathbf{B}$ is

$$(\mathbf{mB}),$$

and this in the usual way represents a diminution in the magnetic potential energy of the field. Thus in the establishment of a distribution of magnetic polarisation of intensity $\mathbf{I}$ the magnetic potential energy increases by

$$-\int (\mathbf{IB})\, dv.$$

This store of potential energy which is associated with the magnetic field and media is created by the combined action of the mechanical forces acting on the media and by the internal elastic or motional forces resisting the setting up of the polarisation. If then in any small virtual displacement the work of the external forces is δW and the internal energy increases by δE_i—so that the work done by the internal forces themselves is $-\delta E_i$—we have

$$\begin{aligned}\delta W - \delta E_i &= -\delta \int (\mathbf{IB})\, dv \\ &= -\int (\mathbf{I}\delta\mathbf{B})\, dv - \int (\mathbf{B}\delta\mathbf{I})\, dv.\end{aligned}$$

The usual argument by the principle of virtual work now shows that

$$\delta W = -\int (\mathbf{I}\delta\mathbf{B})\, dv$$

is the virtual work of the mechanical forces acting on the magnetic media, from which these forces can be derived. In fact this is the part of the total energy variation which is left when the internal constitution defined by the vector **I** is maintained constant, so that the internal energy associated with it must also be constant.

175. Confining attention for the moment to the element δv we see that the work supplied by it to outside mechanical systems which it drives—that is, the reverse of the work which the reaction forces exerted by these outside systems themselves do—in traversing any path is thus

$$\delta v \int (\mathbf{I}\delta\mathbf{B}),$$

the integral being taken along the path. If **I** is a function of **B**, that is, if the magnetism is in part thoroughly permanent and in part induced without hysteresis, so that the operation is reversible, this work must vanish for a complete cycle; otherwise energy would inevitably be created either in the direct path or else in the reversed one of the complete system of which δv is a part. Thus the negation of perpetual motion in that case demands that

$$(\mathbf{I}\delta\mathbf{B}) = d\phi,$$

$d\phi$ being the complete differential of a function ϕ of $(\mathbf{B}_x, \mathbf{B}_y, \mathbf{B}_z)$ involving only even powers. If the polarisation follows a linear law, as it must do if the field is small, ϕ is quadratic or

$$\phi \equiv \tfrac{1}{2}(\mu_{11}'\mathbf{B}_x^2 + \mu_{22}'\mathbf{B}_y^2 + \mu_{33}'\mathbf{B}_z^2 + 2\mu_{12}'\mathbf{B}_x\mathbf{B}_y + \ldots),$$

so that
$$\mathbf{I}_x = \mu_{11}'\mathbf{B}_x + \mu_{12}'\mathbf{B}_y + \mu_{13}'\mathbf{B}_z,$$
$$\mathbf{I}_y = \mu_{21}'\mathbf{B}_x + \mu_{22}'\mathbf{B}_y + \ldots,$$
$$\mathbf{I}_z = \mu_{31}'\mathbf{B}_x + \ldots.$$

In the simple case of isotropic media we have

$$\phi = \tfrac{1}{2}\mu'\mathbf{B}^2,$$

so that
$$\mathbf{I} = \mu'\mathbf{B}.$$

In addition to this energy concerned with the attractions, the magnetic forces in the field expend energy in polarising or orientating the individual molecules against the internal forces of the medium, of aggregate amount in the whole field

$$\delta E_i = \int (\mathbf{B}\,\delta\mathbf{I})\,dv.$$

This part has nothing to do with mechanical forces. It is stored up as internal energy of a purely elastic or thermal character. If the polarisation gradually breaks away some or all of this is lost in heat and the phenomenon of hysteresis results.

When the relation between **I** and **B** is linear and isotropic the two parts of the total energy, the mechanical and molecular parts, are both equal numerically

$$\delta v \int (\mathbf{B}\,\delta\mathbf{I}) = \delta v \int \mu' \frac{\delta\,(\mathbf{B}^2)}{2} = \delta v \int \delta \left(\frac{\mu' \mathbf{B}^2}{2}\right)$$

$$= \delta v \int (\mathbf{I}\,\delta\mathbf{B}),$$

and this is the usual theoretical result. If there is no hysteretic loss of any kind we may take

$$-W = \int \frac{\mu' \mathbf{B}^2}{2}\,dv,$$

where the integral is extended throughout the substance, as the potential energy of the mechanical forces acting on the medium.

176. When there are currents in the field the differential work equations of the previous paragraphs have to be modified by the inclusion of terms depending on these currents.

The energy of an electric current in a magnetic field, that is the mutual energy of the field and the current, can be calculated as the energy of the equivalent magnetic shell in the same field and it is therefore equal to

$$-\frac{1}{c} J \int (\mathbf{n}_1 \mathbf{B})\,df = -\frac{1}{c} J \int \mathbf{B}_n df$$

$$= -\frac{1}{c} NJ,$$

N being the number of unit tubes of magnetic force which thread the circuit.

This fundamental result may also be approached in another more tentative manner. Taking first the case where the magnetic field is due to a simple magnetic pole of strength m, the mutual potential energy of the pole and the current J is the potential of either in the field of the other. In this special case it is therefore

$$-\frac{m}{c} J\omega,$$

where ω is the solid angle subtended by the current circuit at the position of m. But this is equal to

$$-\frac{1}{c}NJ,$$

for $m\omega$ is the number of tubes of force in the field of the pole m which pass through J, or its aperture, in the positive direction corresponding to the circulation. By addition then the general result is seen to be true, whether or not a potential function exists at all points of space.

177. In the case when there are a number of linear current circuits in the field as well as the magnetic media the total magnetic potential energy in the field is

$$-\int (\mathbf{IB})\,dv - \frac{1}{c}\sum_{r=1}^{n} N_r J_r,$$

and now the variational equation is

$$\delta W - \delta E_i = -\int (\mathbf{I}\delta\mathbf{B})\,dv - \frac{1}{c}\sum_{r=1}^{n} J_r \delta N_r - \int (\mathbf{B}\delta\mathbf{I})\,dv - \frac{1}{c}\sum_{r=1}^{n} N_r \delta J_r,$$

in which again the two parts are easily recognised, the currents J serving for each circuit to define an intrinsic property of that circuit. Thus

$$\delta W = -\int (\mathbf{I}\delta\mathbf{B})\,dv - \frac{1}{c}\sum_{r=1}^{n} J_r \delta N_r$$

is the work done in the displacement by the external forces applied to the magnetic media and current circuits, whilst

$$\delta E_i = \int (\mathbf{B}\delta\mathbf{I})\,dv + \frac{1}{c}\sum_{r=1}^{n} N_r \delta J_r$$

represents the total change in the internal energy of the system.

In this more general case the usual argument based on the negation of perpetual motion requires that δE_i should be a complete differential in all cases not involving hysteretic degradation. That is that E_i is a function of the current intensities J_r such that

$$\frac{\partial E_i}{\partial J_r} = \frac{1}{c} N_r.$$

Thus also

$$\frac{\partial N_r}{\partial J_s} = \frac{\partial N_s}{\partial J_r} = c\,\frac{\partial^2 E_i}{\partial J_r \partial J_s},$$

for all values of r and s. In the simplest cases E_i will be quadratic and thus the N's will be linear functions of the J's with constant terms depending possibly on the distribution $\mathbf{I}_0$. The significance of these relations and the coefficients in them—the coefficients of self and mutual induction of the circuits—has already been discussed and need not further detain us; the only additional point brought out in the present method of approach is that the mutual relation

$$a_{rs} = a_{sr}$$

for any two circuits is an essential general consequence of the principle of conservation of energy.

178. Now let us apply these results in a particular case. Suppose the magnetic field arises from a distribution of rigid magnetic polarity of density $\mathbf{I}_0$ at any point in the field together with a system of linear currents. The total energy in the field can then be calculated as the work done in building up the rigid magnetism and the currents gradually in the presence of the magnetisable substances, the induced magnetism simultaneously taking the appropriate value at each stage of this process.

If then at any instant the force intensity in the field is $\mathbf{B}$ the work done by external agency in bringing up an additional small increment of polarity $\delta\mathbf{I}_0$ to each place in the field is equal to

$$-\int (\mathbf{B}\,\delta\mathbf{I}_0)\,dv,$$

integrated over the field. Similarly the work done in increasing the currents by the small amount δJ amounts to

$$-\frac{1}{c}\sum_{r=1}^{n} N_r \delta J_r = -\frac{1}{c}\sum_{r=1}^{n} \delta J_r \int_{f_r} \mathbf{B}_n\,df_r,$$

where N_r is the total flux of the field through the circuit of the typical current and f_r is a barrier surface closing this circuit on which therefore N_r can be counted.

The total amount of energy stored in the field is therefore such that its differential increment is

$$-\int (\mathbf{B}\,\delta\mathbf{I}_0)\,dv - \frac{1}{c}\sum_{r=1}^{n} \delta J_r \int_{f_r} \mathbf{B}_n\,df_r.$$

179. The last part of this differential expression can however be converted into a volume integral over the field. In fact if ϕ is the magnetic potential of the field at the typical stage of its development

$$\frac{1}{c}\,\delta J = \int_s \delta\mathbf{H}_s\,ds = \delta\phi_1 - \delta\phi_2,$$

where s denotes any closed path which encircles once the typical current only and $\delta\phi_1$, $\delta\phi_2$ denote the values of $\delta\phi$ at two infinitely near points on either side of the barrier f, the first being obtained on starting along s where it cuts through f and the second on arriving back at f. Thus our integral for each circuit is

$$-\int_f \mathbf{B}_n\,(\delta\phi_1 - \delta\phi_2)\,df,$$

or if we reckon the two sides of the barrier as different sides of the same closed surface we have this equal to

$$-\int_f \mathbf{B}_n\,\delta\phi\,df.$$

Since each of the currents has a barrier surface the whole space is a singly connected region with the two sides of each barrier and a surface at infinity as boundary. We may therefore apply Green's theorem to the region and convert the surface integral into the volume integral

$$\int (\mathbf{B}\nabla)\,\delta\phi\,dv + \int \delta\phi\,\operatorname{div}\mathbf{B}\,dv,$$

the infinite surface contributing nothing in all regular cases*. Since $\operatorname{div}\mathbf{B} = 0$ and $\nabla\delta\phi = -\,\delta\mathbf{H}$ this is simply

$$-\int (\mathbf{B}\,\delta\mathbf{H})\,dv,$$

and is therefore an integral over the whole field.

The total work put into the field is thus expressible in the form

$$-\int (\mathbf{B}\,\delta\mathbf{H})\,dv - \int (\mathbf{B}\,\delta\mathbf{I}_0)\,dv.$$

Of course if there are no currents in the field we have

$$\int (\mathbf{H}\mathbf{B})\,dv = \int (\mathbf{H}\,\delta\mathbf{B})\,dv = \int (\mathbf{B}\,\delta\mathbf{H})\,dv = 0.$$

* The change of sign occurs because the $\mathbf{B}_n$ is the component of $\mathbf{B}$ along the inward normal at the boundary of our volume.

180. Now if $\mathbf{I}$ is the intensity of the induced magnetism at any stage

$$\mathbf{B} = \mathbf{H} + (\mathbf{I} + \mathbf{I}_0),$$

and thus

$$\delta\mathbf{B} = \delta\mathbf{H} + \delta\mathbf{I}_0 + \delta\mathbf{I},$$

and the total work put into the field is therefore equal to

$$-\int (\mathbf{B}\,\delta\mathbf{B})\,dv + \int (\mathbf{B}\,\delta\mathbf{I})\,dv$$

$$= -\tfrac{1}{2}\delta\int \mathbf{B}^2 dv + \int (\mathbf{B}\,\delta\mathbf{I})\,dv.$$

The second part of this total energy represents the internal elastic energy stored up in the magnetic media on account of the magnetic polarity induced in them; it is stored up in these media as energy of an effectively non-magnetic nature and is mechanically unavailable. The first part therefore represents the true magnetic potential energy of the field and on a tentative theory we could regard it as distributed throughout the field with a density at each place equal to

$$-\tfrac{1}{2}\,\delta\mathbf{B}^2$$

for the infinitesimal change or

$$-\tfrac{1}{2}\mathbf{B}^2$$

for the complete field.

181. In the case of a linear law of induction and when the medium is isotropic we have

$$\mathbf{I} = \mu'\mathbf{B}$$

$$= (1 - \mu)\,\mathbf{B},$$

so that the energy stored in the media on account of the polarisations induced in them appears as the integral of

$$\left(\frac{1-\mu}{2}\right)\mathbf{B}^2.$$

The total energy in the field in this case however appears as the volume integral of

$$-\tfrac{1}{2}\mathbf{B}^2 + \left(\frac{1-\mu}{2}\right)\mathbf{B}^2$$

$$= -\frac{\mu}{2}\,\mathbf{B}^2,$$

so that it can be regarded as distributed throughout the field with this density.

182. Of the total energy of purely magnetic nature stored in the field the part

$$-\int dv \int_0^H (\mathbf{I}_0 d\mathbf{B})$$

is concerned mainly with the rigidly magnetised masses, being in fact the force function of their mechanical interactions and of the interactions of the currents and induced magnetism with them. Of the remainder, the part

$$-\int dv \int_0^H \mathbf{I} d\mathbf{B}$$

is concerned in a similar manner with the induced polarity, being the force function of the mechanical actions on the induced magnets.

The remainder, or

$$-\tfrac{1}{2}\int \mathbf{B}^2 dv + \int dv \int_0^H (\mathbf{I} + \mathbf{I}_0,\ d\mathbf{B}),$$

is therefore concerned solely with the currents and is the force function of their mechanical interactions and the interactions of the magnets with them. Since

$$\mathbf{B} = \mathbf{H} + (\mathbf{I} + \mathbf{I}_0)$$

this part reduces to

$$-\int dv \int_0^H (\mathbf{H}\delta\mathbf{B}),$$

and may therefore be regarded as distributed throughout the field with the density

$$-\int_0^H (\mathbf{H}\delta\mathbf{B})$$

at each place. If, as is generally the case, there are no rigid magnets about, this latter part of the total energy is the only part that is mechanically available, provided it is not prevented from so being by frictional forces tending to degrade it. The part of the energy corresponding to the induced polarisations and arising from their reactions with the currents is in some respects mechanically available, but only in so far as the presence of the magnetic masses increases the available energy associated with the currents giving rise to the field.

A similar transformation to that employed above in the reverse manner shows that

$$-\int dv \int_0^H (\mathbf{H}\delta\mathbf{B}) = -\frac{1}{c}\Sigma \int_0^N J dN,$$

where Σ denotes a sum relative to the linear circuits. This exhibits the energy in its relation to the currents in its more usual form.

183. In some important cases the process of establishing a field with induced polarisation is not a reversible one, that is a gradual reversal of the field from any stage in its establishment will not result in the traverse of the previous states in their exact reverse order, but in an entirely new set of states. In this case in any complete cycle of changes the integral

$$-\delta v\int(\mathbf{I}\,d\mathbf{B})$$

is not zero, so that it represents a loss of energy in the cycle. This is the so-called *hysteresis* loss. It is also represented by

$$-\int(\mathbf{BI})/\delta v+\delta v\int(\mathbf{B}\,d\mathbf{I}),$$

and the first term is zero on the average for a large number of cycles. Thus the energy loss per unit volume is

$$\int(\mathbf{B}\,d\mathbf{I})=\int(\mathbf{B}\,d\mathbf{B})-\int(\mathbf{B}\,d\mathbf{H})-\int(\mathbf{B}\,d\mathbf{I}_0),$$

and the first term vanishes in a complete cycle, so this is

$$-\int(\mathbf{B}\,d\mathbf{H})-\int(\mathbf{B}\,d\mathbf{I}_0),$$

or if there is no rigid magnetism about, this is simply

$$-\int(\mathbf{B}\,d\mathbf{H}),$$

the integral being taken throughout the cycle. This means that if a piece of iron is subjected to the complete cycle of changes involved in magnetising it to any intensity, demagnetising it and taking it through to the same intensity in the opposite direction and then back again to its zero value, then of the energy which has been put into this process the amount

$$L=-\frac{1}{4\pi}\int dv\int(\mathbf{H}\,d\mathbf{B})$$

is converted into heat in the iron and is consequently lost to the system.

Such losses of energy are of course of considerable importance in engineering practice and a good deal of experimental work has been carried out to determine the precise conditions governing them. The energy dissipated in a single cycle in soft iron may amount to 10^4 ergs per cubic centimetre and in hard steel to even ten times this amount. Various empirical formulae have been proposed to allow of its approximate calculation for practical purposes. Until quite recently the best of these appeared to be the Steinmetz formula expressing the loss as a function of the maximum H used in the cycle

$$L = \eta H^{\beta},$$

where $\beta = 1{\cdot}6$ approximately and L is a constant of the material used. Recent work seems however to favour the somewhat simpler binomial formula

$$L = \alpha H + \beta H^2,$$

a form which is in fact suggested by the geometrical shape of the (B-H) curve whose area represents this loss for one cycle.

184. Kinetic or potential energy? Since all our expressions for magnetic energy occur with a negative sign it is reasonable to enquire as to whether it is not possible to alter our conception of these things in such a way as will enable us to use the expressions with the positive sign.

Now it will be remembered that in general dynamics everything is determined by the Lagrangian function L which is the difference between the kinetic energy T and the potential energy W,

$$L = T - W.$$

Thus if a part of the energy of any system be counted as kinetic energy, that is reckoned in T, it must have the opposite sign to what it would have if reckoned in W. At the bottom all potential energy is probably of kinetic origin, but if counted as kinetic its sign must be reversed. Calling it potential energy merely means that we do not want to trouble about its actual constitution*.

The difference of sign in the energies does not of course affect the sign of the forces with which they are associated. In the most general case for steady systems the Lagrangian analysis shows that

* The process of 'ignoring' coordinates in analytical dynamics is practically equivalent to converting the energy in the coordinates from kinetic to potential. Cf. Larmor, *Proc. L. M. S.* 15 (1884).

the internal forces are determined as the positive gradients of the function, so that for instance the force in the θ coordinate is

$$\frac{\partial L}{\partial \theta} = \frac{\partial T}{\partial \theta} - \frac{\partial W}{\partial \theta}.$$

Thus if the energy is reckoned as kinetic the force is determined as its positive gradient, whereas if it is potential the negative gradient has to be taken. The difference of sign in the gradient thus just balances the difference in the sign of the types of energy.

185. The suggestion has therefore been made that all magnetic energy should be treated as kinetic energy. This of course conforms to the modern view as to the mechanism involved.

In the establishment of a current by the direct application of a finite electromotive force in the current circuit we find that the full current is not immediately produced, it rises gradually to its steady value. We may therefore enquire as to what the electromotive force is doing during this time when the opposing resistance is not able to balance it. It is increasing the current of course; but an ordinary force acting on a body in the direction of its motion increases its momentum and communicates kinetic energy to it, or the power of doing work on account of its motion. Thus if, as appears most natural, we assume that a current has motional energy, we may say that the unresisted part of the electromotive force has been employed in increasing the internal motional or kinetic energy of the current. But the internal motional energy of a current is identical with the energy of the magnetic field associated with the current, so that if we regard the energy of a current as of the kinetic type we must also regard the energy in a magnetic field as kinetic energy.

There is of course no essential difference between the magnetic field of a current and that of a magnet, so that we must also assume in that case that *all* magnetic energy is kinetic. This of course fits in with the Ampèrean view which regards magnetism as constituted of minute molecular current whirls.

186. In spite of the slight confusion which may thus arise we shall follow Faraday's suggestion and treat all magnetic energy as of the kinetic type and we shall henceforth denote it by T. It must however be particularly emphasised that we have no definite proof that this energy is kinetic. It is merely a matter of convenient choice so to regard it, and the mechanical relations of our system are entirely independent of such choice, for the Lagrangian function, on which everything depends, is the same in both cases.

In our previous work we have always regarded the energy of magnetism as potential energy and all our results have been deduced on this basis. Thus if we quote in future any of our previous expressions for magnetic energy care must be exercised to see that they are in all cases quoted with the reversed sign. As can be well imagined this sudden change of procedure is of some significance in the theory, the lack of appreciation of which has resulted in some confusion in the usual presentations of this subject.

187. The magnetomotive forces. The forces acting on a magnetically polarised medium can now be obtained either from the energy expressions or by an analysis by forces on the elementary poles similar to that given in the second chapter. But now it is the force on the volume elements alone that is required and therefore the part of the complete force **B** which has to be used is the mechanically effective part **H**, omitting the local term in **I**, the polarisation. This local part of the force will of course act on the medium but the reaction to it is not transmitted by material stress, but is compensated on the spot by molecular action due to change of physical state induced by it. Thus the mechanically effective force is made up of a bodily force **F** and a torque **G**, where

$$\mathbf{F} = (\mathbf{I}\nabla)\,\mathbf{H}, \quad \mathbf{G} = [\mathbf{IH}].$$

Under the usual circumstances these expressions are identical with the ones given by Maxwell in his treatise. The remarkable property is there established, and is the direct analogue of our result for polarised media, that independently of the form of the relation between the magnetic induction and magnetic force in the medium and whether there is permanent magnetism or not, this bodily forcive can be formally represented in explicit terms as equivalent to an imposed stress: viz. it is the same as would arise from (i) a hydrostatic pressure $\frac{1}{2}H^2$, (ii) a tension along the bisector of the angle θ between H and B, equal to $HB\cos^2\theta$, (iii) a pressure along the bisector of the supplementary angle, equal to $BH\sin^2\theta$, together with an outstanding bodily torque turning from **B** towards **H** equal to $HB \sin 2\theta$. When **B** and **H** are in the same direction, the torque vanishes, and a pure stress remains in the form of a tension $(BH - \frac{1}{2}H^2)$ along the lines of force and a pressure $\frac{1}{2}H^2$ in all directions at right angles to them. There is of course no warrant for taking this stress to be other than a mere geometrical representation of the bodily forcive. It is however a convenient one for some purposes. Thus the traction acting on the layer of transition between two media, in which **H** changes very rapidly, which might be directly deduced in the same manner as the electric traction above may also be expressed directly as the resultant of these Maxwellian tractions towards the two sides of the interface. As there cannot be free magnetic surface density the traction on the interface is represented, under the most general circumstances, whatever extraneous magnetic field may there exist, by purely normal pull of intensity $2\pi \mathbf{I}_n{}^2$ towards each side.

188. To obtain the forces on the magnetic media or the rigid magnets we may however—and often with more advantage—proceed to calculate first the energy of the media in terms of the coordinates which determine their general configuration and then to apply the general principle of virtual work.

If the positions of the magnets are determined by generalised coordinates $\theta_1, \theta_2, \ldots$ in the Lagrangian sense and if the external or applied force components corresponding to these coordinates are $\Theta_1, \Theta_2, \ldots$, then the work done by external agency during a displacement of these media is

$$\Theta_1 \delta\theta_1 + \Theta_2 \delta\theta_2 + \ldots,$$

whilst the work done by the internal forces resisting the setting up of the polarisation is

$$-\int_v (\mathbf{B}\,\delta\mathbf{I})\, dv.$$

Thus in the most general virtual change in the configuration of the system—a complete definition of which involves a specification of the polarisation—the magnetic potential of the system increases by

$$\delta W = -\int (\mathbf{B}\,\delta\mathbf{I})\, dv + \Theta_1 \delta\theta_1 + \Theta_2 \delta\theta_2 + \ldots.$$

The usual argument based on the assumption of reversibility and the negation of perpetual motion now requires that δW should be a complete differential of some function W which ultimately measures relative to some standard configuration the potential energy of the system which it possesses in virtue of the magnetisation of its constituent masses. We have seen however in our previous discussions that under the simplest conditions

$$W = -\tfrac{1}{2}\int (\mathbf{BI})\,dv,$$

so that its complete evaluation is a mere matter of integration.

If now the polarisation is maintained constant during the displacement the first part of δW vanishes and we have then simply

$$\delta W = \Theta_1\delta\theta_1 + \Theta_2\delta\theta_2 + \dots,$$

or since δW is a complete differential

$$\Theta_r = \frac{\partial W}{\partial\theta_r}$$

is the force component in the coordinate θ_r which is exerted on the magnetic media by external agency in order to maintain the presumed equilibrium conditions.

189. The determination of the forces exerted on the conductors carrying currents through a magnetic field can be effected in the same general manner. The general equation of virtual change of energy, now including a term due to changes in the current strengths, is

$$\delta W = -\int_v (\mathbf{B}\,\delta\mathbf{I})\,dv - \frac{1}{c}\sum_{r=1}^{n} N\delta J + \Theta_1\delta\theta_1 + \Theta_2\delta\theta_2 + \dots,$$

where also the coordinates θ include those of the conductors: and the value of W is

$$-\tfrac{1}{2}\int \mathbf{B}^2 dv,$$

integrated throughout the field.

The force in any coordinate θ_r is then

$$\Theta_r = \frac{\partial W}{\partial\theta_r},$$

and can be found as soon as the value of W can be determined in the appropriate form. This general method of procedure for finding the forces, to which we shall return at a later stage, is not however always the most convenient. It is sometimes simpler to examine the matter in closer detail.

190. The interaction between a steady current and a single magnetic pole is first analysed. The potential of the field of the current at any point is

$$\psi = -\frac{J\omega}{4\pi c} = -\frac{J}{4\pi c}\int_f \frac{\cos\theta\, df}{r^2},$$

the integral being taken over any barrier surface f bounded by the circuit, r denoting the distance of the point from the typical element df and θ the angle between the normal to df and r. This can be written in the form*

$$\psi = \frac{J}{4\pi c}\int_f (\mathbf{n}_1\nabla)\frac{1}{r}\, df.$$

The force in the field is therefore the vector

$$-\nabla_P\psi = \nabla\psi = \frac{J}{4\pi c}\int_f (\mathbf{n}_1\nabla)\nabla\left(\frac{1}{r}\right) df,$$

which transforms by Stokes' theorem to the vector

$$\nabla\psi = \frac{J}{4\pi c}\int_s [d\mathbf{s},\ \nabla]\frac{1}{r},$$

the integral now being taken round the circuit s bounding the surface f.

This result admits of interpretation as the sum of effects due to each element of the current. If we take a vector $Jd\mathbf{s}$ in the direction of the element ds of the current and another $\mathbf{r}$ from ds to the point in the field, the two form the sides of a triangle of area

$$Jds\,.\,r\sin(\widehat{r, ds});$$

thus the part of the force due to ds is

$$\frac{Jds\,.\,\sin \widehat{r, ds}}{4\pi r^2},$$

and is at right angles to both vectors so as to turn positively round the tangent to the circuit when it is taken positively onwards†.

191. The action of the single magnetic pole on the circuit has a resultant equal and opposite to this force. The constituent forces

* In these expressions the operator ∇ refers to differentiation at the point on the surface or curve, these being equal but opposite in sign to the same differentiations ∇_p at the field-point.

† The law of Biot and Savarts, *Jour. de Savants* (Paris, 1821), p. 221.

may however be applied at the elements of the circuit; the reason being that the couples

$$\frac{J ds \sin \widehat{r ds}}{4\pi c r}$$

so introduced are in equilibrium*. Thus the action of a pole m on a current J may be calculated by combining the forces

$$-\frac{1}{4\pi c}[J d\mathbf{s}, \nabla]\frac{m}{r}$$

at each element of the current.

This result admits of immediate generalisation: in fact for any magnetic system we see that the force on the current element in ds is

$$-\frac{1}{4\pi c}\Sigma\,[J d\mathbf{s}, \nabla]\frac{m}{r}$$

$$= -\frac{1}{4\pi c}[J d\mathbf{s}, \nabla]\Sigma\frac{m}{r},$$

the sum Σ being taken over all the elementary poles of the system; but

$$\mathbf{B} = -\frac{1}{4\pi}\nabla\Sigma\frac{m}{r}$$

is the magnetic force due to the whole magnetic system at the position of ds. Thus the force on the current in any field is

$$\frac{1}{4\pi c}[J d\mathbf{s}, \mathbf{B}],$$

it is perpendicular to the plane of $\mathbf{B}$ and $d\mathbf{s}$ and its actual amount is

$$\frac{J\,.\,B}{4\pi c} ds \sin \widehat{B ds}.$$

This argument is however restricted to the case where the magnetic force is derivable from a potential which is only the case when the current elements under investigation are in parts of space devoid of magnetic matter, where the force is identical with the induction which is always so derivable under static conditions. The result is however quite general as can be seen in another way

* The couple associated with any element of the circuit is completely represented in the vector sense by the chord of the projection of this element on a unit sphere round the field-point, this point being the centre of projection. The circuit being closed the vector polygon of the couples is closed so that they are in equilibrium.

by an application of the other form for the potential of a current in a field, viz.

$$W = -\frac{1}{c} NJ.$$

This expression was of course deduced as the potential of the field system in the presence of the current, but it is of necessity a mutual affair, that is, it is the energy of either system in the field of the other. It is a function containing the coordinates of both bodies and by varying one set we get the actions and by varying the other set the equal and opposite reactions. The principle of equality of action and reaction which in its simplest form is Newton's third law of motion, is in its generalised form a consequence of the existence of a potential energy function.

192. Suppose now that the variation in the configuration of the system is produced by the element AB of the wire of length ds moving out into a near parallel position $A'B'$ at a vectorial distance $\delta\mathbf{s}$ from its original position. The change in W due to the displacement of this bit alone is

$$\delta W = -\frac{J\delta N}{c}$$

$$= -J \text{ (number of tubes of induction through } ABB'A')/c.$$

We must now estimate the number of tubes through this small area $ABA'B'$. It is obviously equal to the product of the component of the magnetic induction perpendicular to the area by the area, and this is

$$\delta N = ds \,.\, \delta s' \,\mathbf{B}_p \sin \widehat{ds\delta s'} = ([\delta\mathbf{s}' \,.\, d\mathbf{s}]\, \mathbf{B}).$$

But if F is this resultant force

$$(\mathbf{F}\,\delta\mathbf{s}') = -\delta W = \frac{J}{c}(\mathbf{B}\,.\,[\delta\mathbf{s}'\,.\,d\mathbf{s}])$$

$$= \frac{J}{c}([d\mathbf{s}\,.\,\mathbf{B}]\,.\,\delta\mathbf{s}'),$$

and thus the resultant force on the current element is

$$\mathbf{F} = \frac{J}{c}[d\mathbf{s}\,.\,\mathbf{B}],$$

its direction being perpendicular to the directions of $\mathbf{B}$ and $d\mathbf{s}$ and the actual magnitude

$$\frac{Jds}{c}\,.\,\mathbf{B}\,.\sin \widehat{Bds}.$$

This agrees with the former result deduced from elementary principles.

The whole force exerted by the magnetic system on the current is thus compounded of all these elementary forces applied at the corresponding elements of the circuit. Their composition is effected in the usual way.

193. Electromagnetic induction—Faraday's Law. If we push the arguments of the previous paragraph further and examine the effects of changes in position on the electrical conditions of the conductors we shall be led to another important law. We again confine our attention to the simple case where a single conducting circuit of resistance R and in which an electromotive force E, arising from voltaic action, exists is under the action of a distribution of rigid magnetism: more general cases will be examined at a later stage.

If the magnets are allowed to move the external forces necessary to maintain the equilibrium conditions do work during the small time-interval δt of total amount

$$+\int (\mathbf{I}\delta\mathbf{B})\,dv = +\,\delta t \int_v \left(\mathbf{I}\,\frac{d\mathbf{B}}{dt}\right) dv.$$

During this time the electromotive forces from the battery also provide an amount of work $EJ\delta t$ in driving the current so that the total work provided to the system is

$$EJ\delta t + \delta t \int_v \left(\mathbf{I}\,\frac{d\mathbf{B}}{dt}\right) dv.$$

This work is accounted for in two ways (i) heat is generated in the circuit of amount $J^2R\,\delta t$, and (ii) the magnetic (kinetic) energy of the field has been increased by

$$\frac{\delta t}{2}\int_v \frac{d\mathbf{B}^2}{dt}\,dv = \delta t \int_v \left(\mathbf{B}\,\frac{d\mathbf{B}}{dt}\right) dv.$$

The principle of energy thus requires that

$$EJ + \int_v \left(\mathbf{I}\,\frac{d\mathbf{B}}{dt}\right) dv = J^2R + \int_v \left(\mathbf{B}\,\frac{d\mathbf{B}}{dt}\right) dv,$$

or

$$J^2R = EJ - \int_v \left(\mathbf{H}\,\frac{d\mathbf{B}}{dt}\right) dv.$$

The integral in the last equation is however, in the usual way, equivalent to

$$\frac{1}{c} J \frac{dN}{dt},$$

where N represents the number of lines of force through the circuit due to the magnetic field. Thus we have

$$JR = E - \frac{1}{c}\frac{dN}{dt}.$$

In other words in addition to the electromotive force of the battery there is an additional apparent electromotive force in the circuit which is proportional to the time rate of reduction of the magnetic force flux through the circuit. This is precisely Faraday's result.

194. Faraday's* notion of representing a magnetic field was by lines of force, as a field of flux in tubes. He then discovered that the essence of the electromotive force necessary to drive the current observed in the near circuits was the change in the flux through the circuit. He deduced experimentally the following rule:

Whenever the total flux of magnetic force through any circuit varies there is an electromotive force induced in the circuit whose magnitude is proportional to the rate of diminution of the total number of tubes of induction threading the circuit.

This law has been found to be true universally when either or both the magnetic system and current circuit are moving, and it has in consequence become the foundation stone of electromagnetic theory, with the various developments of which the remainder of our discussions are concerned. The deduction we have given shows the connection of this law with the mechanical principle of energy, a connection first pointed out by Helmholtz and Kelvin.

195. If $\mathbf{E}$ denotes the electric force measured in electrostatic units the electromotive force in any closed circuit is

$$\int_s \mathbf{E}_s ds,$$

and according to Faraday this is equal to

$$-\frac{1}{c}\frac{d}{dt}\int_f \mathbf{B}_n df,$$

* *Exp. Res.* II. p. 127.

where **B** denotes the vector of magnetic induction, the integral being extended over any surface f bounded by the circuit s, and c is a constant depending on the units adopted. We thus have

$$\int_s \mathbf{E}_s ds = -\frac{1}{c}\frac{d}{dt}\int_f \mathbf{B}_n df,$$

another circuital relation forming the second fundamental equation of our subject. Faraday's name is usually attached to it.

This relation of course implies that there is a definite value for the magnetic force flux through the circuit, i.e. it must be the same on whatever barrier surface f it is calculated. This implies of course that **B** is a stream vector or that

$$\operatorname{div}\mathbf{B} = 0,$$

which is of course a relation always satisfied by this vector; so that the equation so interpreted is quite consistent with our former ideas and needs no extension as in the previous case.

196. Faraday's law can easily be expressed in differential form in a similar manner to that employed in paragraph 158 for Ampère's law. Consider for example any closed circuit drawn in the conducting material and count the number of lines of magnetic induction which pass through it. Then according to Faraday's law there will be an electromotive force in the circuit which is proportional to the rate of diminution of this number of lines of induction. The electromotive force is

$$\int_s \mathbf{E}_s ds,$$

and according to Faraday this is

$$= -\frac{1}{c}\frac{dN}{dt},$$

where $N = \int_f \mathbf{B}_n df$, so that we have

$$\int_s \mathbf{E}_s ds = -\frac{1}{c}\frac{d}{dt}\int_f \mathbf{B}_n df,$$

a form already quoted for the case of linear conductors. It is here regarded as applying to any circuit drawn in the conducting substance, whether that circuit is the path of one of the elementary currents or not. By again using Stokes' theorem we can write this in the form

$$\int_f (\operatorname{curl} \mathbf{E})_n \, df = -\frac{1}{c}\frac{d}{dt}\int_f \mathbf{B}_n \, df,$$

so that if the circuit is not moving we have

$$\int_f \left(\operatorname{curl} \mathbf{E} + \frac{1}{c}\frac{d\mathbf{B}}{dt}\right)_n df = 0,$$

or owing to the arbitrary nature of the circuit

$$-\frac{1}{c}\frac{d\mathbf{B}}{dt} = \operatorname{curl} \mathbf{E},$$

which is the required form.

It can also be expressed in a very simple form in terms of the vector potential introduced above so that

$$\mathbf{B} = \operatorname{curl} \mathbf{A}.$$

From the fundamental solenoidal property of $\mathbf{B}$ it follows that

$$\int_f \mathbf{B}_n \, df$$

taken over any surface must be expressible in terms of the boundary curve of that surface. But Stokes' theorem shows that

$$\int_f \mathbf{B}_n \, df = \int_s \mathbf{A}_s \, ds,$$

where $\mathbf{A}$ is the vector defined above. Thus for any surface

$$-\frac{1}{c}\frac{dN}{dt} = -\frac{1}{c}\frac{d}{dt}\int_s \mathbf{A}_s \, ds,$$

and if the circuit is fixed this is

$$= -\frac{1}{c}\int_s \frac{d\mathbf{A}_s}{dt} \, ds.$$

But by Faraday's rule this is the same as

$$\int_s \mathbf{E}_s \, ds,$$

so that

$$\mathbf{E} = -\frac{1}{c}\frac{d\mathbf{A}}{dt}$$

is a particular solution for the electric force at a point: to generalise it we must add the gradient of any acyclic function ϕ which would give nothing on integration round the closed curve

$$\mathbf{E} = -\frac{1}{c}\frac{d\mathbf{A}}{dt} - \nabla\phi.$$

This is an expression for the electric force which is in fact equivalent to Faraday's law.

197. A current possesses a magnetic field of its own and if this field is varying, as it will do if the current is changing, the flux through the circuit will vary and there will then be an electromotive force in the circuit. The flux through the circuit is at each instant proportional to the current and increases with it, so that the induced electromotive force in the circuit is negative in the case of increasing current and positive in the case of decreasing current; it therefore always acts to impede the change of current. It is as though the current possessed inertia and momentum proportional to the flux through it, which resists any tendency to change. This is the idea of the *self-induction* in a circuit.

If a is the coefficient of self-induction in a circuit, the flux of force through the circuit when a current J flows in it is

$$aJ.$$

The induced electromotive force for changing J is therefore

$$-\frac{a}{c}\frac{dJ}{dt},$$

so that Ohm's law for the circuit is now

$$JR = E - \frac{a}{c}\frac{dJ}{dt}.$$

The self-induction of the circuit will not therefore show itself whilst the current is steady but will have considerable effect on the flow of a variable current or even on establishment of a steady current which for a short period is a variable affair.

198. The general electrodynamic theory of currents. The developments in the previous paragraph illustrate in certain cases the application of elementary dynamical principles to the solution of electrodynamic problems concerning the action and interaction of linear current systems. We have however so far only applied the energy principle and this, although sufficient in statical cases, is not wide enough for the general case when there is more than one degree of freedom. We shall now discuss an application of the general results of analytical mechanics to a scheme which will enable us to deduce from the energy expressions already obtained all the mechanical and electrical equations of a system of linear currents with any number of degrees of freedom. It was found by Maxwell that the application of Lagrange's equations to the most general problem expressed in this way leads to results which are in entire agreement with the experimental facts.

The palpable coordinates of such a system are evidently the current strengths and the geometrical coordinates of the system, these being the only quantities directly accessible to measurement. The motions which define the currents are necessarily of the type which Helmholtz described as cyclic. The parameters which determine the instantaneous position of such a system are of two kinds: (i) the parameters of the first kind which are of the general type of coordinates in mechanics. These parameters occur in general in the expression of the kinetic energy with their differentials with respect to the time. To this group belong, in the present instance, those geometrical parameters which determine the position of the circuits: (ii) the parameters of the second group on the other hand, which are the coordinates of the cyclic motion, do not themselves appear in the expression for the kinetic energy, only their time rates of change being involved. If the kinetic energy is known to be correctly expressed in terms of the coordinates and velocities explicitly we have therefore no difficulty in separating the coordinates into their respective classes. But in systems where the internal connections are only partially known, a difficulty may occur, in so far as in obtaining the expression for the kinetic energy it may have been convenient or even necessary to introduce the generalised momenta in the cyclic motions, in order to obtain a usable expression. For example in the hydrodynamical analogue, in determining the forcives between cores in problems of cyclic motion, the circulations in terms of which the energy is usually expressed must be treated as generalised momenta. It is therefore necessary and essential to have a clear view of the circumstances which determine whether the various quantities which enter into the specification of the energy are to be classed as velocities or momenta. The basis of the distinction between these two classes of quantities is of course fundamental; it is to be found in the way in which they occur in the Hamiltonian analysis of the dynamical problem. The essential property of a velocity is that it is a perfect differential coefficient with respect to the time; any function involving rate of change of configuration, which enjoys this property, so that its time integral is a function of position only may be taken to be a velocity; provided we, if need be, contemplate also a corresponding force. On the other hand, any such function of the rate of change of configuration, even though it be a perfect differential with respect to the time, must be treated as a momentum, if it is known to remain constant with time while no external forces are applied to it; for if it were a velocity, linked up with other velocities, its constancy in the free motion could not usually fit in with the analytical theory.

199. In the theory of cyclic fluid motion, the circulations being constant must thus be taken as momenta, and, when the energy is expressed in terms of them, it must be modified before the forcives can be derived from it in the manner of Lagrange and Hamilton. In the theory of electrodynamics, on the other hand, the electric currents are not unalterable with the time, even if no applied electromotive forces are applied, and as they are the differential coefficients with respect to the time of definite physical quantities, the charges of electricity, they may be taken as the velocities, provided we recognise the play of the corresponding (electromotive) forces.

In the electrodynamics of complete circuits however there is no reason, in that theory taken by itself, why the functions defined hereinafter as the electrokinetic momenta should not be taken as the velocities instead, if so desired, for they satisfy all the above conditions, though of course the corresponding forcives would be of quite different types from the usual ones. This remark is in illustration of the fact that the distinction between momenta and velocities is to a certain extent one of convenience. We shall however adopt the method indicated on account of the enormous difficulties underlying this suggested alternative.

200. The system we shall treat will consist of n-conducting circuits carrying currents $J_1, J_2, \dots J_n$. We have therefore n-cyclic coordinates in which the velocities are $J_1, J_2, \dots$, in addition to a certain number of ordinary geometrical coordinates $\theta_1, \theta_2, \dots \theta_m$ which determine the relative configuration of the system. The actual cyclic coordinates may be taken as the integrals of the currents with respect to the time reckoned from a definite instant, i.e. the quantities $Q_1, Q_2, \ldots Q_n$ of electricity which since that instant have crossed any cross-section of the respective conductors

$$J_1 = \frac{dQ_1}{dt}, \quad J_2 = \frac{dQ_2}{dt}, \ldots.$$

The generalised force components corresponding to the cyclic coordinates are the electromotive forces which work on the currents flowing. The work function of the forces applied in these coordinates would thus be

$$\delta W_Q = \sum_{r=1}^{n} E_r \delta Q_r,$$

if we assume impressed electromotive forces $E_1, E_2, \ldots E_n$ in each circuit respectively, since the work of the electromotive force E in any virtual increase of the coordinate Q defined as above is $E\,\delta Q$.

We have also to include the virtual work of the applied mechanical forces if there are any. In the general Lagrangian method the generalised force component Θ which corresponds to the coordinate θ is the coefficient in the work done when that coordinate is alone altered. Thus

$$\delta W_\theta = \Sigma \Theta\, \delta\theta.$$

If the forces are applied from without δW would represent energy added to the system. If they are merely forces exerted by one part of the system on another or against the external system the work in them would come from the energy of the system and must therefore be taken as $-\,\delta W_\theta$.

201. We have to take into account the resistances to the flow of the currents. These may be introduced either by including them in the generalised impressed or external force components corresponding to the cyclic coordinates, or by the more general method involving the introduction of a dissipation function. If the resistances in the circuits are $R_1, R_2, \ldots$ then on the first method the impressed electromotive forces in the separate circuits would have been respectively diminished by $R_1J_1, R_2J_2, \ldots R_nJ_n$, in order to obtain the resultant force components. The more general method consists in the introduction of the function

$$F = \tfrac{1}{2}\left(R_1J_1^2 + R_2J_2^2 + \ldots\right)$$

into the general dynamical scheme.

In order to obtain complete generality we shall assume that each circuit has included in it a condenser, or an appreciable capacity for storing energy. It is only in this case that the potential electric energy is at all comparable with the magnetic kinetic energy. If the original charges in these condensers were $Q_{0_1}, Q_{0_2}, \ldots$ then the general potential energy function for the system would be

$$W = \tfrac{1}{2}\Sigma b_{rr}(Q_{0_r} - Q_r)^2 + \Sigma b_{rs}(Q_{s_r} - Q_r)(Q_{0_s} - Q_s),$$

and the product terms would be negligible in most cases if the condenser in each circuit is of such a form as to concentrate its field sufficiently to prevent mutual influence with the others.

We shall leave out of account any so-called permanent magnets or magnetisable substances in which the magnetisation is not capable of following the field without hysteretic loss. 'Permanent' magnets are in reality far from permanent and are indeed very erratic things; their properties are very indefinite and a theory including them becomes largely an empirical subject. We may thus regard the coefficients of induction of the circuits to be dependent merely on the geometrical configurations in the circuits.

202. The electrokinetic energy of the system is, as before, given by

$$2T = \Sigma a_{rr}\left(\frac{dQ_r}{dt}\right)^2 + 2\Sigma a_{rs}\frac{dQ_r}{dt}\cdot\frac{dQ_s}{dt}.$$

This is kinetic energy of some kind; we do not as before need to know of what kind. We only want its amount in suitable terms to enable us to apply general dynamical methods; this is the great advantage of the present line of attack.

We must now also include the kinetic energy of the movement of the material conductors because the material conductors involved may possess very considerable masses. The positions and general configurations of these masses are as before specified by the generalised coordinates $\theta_1, \theta_2, \ldots \theta_m$ and the kinetic energy corresponding to them will be denoted by T_1 so that the total kinetic energy is given by

$$T + T_1;$$

T_1 is of course a quadratic function of $\dot{\theta}_1, \dot{\theta}_2, \ldots \dot{\theta}_m$ in which the coefficients are functions of $\theta_1, \theta_2, \ldots \theta_m$.

For absolute generality we should include in the complete expression for the kinetic energy terms involving such things as

$(\dot{Q}_r\dot{\theta}_s)$ but this would be getting beyond our theory. In all realisable cases and certainly in all those cases where an equilibrium theory is applicable the electric changes adjust themselves so quickly compared with the slow motions of ordinary matter that the general electromagnetic system is at each moment sensibly in an equilibrium condition; so that there is practically no interaction between the kinetic energies of the electromotive and material systems such as would arise from mixed terms in the energy function involving both their velocities—a fact verified experimentally by Maxwell. The expression for T thus represents completely the energy of the system as far as electromotive disturbances are concerned, whether the system is in motion or not. It is therefore sufficient for the determination of the electrical conditions. The other part T_1 is solely the ordinary kinetic energy of motion of the conductors and is alone necessary for the determination of the mechanical relations of the electrodynamic field.

203. We can now proceed to apply any of the usual methods of obtaining the equations of the motion in the various coordinates, electrical and geometrical. The most general method involves a use of the principle of least action which represents the most general principles of dynamics in their most condensed form. If the system is governed by dynamical laws at all, we have merely to obtain the energies in their most compact form and then to substitute them in this principle. Lagrange's principle of least action is in fact the infallible method to apply to all dynamical systems and is the one which avoids the investigation every time of the conditions peculiar to each case. We shall however find it more convenient for the present case, in which we have actually determined the form of the functions, to take the slightly less general form of the principle which is contained by the expression that the motion is determined by the ordinary Lagrangian equations in dynamics.

If we apply the general dynamical equations to the geometrical coordinates which determine the configuration of the system we find that there is an equation for each such coordinate θ of the type

$$\frac{d}{dt}\frac{\partial (T+T_1)}{\partial\dot{\theta}} - \frac{\partial (T+T_1)}{\partial\theta} + \frac{\partial W}{\partial\theta} - \frac{\partial F}{\partial\dot{\theta}} = \Theta,$$

which since $\frac{\partial T}{\partial \dot{\theta}_r} = 0$ can be put in the form

$$\frac{d}{dt}\left(\frac{\partial T_1}{\partial \dot{\theta}}\right) - \frac{\partial T_1}{\partial \theta} = \Theta - \frac{\partial W}{\partial \theta} + \frac{\partial T}{\partial \theta} + \frac{\partial F}{\partial \dot{\theta}}.$$

The two terms on the left represent with signs changed the kinetic reaction of the circuits to ordinary motions. The terms on the right,

$$-\frac{\partial W}{\partial \theta} + \frac{\partial T}{\partial \theta},$$

represent the mechanical force exerted by the system in the coordinate θ. The second part arises from the electrokinetic energy of the system and the first from the electrostatic part. We call the former the electrodynamic forces and the others the electrostatic ones.

We thus see that the electrodynamic forces of stationary currents always tend to increase the electrokinetic energy T. We then speak of $-T$ in this sense as the *electrodynamic potential* of the system; it plays the part of the force function of the electrodynamic forces. This is Neumann's result*, which is in agreement with the result of our discussion of the steady case in paragraph 189.

204. If we apply the general dynamical equations to the electrical coordinates we find that there is an equation for each circuit of a type

$$\frac{d}{dt}\frac{\partial (T + T_1)}{\partial \dot{Q}_r} - \frac{\partial (T + T_1)}{\partial Q_r} + \frac{\partial W}{\partial Q_r} - \frac{\partial F}{\partial Q_r} = E.$$

But since T_1 is a function of the θ's and $\dot{\theta}$'s only and does not contain either Q_r or $\dot{Q}_r$, this reduces to

$$\frac{d}{dt}\left(\frac{\partial T}{\partial \dot{Q}_r}\right) - \frac{\partial T}{\partial Q_r} + \frac{\partial W}{\partial Q_r} - \frac{\partial F}{\partial \dot{Q}_r} = E_r,$$

and there are as many of these equations as there are circuits, so they are sufficient to determine the electrical motions in the circuits.

On the analogy with the ordinary Lagrangian equations the term

$$E_r - \frac{\partial W}{\partial Q_r} + \frac{\partial F}{\partial \dot{Q}_r}$$

* "Ueber ein allgemeines Prinzip der mathematischen Theorie der induzierten Ströme," *Berlin. Abhdg.* (1848).

is the applied force acting in the coordinate. The term on the other side,

$$\frac{d}{dt}\left(\frac{\partial T}{\partial \dot{Q}_r}\right) - \frac{\partial T}{\partial Q_r},$$

which, since the actual coordinates Q_1, Q_2, ... corresponding to the cyclic velocities do not explicitly appear in any of the functions involved, reduces to

$$\frac{d}{dt}\frac{\partial T}{\partial \dot{Q}_r},$$

and with sign changed it represents what Kelvin calls the kinetic reaction of the system. In fact $\frac{\partial T}{\partial \dot{Q}_r}$, that is the differential coefficient of T with respect to the cyclic velocity Q_r, is a sort of momentum in the circuit—it is called the *electrokinetic momentum*—and then on the ordinary analogy there is a kinetic reaction to the variation of this momentum in each circuit, which in the present case acts simply as a sort of additional electromotive force in the circuit.

If we write $J_r = \dot{Q}_r$ then

$$\frac{dT}{dJ_r} = \sum_{s=1}^{n} a_{rs} J_s,$$

so that the momentum corresponding to each circuit is identical with the instantaneous magnetic induction flux through the circuit. Thus there is an additional electromotive force in each circuit which is proportional to the time rate of reduction of the magnetic force flux through that circuit. This is the general form of Faraday's result applicable for each circuit when the field is of any variable type.

205. Before closing this discussion it may be as well to add further details of two special cases illustrating certain general principles. We have the kinetic and potential energies and the dissipation function in the form

$$2T = \Sigma a_{rr}\left(\frac{dQ_r}{dt}\right)^2 + 2\Sigma a_{rs}\frac{dQ_r}{dt}\frac{dQ_s}{dt},$$

$$2W = f(Q_1, Q_2, \dots Q_n),$$

$$2F = \Sigma R_r\left(\frac{dQ_2}{dt}\right)^2,$$

and in addition there are the impressed electromotive forces $E_1, E_2, \ldots E_n$ in the respective circuits.

The general equations are then of the type

$$\frac{d}{dt}\left(\frac{\partial T}{\partial \dot{Q}_r}\right) + \frac{\partial W}{\partial Q_r} - \frac{\partial F}{\partial \dot{Q}_r} = E_r .$$

If the solution shows that the currents are periodic forced vibrations with a frequency n then all the quantities may be treated as dependent on the time by the factor e^{int} and thus we can have

$$\frac{d}{dt}\left(\frac{\partial T}{\partial \dot{Q}_r}\right) = in \frac{\partial T}{\partial \dot{Q}_r},$$

and when n increases indefinitely we must have

$$\frac{\partial T}{\partial \dot{Q}_r} = 0$$

in each circuit. Thus in the case of enormously rapid vibrations of the currents, their distribution in the various conductors is independent of the resistances and is determined by the fact that the kinetic energy (and not the dissipation) function is a minimum.

A similar remark applies when the question under consideration is one of initial impulse effects.

This explains why it is that when a rapidly alternating current is sent along a wire, the current really only travels in the outer layer of the wire; the mean distance between the various filaments in the current being thereby increased and their mutual inductance and the kinetic energy of the field decreased to their minimum values.

When $W = 0$ it is convenient to express everything in terms of the currents $J_r = \frac{dQ_r}{dt}$. Thus in the problem of steady electric flow when all the quantities E_r representing impressed electromotive forces are constant, the currents are determined directly by the linear equations

$$\frac{\partial F}{\partial J_r} = E_r,$$

which express the condition that the function

$$(F - \Sigma E_r J_r)$$

is a minimum. If all the E's are zero this is Joule's law of minimum dissipation for steady currents. The above is the more generalised form including impressed electromotive forces.

206. Equilibrium theory. In the usual treatment of the dynamics of rigid bodies we express the energy functions, and dynamical relations generally, in terms solely of the palpable coordinates of the body and the velocities in them. This however implies certain restrictions on the nature of the body and its motion which are in practice never actually realised. In fact we know that when we hit a body so as to start it off suddenly, the actual observed motion is set up by a most complicated process. Firstly a wave of compression is sent off through the body from the point of application of the blow, and the further parts of the body take up their motion only when this wave reaches them and imparts the necessary momentum and energy. The uniform motion of the body is not therefore attained until this wave disturbance has been smoothed out over the whole body; in most cases of practical importance however the time occupied by this process is so exceedingly small that we can neglect it altogether and regard the observed conditions as instantaneously established, and this is implied in the term 'rigid body.' Of course there is a slight dissipation of the energy of the wave disturbance, and it is not all transformed into the energy of the observable motion, but some goes into heat; but this quantity is usually so small that we are fully justified in treating it as non-existent or, again, the body as rigid. Of course there is the theory at the other end where we investigate these waves of compression, but the two cases are extremes and as a rule we need not mix the one with the other.

There is just the same point in the usual electrodynamic theories. We shall soon see how Maxwell was induced to his theoretical explanation of all observed electrodynamic actions as being transmitted through and by that hypothetical medium, the aether, which occupies the whole of space; an essential point being that the actions are transmitted by this medium with a finite velocity. It therefore follows that any disturbance produced in the conditions at any part of an electromagnetic field will smooth itself over the whole field by sending out an electromagnetic wave through the field in a manner exactly analogous to that described above. If however the variations arbitrarily produced in the conditions at any point of an electromagnetic field are slow enough, we can consider the adjustment of the corresponding new conditions throughout the whole of the field to take place so quickly as to be almost instantaneous, so that the whole field is at each instant practically in the equilibrium condition under the circumstances pertaining at that instant.

Thus if we confine ourselves to the discussion of phenomena the period of any change of which is large compared with the time taken by radiation to get across the system and adjust any new conditions, we may adopt an equilibrium theory and use the results obtained for absolutely stationary states as applicable at each instant to the slowly varying motion (quasi-stationary state). That is, in the present case if the changes in the values of the current strengths or the positions of the circuits are very small and insignificant in the time taken to adjust an equilibrium condition throughout the field, then we may treat the currents in each circuit as definite quantities, i.e. the same all round the circuit, and the field at each instant can be determined in terms of their instantaneous values.

The velocity of adjustment of electrical conditions in air turns out to be 3×10^{10} cms. per second, so that for any ordinary sized system the condition that changes in its configuration should be small in the time occupied by radiation in getting across the system hardly restricts the application of the results obtained except in the case of very rapid oscillations, of the type in fact used to start electric waves. For the present therefore we have implied this restriction to an 'equilibrium' theory and have assumed that the state of the motion actually observed adjusts itself so quickly that at each instant it is practically in equilibrium under the conditions pertaining at that instant.

207. Magnetostriction. As paramagnetism is in general a constitutive phenomenon we should expect relations to exist between the magnetic and mechanical deformations of media exhibiting this type of magnetisation. Such relations are however observed only in the case of the ferromagnetic substances: in general a magnetic field causes a change in dimensions of such bodies while, reciprocally, mechanical deformations produce changes in the magnetic properties. The various effects of this kind that have been observed are:

(i) *Joule effect**. If a rod of iron or steel is subjected to a magnetic field which may be varied continuously from zero upwards it will be found that the rod first increases in length up to a maximum at a certain field strength after which it again begins to shorten and may even become shorter than its original length. For very large fields however it appears to approximate again to its natural length.

Accompanying this change of length there is also a sideways contraction or expansion but Joule's original idea that these are just what are required to balance the extension so that the volume remains constant does not appear to be justified. In the case of iron the volume is increased by magnetisation whilst in the case of cobalt there is actually a decrease of the volume.

(ii) *Wiedemann effect*†. When a rod of ferromagnetic substance is magnetised longitudinally and then simultaneously a current is sent along the rod producing a circular magnetic field, the superposition of the two fields causes the two ends of the rod to rotate in opposite directions.

This is probably only a complex Joule effect and it is accompanied by the inverse effects, for when a rod of iron is magnetised longitudinally in a solenoid a twist imparted to the rod establishes a circular magnetic field in it, or if it is magnetised circularly the twist will give rise to a longitudinal field.

(iii) *Villari effect*‡. Thomson§ has pointed out that there are certain reciprocal relations in magnetism where, if the changes in length were known, it might be predicted with certainty the effect which a longitudinal pull or compression would produce in the same specimen in the way of changing its magnetic properties. If a ferromagnetic rod shows an increase in length due to a magnetic field, the same rod will show an increase in magnetisation when stretched or a decrease in magnetisation when compressed longitudinally. For substances which show an increase in magnetisation for weak fields and a decrease for strong fields, there is a certain

* *Phil. Mag.* 30 (1847), pp. 76 and 225. Cf. also Bidwell, *Proc. R. S.* 55 (1894), p. 228; 56 (1894), p. 94; Williams, *Phys. Rev.* 34 (1912), p. 258.

† *Elektrizität*, Bd. III. p. 689.

‡ *Pogg. Ann.* 126 (1868), p. 87.

§ *Applications of Dynamics to Phys. and Chem.* (1888), p. 47.

critical field strength where the intensity is unaffected by extension or compression. This is known as the Villari reversal point. Substances which show this reversal effect also show an analogous reversal of the Joule effect as Bidwell demonstrated for iron.

A theory of these effects has been attempted by various writers but perhaps the most generally successful one is that given by Thomson* who makes use of the energy principle to obtain the reciprocal relationships observed to exist between strain and magnetisation. The first step in the deduction is to set up the potential energy function for a magnetised medium in terms of suitable magnetic and strain coordinates. For the former we may take the magnetic field $\mathbf{B}$ and the intensity of magnetisation $\mathbf{I}$ where $\mathbf{I} = \kappa\mathbf{B}$ so that the energy corresponding to this part is potential and of amount

$$+\int (\mathbf{I}\,d\mathbf{B})$$

per unit volume. The elastic energy per unit volume is as usual of the general form

$$\tfrac{1}{2}m\,(e^2+f^2+g^2)+\tfrac{1}{2}n\,(e^2+f^2+g^2-2ef-2eg-2fg),$$

e, f, g being in the usual notation of elastic solid theory the component strains, n the coefficient of rigidity and $m - n/3$ the bulk modulus. The total potential energy is therefore

$$\int W dv = \int dv\left[\int (\mathbf{I}\,d\mathbf{B}) + \tfrac{1}{2}m\,(e^2+f^2+g^2) + \tfrac{1}{2}n\,(e^2+f^2+g^2-2ef-2eg-2fg)\right].$$

Now by the principle of energy this energy function must be a minimum in the positions of equilibrium so that its variation with respect to any coordinate is zero. In this way equations are obtained which appear to be adequate to account qualitatively for all the observed magnetostriction effects, including the reciprocal relationships between torsion and magnetisation; but apparently no attempt has been made to apply them quantitatively.

* *Applications of Dynamics, etc.* p. 47. Cf. also Larmor, *Phil. Trans.* A, 190 (1897), pp. 283–299.

CHAPTER VI

MAXWELL'S ELECTROMAGNETIC THEORY

208. Maxwell's generalisation of Faraday's theory. The previous discussions refer to electric and magnetic phenomena which are either actually stationary or at least so slowly variable that they admit of treatment along similar lines. For cases in which this restriction is satisfied the laws of Ampère and Faraday, applicable in reality only to closed currents, provide a sufficient and satisfactory foundation. To obtain a theory applicable under more general conditions Maxwell assumed that all electric discharges and motions are effectively of the nature and possess the properties of systems of closed currents, being completed when necessary by the so-called displacement current in the aether and in the dielectric media and then it was merely necessary to complete the scheme to assume that the fundamental relations in their simplest and differential form are applicable under all circumstances.

Maxwell, with Faraday, saw all the obvious phenomena of electromagnetics merely as the terminal aspects of a variation of condition in the space (or field) surrounding the apparatus. The observed actions are then to be imagined as transmitted through and by a something, the aether, in this field, this aether being capable of varying its condition. In reality such an aether is merely a definite framework to which we can attach the vector quantities of the electromagnetic field, although it is frequently convenient to have a definite representation of its mode of action, that is a description of its mode of activity in dynamical or analytical terms. The essential characteristics of a description of this kind is of course that the values of the vectors at a definite point of space are directly connected only with their values at infinitely near points and only indirectly with the conditions at finitely distant points. This is why the differential relations of the field are likely to be of more fundamental physical importance than any integral or other equivalent forms of such relations.

209. The vectors necessary for a complete specification of the conditions in the electromagnetic field at any point are:

(i) $\mathbf{E}$, the intensity of the electrostatic field measured generally by the ratio of the resultant force to the charge on a small conductor placed at the corresponding point, a reservation being made if the point is inside polarised matter.

(ii) $\mathbf{P}$, the polarisation intensity of the dielectric media, which with the aethereal displacement $\mathbf{E}$ forms the composite vector $\mathbf{D}$

$$\mathbf{D} = \mathbf{E} + \mathbf{P}.$$

(iii) $\mathbf{B}$, the magnetic force intensity, this being defined so that under all conditions

$$\operatorname{div} \mathbf{B} = 0$$

everywhere, there being no free magnetism.

(iv) $\mathbf{I}$, the magnetic polarisation intensity of the magnetisable media. This vector also gives rise to an auxiliary vector $\mathbf{H}$ defined at each point so that

$$\mathbf{B} = \mathbf{H} + \mathbf{I};$$

this vector $\mathbf{H}$ we call the *magnetic induction*, but add a warning that it is the induction of *mechanically effective force* that is to be understood.

(v) $\mathbf{C}$, the total current density of Maxwell's theory, which in the most general case contains terms representing all possible types of coordinated or average motions of electrons. There is

(*a*) $\mathbf{C}_1$, the current of conduction, is, as we have already seen, made up of a drift of electrons or ions, the positive ones travelling in one direction, the negative ones in the opposite direction, under the influence of the electric force.

(*b*) $\mathbf{C}_2$, the polarisation current associated with the material medium. We have already seen how the establishment of a condition of polarisation in a dielectric medium is accomplished on the Larmor-Lorentz view of the subject, by an electric displacement which may be caused either by turning round the little bi-polar molecules, or by an actual displacement of the charges relative to one another in the molecule. A time variation of such a state of affairs would therefore involve an electric current in the dielectric. The displacement in any position is the same as if the positive pole started from the final position of the negative pole and moved up to its final position. The total amount can thus be estimated and is such that

$$\mathbf{C}_2 = \frac{d\mathbf{P}}{dt}.$$

(*c*) A material medium moving with the velocity equal at the (x, y, z) point to $\mathbf{u}$ and having in the neighbourhood of that point a charge of amount ρ per unit volume, clearly contributes a convection current of density $\rho\mathbf{u}$. The elements of the moving medium may be the molecules or ions as in electrolysis, so that conduction currents are merely particular examples of convection currents. The convection current may however also consist simply of a convection of free electrons so that it is difficult to distinguish between conduction and convection currents, although it is usual and convenient to use the distinction provided by the applicability of Joule's and Ohm's laws.

(*d*) The convection of a material medium merely polarised to intensity $\mathbf{P}$ also supplies a part to the volume distribution of electric currents: but its determination requires more refined analysis. We first notice that except in so far as it is equivalent to a volume distribution of charge of density $-\operatorname{div}\mathbf{P}$, the polarisation cannot give rise to a current unless it is changing and that then its contribution to the volume density of the current is increased by

$$\frac{\delta\mathbf{P}}{dt},$$

its rate of change at a point fixed relatively in the moving dielectric. But if $\frac{d\mathbf{P}}{dt}$ denote the rate of variation of $\mathbf{P}$ at a fixed point of space we know from a well-known theorem in vector analysis that the rate of change of the polarisation flux $\mathbf{P}_n df$ through any surface df moving with the dielectric is

$$\frac{\delta\mathbf{P}_n}{dt}df = \left(\frac{d\mathbf{P}}{dt} + \operatorname{curl}[\mathbf{Pu}] + \mathbf{u}\operatorname{div}\mathbf{P}\right)_n df,$$

so that
$$\frac{\delta\mathbf{P}}{dt} = \frac{d\mathbf{P}}{dt} + \operatorname{curl}[\mathbf{Pu}] + \mathbf{u}\operatorname{div}\mathbf{P}.$$

The total convection current due to the polarisation is therefore simply

$$\frac{d\mathbf{P}}{dt} + \operatorname{curl}[\mathbf{Pu}],$$

the first term representing of course the current due to the varying polarisation mentioned in (*b*) above.

210. The significance of the various terms in the complete expression for the true electric flux density which depend on the electric [and magnetic] polarisations can also be exhibited in a more analytical form, if we assume that all such polarisations result entirely as the average aspects of a more or less complex distribution of electrons or point charges of both signs. In such a case the part of the current density depending on these electrons at any point in the medium is the limiting ratio of the volume of the physically small element δv to the sum

$$\Sigma q\mathbf{v},$$

extended over all the electrons inside it; $\mathbf{v}$ is the velocity of the typical electron. In effecting the summation care must be taken to refer each electron to its proper position in the matter. Having fixed on a definite point in the medium moving with the velocity $\mathbf{u}$ we notice that the electron attached to it is displaced from that point through a distance $\mathbf{r}$ on account of the polarisation, and has therefore a velocity

$$\mathbf{u} + \dot{\mathbf{r}} = \mathbf{u} + \frac{d\mathbf{r}}{dt} + (\mathbf{u}\nabla)\,\mathbf{r}.$$

The electron which is actually at the point in the matter under review and which properly belongs to some other point has therefore the velocity

$$\mathbf{u} + \dot{\mathbf{r}} - (\mathbf{r}\nabla)\,\mathbf{u} - (\mathbf{r}\nabla)\,\dot{\mathbf{r}},$$

it being assumed that the velocity of an electron is a continuous function of its position in the matter. This is the value that has to be taken for $\mathbf{v}$ in effecting the summation as above.

Again we must notice that the summation is to be taken only over those electrons actually in the volume element under consideration, the number of which is a function of their displacements. In fact in setting up the displacements typified by $\mathbf{r}$ the charge displaced across any surface is

$$\int_f \Sigma q\,(\mathbf{r}_n\,df),$$

Σ denoting a sum for the electrons in the volume element $\mathbf{r}_n df$. A simple application of Green's lemma shows that the statistical effect of this displacement is the same as if each electronic charge inside the surface were reduced in the ratio

$$1 - \operatorname{div}\mathbf{r} : 1,$$

the number remaining unaltered.

It is thus the summation of

$$\Sigma q\,\{1-(\nabla\mathbf{r})\}\left[\mathbf{u}+\frac{d\mathbf{r}}{dt}+(\mathbf{u}\nabla)\,\mathbf{r}-(\mathbf{r}\nabla)\,\mathbf{u}-(\mathbf{r}\nabla)\frac{d\mathbf{r}}{dt}-(\mathbf{r}\nabla)\,(\mathbf{u}\nabla)\,\mathbf{r}\right],$$

or, neglecting quantities of higher order in the displacement, of

$$\Sigma q\left[\mathbf{u}-\mathbf{u}\,(\nabla\mathbf{r})+(\mathbf{u}\nabla)\,\mathbf{r}-(\mathbf{r}\nabla)\,\mathbf{u}+\frac{d\mathbf{r}}{dt}-\frac{d\mathbf{r}}{dt}(\nabla\mathbf{r})-(\mathbf{r}\nabla)\frac{d\mathbf{r}}{dt}\right],$$

that is to be effected. Owing to the smallness of the volume element δv we may assume that $\mathbf{u}$ and its space gradients are constant throughout, whilst

$$\Sigma q=\rho\,\delta v,\quad \Sigma q\mathbf{r}=\mathbf{P}\,\delta v,$$

where ρ is the density of the free charge and $\mathbf{P}$ the polarisation intensity of the element. Thus

$$\Sigma q\,\frac{d\mathbf{r}}{dt}=\frac{d}{dt}(\mathbf{P}\delta v)=\left\{\frac{d\mathbf{P}}{dt}+\mathbf{P}\,(\nabla\mathbf{u})\right\}\delta v,$$

$$\Sigma q\,(\mathbf{u}\nabla)\,\mathbf{r}=(\mathbf{u}\nabla)\,\Sigma q\mathbf{r}=(\mathbf{u}\nabla)\,\mathbf{P}\delta v,$$

$$\Sigma q\mathbf{u}\,(\nabla\mathbf{r})=\mathbf{u}\,(\nabla,\Sigma q\mathbf{r})=\mathbf{u}\,(\nabla\mathbf{P})\,\delta v,$$

$$\Sigma q\,(\mathbf{r}\nabla)\,\mathbf{u}=(\Sigma q\mathbf{r},\nabla)\,\mathbf{u}=(\mathbf{P}\nabla)\,\mathbf{u}\,\delta v.$$

Due to these terms we have therefore a current of density

$$\rho\mathbf{u}+\frac{d\mathbf{P}}{dt}+(\mathbf{u}\nabla)\,\mathbf{P}-(\mathbf{P}\nabla)\,\mathbf{u}-\mathbf{u}\,(\nabla\mathbf{P})+\mathbf{P}\,(\nabla\mathbf{u})$$
$$=\rho\mathbf{u}+\frac{d\mathbf{P}}{dt}+\operatorname{curl}\,[\mathbf{P}\mathbf{u}],$$

which agrees with the result obtained above.

The remaining terms have another significance. They are

$$-\Sigma q\left\{\frac{d\mathbf{r}}{dt}(\nabla\mathbf{r})+(\mathbf{r}\nabla)\frac{d\mathbf{r}}{dt}\right\}=-\Sigma q\,(\nabla\mathbf{r})\,\frac{d\mathbf{r}}{dt}$$
$$=-\tfrac{1}{2}\Sigma q\,\frac{d}{dt}\{(\nabla\mathbf{r})\,\mathbf{r}\}+\operatorname{curl}\tfrac{1}{2}\left\{\Sigma q\left[\mathbf{r},\frac{d\mathbf{r}}{dt}\right]\right\},$$

where in these equations the operator ∇ is presumed to affect all quantities immediately following it.

The second term on the right-hand side of the last equation receives its main contribution from those electrons which are executing rapid orbital motions about their mean positions in the matter, the contribution of any one electron being proportional to the moment of its momentum about the equilibrium position.

Moreover if the orbital motions are to any extent permanent the sum

$$\Sigma q\,(\nabla \mathbf{r})\,\mathbf{r}$$

for these electrons will be practically independent of the time. Thus when some or all of the electrons are executing motions of the type considered there is an additional term in the complete expression for the current density which to the first approximation is equal to

$$c \operatorname{curl} \mathbf{I},$$

where

$$\mathbf{I}\,\delta v = \frac{1}{2c}\Sigma q \left[\mathbf{r}, \frac{d\mathbf{r}}{dt}\right].$$

In this form we recognise that the additional term in the current is the same as would be contributed by the medium magnetically polarised to intensity **I**, if the magnetism is regarded as equivalent to a distribution of minute current whirls. This is a tentative suggestion of an electron theory of magnetism, which will be further discussed in the sequel. For the present we shall usually disregard terms of this type in the current expression.

211. In the whole of the above discussion it has been assumed that the matter extends continuously throughout and beyond the element under direct observation. If as is sometimes the case it is necessary to include the effects of discontinuities in the material distribution these can always be estimated by regarding such discontinuities as continuous rapid transition regions throughout which the definitions given above remain effective. In this way it is easily seen that a surface of discontinuity in the material medium is to be regarded as the seat of a current sheet of density

$$\left| c\,[\mathbf{n}_1 \mathbf{I}_1] \right|_1^2,$$

where

$$\mathbf{I}_1 = \mathbf{I} + \frac{1}{c}[\mathbf{P}\mathbf{u}],$$

and $\mathbf{n}_1$ is the unit vector normal to the surface in the direction from the side 1 to the side 2. The notation implies that it is the difference of the values of the function on the two sides that is to be taken as the current density.

Thus, for example, in any continuous piece of matter with a definite boundary the electronic motions in it may be specified in their average aspect as being equivalent to a distribution of body currents of density

$$\mathbf{C}_1 + \rho\mathbf{u} + \frac{d\mathbf{P}}{dt} + c \operatorname{curl} \mathbf{I}_1$$

throughout its mass together with the surface current sheet of density

$$-c\,[\mathbf{n}_1\mathbf{I}_1] = c\,[\mathbf{I}_1\mathbf{n}_1]$$

at any point of its outer boundary.

These surface currents are important when we consider the dynamical aspects of electromagnetic phenomena.

212. Maxwell's theory may now be stated in the general form. It states that under any of the most complex conditions the electric and magnetic vectors at any point of the field always satisfy the two fundamental differential vector relations:

(i) Faraday's Law $$-\frac{1}{c}\frac{d\mathbf{B}}{dt} = \operatorname{curl}\mathbf{E}.$$

(ii) Ampère's Law $$\frac{1}{c}\mathbf{C} = \operatorname{curl}\mathbf{H},$$

the vector **B** being however restricted to satisfy the condition

$$\operatorname{div}\mathbf{B} = 0.$$

These relations are of course not sufficient in themselves to completely determine the correlations of the field. We know however that the vectors are connected by certain constitutional relations which we have examined in some detail on previous occasions. There will be some relation connecting the electric force with the conduction current density—a general form of Ohm's law—and also relations between the polarisations and the inducing forces. These constitutive relations will of course depend essentially on the nature of the ponderable matter involved and as they can only be obtained experimentally they can only be regarded as more or less approximate. The first two relations will be of the nature of dynamical principles and must be regarded as exact.

213. In any given material medium devoid of hysteretic quality, the intensity of electric polarisation **P** must be a mathematical function of the electric force **E** which excites it. In ordinary cases, certainly in all cases in which the exciting force is small, the relation between **P** and **E** is a linear one: thus in the general problem of an aeolotropic medium there will be nine static dielectric coefficients. The principle of negation of perpetual motions requires this linear relation to be self-conjugate and so reduces the nine coefficients to six. In the special case of isotropy there is only one coefficient and the relation may be expressed in the usual form

$$\mathbf{P} = \frac{\epsilon - 1}{4\pi}\mathbf{E},$$

where ϵ is the single dielectric constant of the medium.

In problems relating to moving material media the question may naturally arise whether the value of ϵ for the medium is sensibly affected by its movement through the aether. When it is considered that each molecule that is polarised by the electric force has effectively two precisely complementary poles, positive and negative, it becomes clear that a reversal of the motion of the material medium cannot alter the polarity induced: hence the influence of the motion on ϵ can only depend on square and higher even powers of the velocity.

In cases in which the magnetisation induced in the medium is of sufficient magnitude to be taken into account, similar statements will apply to it. In the general crystalline medium there are six independent coefficients of magnetisation: these reduce for an isotropic medium to a single coefficient and the relation between induction and force is

$$\mathbf{H} = \mu\mathbf{B}.$$

A simple equation of this kind, representing linear and reversible magnetisation, applies to substances such as iron only when the field is of small intensity.

Finally the relation between the current of conduction $\mathbf{C}_1$ and the electric force may be taken as a linear one involving nine independent coefficients of conductivity: in the case of isotropy these reduce to a single one $\mathbf{C}_1 = \kappa\mathbf{E}$.

It has been found by experiment that coefficients of electric conduction, unlike the other coefficients above considered, remain constant for all intensities of the current up to very high limits, so long as the temperature and physical condition of the conducting substance are not altered. This is what was perhaps to be anticipated from the circumstance that conduction arises from the filtering of the simple non-polar electrons or ions through the conducting medium under the directing action of the electric force, not from orientation of polar complex molecules which may originate hysteretic changes in their cohesive grouping in the substance.

For a body of compound nature at rest and in which both polarisation and conduction currents can coexist the relation between the total current of Maxwell's scheme and the electric force is more complicated than those given above. In fact

$$\mathbf{C} = \mathbf{C}_1 + \dot{\mathbf{D}}$$

$$= \mathbf{E} + \epsilon\dot{\mathbf{E}},$$

and this and the relation

$$\mathbf{H} = \mu\mathbf{B}$$

are the two directly required in the theory involving only media at rest.

With these four relations we have then a complete electromagnetic scheme which, if presumed to apply in the whole range of electrodynamic phenomena, is a sufficient basis for the mathematical development of the subject with regard to media at rest.

214. Finally we must notice that the circuital relation of Ampère requires one further condition to be satisfied by the current vectors of the general theory. In fact since

$$\text{curl}\,\mathbf{B} = \frac{4\pi}{c}\mathbf{C}$$

it follows at once that under all circumstances

$$\text{div}\,\mathbf{C} = 0,$$

or the current of the theory must be circuital. This is the same as saying that all currents must follow in complete circuits.

If we define a current as a flow of electricity (or electrons) the discharge of a condenser with a vacuum dielectric must constitute a current with a break in it, so that in a case of this kind the current is an open one and our theory could not possibly apply to it. However Faraday insisted from the beginning that all currents however produced are circuital. Maxwell, following this hint, soon discovered the wonderful simplicity introduced into the theory by the assumption and he therefore formally included it as an integral part of his theory. In cases like the discharge of a condenser it involves the hypothesis that there is in the aether gap some sort of release of strain taking place which possesses the electrodynamic properties of a current or a movement of electricity.

215. The point here involved has been discussed at length in a previous chapter*, but the following example will provide the general analytical form of the hypothesis. Consider the process involved in charging a conductor existing alone in the field. As we charge the conductor a state of 'polarisation' is gradually established in the surrounding dielectric medium (and aether) being accompanied by a 'displacement' in the Maxwellian sense away from the conductor. It is the essence of Maxwell's hypothesis that, in addition to the true electric displacement involved in the polarisation of the dielectric medium, there is some effect in the aether, an aethereal displacement he calls it, which is *not* an electrical displacement but for some reason or other its rate of change has the properties of a true electric current. The actual measure of this current is obtained as follows: if $\mathbf{C}'$ denote the current intensity of the true electric flow supplying the charge to the conductor then the integral

$$-\int_f \mathbf{C}_n' df,$$

taken over any closed surface f enclosing the conductor, indicates the rate at which the charge on the conductor is increasing; or if we use Q for the charge on the conductor at any time t

$$-\int_f \mathbf{C}_n' df = \frac{dQ}{dt}.$$

When however there is a charge Q on the conductor the conditions in the surrounding field are such that if $\mathbf{D}$ denotes the totality of displacement in the dielectric (aethereal and true electrical polarisation) at any point, then

$$\int_f \mathbf{D}_n df = Q,$$

so that we have

$$-\int_f \mathbf{C}_n' df = \frac{d}{dt}\int_f \mathbf{D}_n df,$$

or

$$\int_f (\mathbf{C}_n' + \dot{\mathbf{D}}_n)\, df = 0,$$

where we use $\dot{\mathbf{D}}_n$ for the time rate of change of $\mathbf{D}_n$. This indicates that the vector

$$\mathbf{C}' + \dot{\mathbf{D}}$$

is always circuital, that is, always flows in closed circuits. Thus if we add to the true current of electrons the time rate of change of

* Page 97.

the total displacement we obtain a total current vector which is always circuital. But

$$\mathbf{D} = \mathbf{P} + \mathbf{E},$$

and thus the total displacement current

$$\dot{\mathbf{D}} = \dot{\mathbf{P}} + \dot{\mathbf{E}}$$

consists of a part $\dot{\mathbf{P}}$ depending on the presence of the dielectric which is a real displacement of electric charge in the molecules of the medium. The part $\dot{\mathbf{E}}$ is the part of the displacement current which must be ascribed to some action in the aether; it is Maxwell's aethereal current. The only way of explaining the existence of this current is by a theory of the constitution of the aethereal medium, about which however we know so little. A dynamical theory of its mode of action can however be described which gives some idea of what sort of thing this displacement current is and of how it simulates an electric flow*.

We must make a distinction between the true current

$$\mathbf{C}' + \dot{\mathbf{P}},$$

which is a true flow of electricity, and this fictitious current in the aether. The contrast is in reality between the total current

$$\mathbf{C}' + \dot{\mathbf{P}} + \dot{\mathbf{E}}$$

and the true current. The total current always contains a part ($\dot{\mathbf{E}}$) which is not electric flow at all but is a something possessing the electrodynamic properties of a true electric flow.

The important thing is that now every current is effectively circuital. Looking at it from the practical side the only case where the distinction comes in is that of the electrostatic discharge. All currents of conduction are in themselves complete. In technical electrodynamics electrostatic discharges are of little account although of late years the phenomena of wireless telegraphy has imparted a technical aspect even to this side of the subject.

It took over thirty years before anything like a decisive test of this hypothesis of Maxwell's was obtained. At the beginning there were no phenomena in which the dynamical relations of an electric discharge could be investigated. Hertz however finally succeeded in discovering the electric waves whose existence and theoretical relations had been deduced by Maxwell as an essential consequence

* Cf. Larmor, *Aether and Matter*.

of his theory and thereby provided the necessary experimental proof of the hypothesis on which the theory is based.

The total current density of the general theory is therefore

$$\mathbf{C} = \mathbf{C}_1 + \frac{d\mathbf{P}}{dt} + \frac{d\mathbf{E}}{dt} + \rho\mathbf{u} + \text{curl}\,[\mathbf{Pu}]$$

and it is always circuital.

216. The units in the electromagnetic equations. In the process here employed of the construction of a mathematical theory of electricity and magnetism definite sets of units have been introduced as associated with the different classes of phenomena. As we now see that all the various phenomena are in the most general case correlated with one another there must be some definite relation between the various systems of units thus adopted: the theory would otherwise not be consistent with itself.

The chief object in the choice of a system of units is to obtain the simplest expression for our purposes of the fundamental physical relations on which the particular theory is based.

In any relation connecting physical quantities of fundamentally different characters, certain constants must occur in order to secure that the dimensions of the fundamental quantities correlated are the same. For example in the expression of the mechanical action between two point charges q_1 and q_2 concentrated at a distance r apart we say that the force between them is proportional to $\frac{q_1 q_2}{r^2}$ a quantity of a fundamentally different kind. We therefore write

$$F = \gamma \frac{q_1 q_2}{r^2},$$

and choose the dimensions of γ so as to make the dimensions of both sides of this equation the same. This constant γ is then an absolute constant of the theory, whose value and dimensions depend however on the choice of units for the other quantities involved in the relation. In the simple electrostatic theory as we have developed it we choose to measure a quantity of electricity so that the constant γ in this expression is a simple number (without dimensions) numerically equal to $\frac{1}{4}\pi$. This means that with the same notation as before the dimensions of a quantity of electricity are given by the symbolical equation

$$[Q] = [m^{\frac{1}{2}} l^{\frac{3}{2}} t^{-1}].$$

The dimensions of the electric displacement D then follow as the quotient of charge by a surface; those of the force intensity E as the quotient of force by charge; those of the potential being then the product of the dimensions of E and a length; and finally the dimensions of a current density are those of a charge divided by a time and a surface. In symbols

$$\begin{aligned}
[Q] &= [m^{\frac{1}{2}} l^{\frac{3}{2}} t^{-1}], \\
[E] &= [m^{\frac{1}{2}} l^{-\frac{1}{2}} t^{-1}], \\
[D] &= [m^{\frac{1}{2}} l^{-\frac{1}{2}} t^{-1}], \\
[\phi] &= [m^{\frac{1}{2}} l^{\frac{1}{2}} t^{-1}], \\
[C] &= [m^{\frac{1}{2}} l^{-\frac{1}{2}} t^{-1}], \\
[J] &= [m^{\frac{1}{2}} l^{\frac{3}{2}} t^{-2}].
\end{aligned}$$

These are the quantities that occur in the specification of the electric part of the general scheme. The magnetic quantities can be similarly examined and the results are identically the same as those just given for the analogous electric quantities or in symbols

$$\begin{aligned}
[H] &= [m^{\frac{1}{2}} l^{-\frac{1}{2}} t^{-1}], \\
[B] &= [m^{\frac{1}{2}} l^{-\frac{1}{2}} t^{-1}].
\end{aligned}$$

217. But we have introduced relations connecting these two classes of quantities here involved: these relations however both contain a constant c so that as usual a definite choice of units is not implied in the form of their expression adopted. If however we interpret the equations in terms of units as defined in the earlier part of this work it is found that the constants have the dimensions of a velocity. Their actual magnitudes could then be obtained by measurements of the relative magnitudes of the quantities involved. We could for instance determine the constant c in the relation

$$\int_s \mathbf{H}_s ds = \frac{J}{c},$$

connecting the magnetic field of a linear current with the strength of that current by examining the field of a given current and evaluating directly the integral on the left. Again, the constant c in the relation

$$-\frac{1}{c}\frac{d}{dt}\int_f \mathbf{B}_n df = \int_s \mathbf{E}_s ds,$$

connecting the electromotive force in a circuit with the rate of change of induction through it, might be deduced by moving a circuit in a known magnetic field and determining the current produced in it. If measurements of this type, or others involving the same principles, are made it is found that the constant c has the same value in both equations, viz.:

$$c = 3 \,.\, 10^{10} \text{ cm./sec.}^*$$

We shall therefore continue to use the equations of Ampère and Faraday with the same constant c and imply the value just given.

218. For practical purposes however the units here introduced are many of them of inconvenient magnitude so that other units, fractions or multiples of these theoretical units, have been introduced. The chief of these practical units are given below with their definitions.

1 *Ampère* is $3 \,.\, 10^9 \,.\, \sqrt{4\pi}$ absolute units of current.

1 *Ohm* is $\frac{1}{36\pi} \,.\, 10^{-11}$,, ,, resistance.

1 *Volt* is $\frac{1}{3\sqrt{4\pi}} \,.\, 10^{-2}$,, ,, potential.

These units are arranged so that the equation

$$\mathbf{C} = \kappa \mathbf{E}$$

still holds when all quantities are measured in practical units.

1 *Coulomb* is $10^{-1}\sqrt{4\pi}$ absolute units of electric quantity, so that 1 ampère is the current which carries 1 coulomb across each section of the wire per unit time.

1 *Watt* is 10^7 ergs.

1 *Joule* is 10^7 ergs per second.

1 *Farad* is $36\pi \,.\, 10^{11}$ absolute units of capacity.

1 *Microfarad* is $36\pi \,.\, 10^5$ absolute units of capacity.

* The most recent and accurate values give $c = 2{\cdot}9980 \times 10^{10}$ and Weinberg's determination for the velocity of light gives $2{\cdot}99852 \pm 0{\cdot}00024$.

The practical units of magnetic induction and force (called the *Gauss* and *Maxwell* respectively) are identical with the theoretical units multiplied by $\sqrt{4\pi}$. In addition there is a unit for what is called the *inductance*, which is the induction through a circuit when the electromotive force in it is 1 volt and the inducing current varies at the rate of 1 ampère per second; this unit is called the *Henry*, and

1 *Henry* is 10^9 absolute units of inductance $= 10^9$ cms.

219. The electromagnetic potentials. It is often convenient to interpret the fundamental equations of Faraday and Ampère in a different mathematical form. We have already introduced the vector potential $\mathbf{A}$ which is defined by the relation

$$\mathbf{B} = \text{curl}\,\mathbf{A}.$$

In terms of this potential Faraday's equation becomes

$$\text{curl}\,\mathbf{E} = -\frac{1}{c}\,\text{curl}\,\frac{d\mathbf{A}}{dt},$$

so that a particular solution for $\mathbf{E}$ is

$$\mathbf{E} = -\frac{1}{c}\frac{d\mathbf{A}}{dt}.$$

The general solution for $\mathbf{E}$ is now simply derived by the addition of the gradient of an acyclic potential ϕ, such a term having always a zero *curl*. Thus generally we may write

$$\mathbf{E} = -\frac{1}{c}\frac{d\mathbf{A}}{dt} - \text{grad}\,\phi$$

and it consists of two parts, the first of which is of electrodynamic origin and the second of static origin.

These formulae have been derived independently in connection with the integral forms of the fundamental relations where additional characteristics of the potentials are brought out. To distinguish it from the *vector potential* $\mathbf{A}$, the static potential ϕ is often called the *scalar potential*.

The three differential equations involved in the vector relation

$$\mathbf{B} = \text{curl}\,\mathbf{A}$$

are however not sufficient to determine $\mathbf{A}$ because if $\mathbf{A}$ is one solution then obviously

$$\mathbf{A} + \text{grad}\,\chi$$

is another, χ being any function of the coordinates. To define $\mathbf{A}$

more completely we may therefore impose another condition. Maxwell takes

$$\operatorname{div} \mathbf{A} = 0,$$

and this appears to be the most convenient although it still leaves a certain amount of indefiniteness. From this definition of **A** we deduce at once that

$$\begin{aligned}\operatorname{curl} \mathbf{B} &= \operatorname{curl} . \operatorname{curl} \mathbf{A} \\ &= -\nabla^2 \mathbf{A} + \operatorname{grad} \operatorname{div} \mathbf{A},\end{aligned}$$

and if we take $\operatorname{div} \mathbf{A} = 0$ this gives

$$\operatorname{curl} \mathbf{B} = -\nabla^2 \mathbf{A}.$$

This equation being analogous to that of Poisson we may consider **A** to be the potential of a distribution of matter of density curl **B**, attracting according to the inverse squares law. We thus see that

$$\mathbf{A} = \frac{1}{4\pi} \int \operatorname{curl} \mathbf{B} \frac{dv}{r},$$

the integral extending to the whole of space occupied by the electromagnetic field.

Now $$\mathbf{B} = \mathbf{H} + \mathbf{I},$$

and $$\operatorname{curl} \mathbf{H} = \frac{1}{c} \mathbf{C},$$

so that $$\mathbf{A} = \frac{1}{4\pi c} \int (\mathbf{C} + c \operatorname{curl} \mathbf{I}) \frac{dv}{r}.$$

In the case of a linear conductor carrying a current of strength J and when there is no magnetic matter about this reduces to

$$\mathbf{A} = \frac{J}{4\pi c} \int \frac{d\mathbf{s}}{r},$$

where $d\mathbf{s}$ is the vectorial element of the conducting line along which the integral is taken.

220. In the general case we have also

$$\rho = \operatorname{div} \mathbf{D} = \operatorname{div} \mathbf{E} + \operatorname{div} \mathbf{P},$$

or on inserting the value of **E** in terms of **A** and ϕ and noticing that $\operatorname{div} \mathbf{A} = 0$ we find that

$$\begin{aligned}\nabla^2 \phi &= -(\rho - \operatorname{div} \mathbf{P}) \\ &= -(\rho + \rho'),\end{aligned}$$

where ρ' is the Poisson density of ideal electrification which is equivalent for some purposes to the electric polarisation. It follows then that

$$\phi = \frac{1}{4\pi}\int \frac{\rho + \rho'}{r}\, dv,$$

so that ϕ is the static electric potential of a distribution of density $(\rho + \rho')$.

221. Most subsequent writers have adopted a slightly different definition of the vector potential which has certain advantages over that given by Maxwell. Starting from the definition of the magnetic induction in terms of a vector potential and the consequent derivation of the electric force in terms of this same potential and a scalar potential ϕ as above, it is assumed that these two potentials are connected by the equation

$$\operatorname{div} \mathbf{A} + \frac{1}{c}\frac{\partial \phi}{\partial t} = 0.$$

It is then easy to verify that $\mathbf{A}$ and ϕ satisfy respectively the equations

$$\nabla^2 \mathbf{A} = \frac{1}{c^2}\frac{d^2\mathbf{A}}{dt^2} - \frac{1}{c}(\mathbf{C}_1 + c \operatorname{curl} \mathbf{I}),$$

and

$$\nabla^2 \phi = \frac{1}{c^2}\frac{d^2\phi}{dt^2} - (\rho + \rho'),$$

where in the former equation $\mathbf{C}_1$ is used to denote the density vector of true electric flux.

Under ordinary circumstances and in finite regular fields the appropriate solutions of these equations follow immediately from the usual solution of the generalised wave-potential equation*. When the conditions of the field vary in both space and time we have in fact

$$\phi = \frac{1}{4\pi}\int [\rho + \rho'] \frac{dv}{r},$$

and

$$\mathbf{A} = \frac{1}{4\pi c}\int [\mathbf{C}_1 + c \operatorname{curl} \mathbf{I}] \frac{dv}{r},$$

the integrals being extended over the whole of the field; r as usual denotes the distance from the typical element of integration to the field-point at which the functions are calculated and the square brackets serve to indicate that the values of the functions affected

* The solution is given by Jeans in *Electricity and Magnetism*, Ch. VII, and by Lorentz in *Theory of Electrons*, p. 233.

are to be taken for the instant $t - \frac{r}{c}$, t being the instant at which the functions themselves are evaluated.

The advantage of this form of definition is that it expresses both potentials directly in terms of the positions and motions of the actual charge elements. The potentials themselves are usually called the *retarded potentials* because the contributions to them due to charges at a distance r away is not due to the instantaneous value of these charges but to their values at the previous time $\left(t - \frac{r}{c}\right)$. This means of course that the effect of any change in the charge distribution is not felt at points a distance r away until a time r/c after it has occurred, which is interpreted as implying that effects of electric changes are propagated outwards through space with the uniform velocity c in all directions.

The retarded potentials* are to be strongly contrasted with the instantaneous potentials of Maxwell's theory, and although they are perhaps more consistent with a propagation theory it is not to be inferred that the instantaneous potentials of Maxwell's theory necessarily imply the instantaneous propagation of effects from all parts of the field. It must be remembered that on both forms of the theory the potentials have been introduced primarily for analytical simplification and they do not necessarily represent directly definite physical quantities, although it may in certain circumstances be convenient to regard them as so doing. The ultimate procedure in either case involves the elimination of these potentials and the expression of all necessary relations in terms of the physical quantities that are propagated, without the aid of any auxiliary mathematical conceptions.

222. The retarded potentials are the most useful for the determination of the field of specified charge and current distributions but in their above form they are not directly suitable for numerical calculation, inasmuch as the elements of charge $[\rho]\,dv$ or current $[\mathbf{C}_1]\,dv$† which enter in their expression are not all present in the

* Cf. Larmor, *Aether and Matter*, pp. 111–112.

† We shall for the present drop the term in the vector potential due to the magnetisation. It can be replaced quite easily at any stage or may be included in the current $\mathbf{C}_1$.

volume element dv at the same effective instant. To render them more useful we must express the integrals explicitly as functions of the instantaneous distributions of the charge and current elements, which are the data usually specified in any problem.

Regarded as functions of t the densities ρ and $\mathbf{C}_1$ for a given point of space may in the limit be discontinuous, as for instance when the boundary of a charged conductor crosses the point; but from the nature of the case the number of discontinuities or infinities which occur during any finite interval of time is necessarily finite, and for each such irregularity the aggregate variation is also finite. Hence the quantities $[\rho]$ and $[\mathbf{C}_1]$ which occur in the integrals for the potential functions can always be expressed as Fourier integrals. Doing this and supposing that the values of both functions at any point (x_e, y_e, z_e) are prescribed for all values of t we get

$$4\pi\phi = \frac{1}{2\pi}\int dv \int_{-\infty}^{+\infty}\int_{-\infty}^{+\infty} e^{i\mu\left(t-\frac{r}{c}-\tau\right)}\rho\frac{d\tau\, d\mu}{r},$$

and
$$4\pi\mathbf{A} = \frac{1}{2\pi c}\int dv \int_{-\infty}^{+\infty}\int_{-\infty}^{+\infty} e^{i\mu\left(t-\frac{r}{c}-\tau\right)}\mathbf{C}_1\frac{d\tau\, d\mu}{r^2},$$

where in both integrals dv is the typical element of volume round the point (x_e, y_e, z_e) and

$$r^2 = (x - x_e)^2 + (y - y_e)^2 + (z - z_e)^2,$$

and ρ and $\mathbf{C}_1$ are now regarded as functions of τ with (x_e, y_e, z_e) as parameters.

Although these integrals appear to require a complete knowledge of the future history of the field for its present determination, they in reality do effectively define the field at the present instant independently of such knowledge. We may in fact choose ρ and $\mathbf{C}_1$ quite arbitrarily as far as future time is concerned; but when these values have been chosen the values of $[\rho]$ and $[\mathbf{C}_1]$ and hence also of ϕ and $\mathbf{A}$ are quite determinate for all time: they have the proper value for all past time whatever values may be selected for the future. If it is desired to express the integrals explicitly in terms of specified quantities only then the sine or cosine integrals must be used, but this seems to be unnecessary in the present instance.

We may now in these expressions rearrange the order of integration in the triple integrals; for this only amounts to a rearrangement of the terms of a triple sum which is for physical reasons known to be absolutely convergent. We may therefore effect the integration with respect to v first and then the difficulties of the kind mentioned above do not present themselves, because the summation is that of instantaneous contributions at the time τ from elements all over the field.

The whole of the circumstances in the surrounding field can now be determined from these forms of the potentials. To determine the electric force and magnetic induction vectors we use the relations

$$\mathbf{E} = -\frac{1}{c}\frac{d\mathbf{A}}{dt} - \operatorname{grad}\phi,$$

$$\mathbf{B} = \operatorname{curl}\mathbf{A},$$

so that
$$4\pi\mathbf{E} = \frac{1}{2\pi c}\frac{\partial}{\partial t}\int dv \int_{-\infty}^{+\infty}\int_{-\infty}^{+\infty} \frac{e^{i\mu\left(t-\frac{r}{c}-\tau\right)}}{r}\left(\rho\mathbf{r}_1 - \frac{1}{c}\mathbf{C}_1\right) d\tau\, d\mu$$
$$+ \frac{1}{2\pi}\int dv \int_{-\infty}^{+\infty}\int_{-\infty}^{+\infty} \frac{e^{i\mu\left(t-\frac{r}{c}-\tau\right)}}{r^2}\rho\mathbf{r}_1\, d\tau\, d\mu,$$

and
$$4\pi\mathbf{B} = \frac{1}{2\pi c^2}\frac{\partial}{\partial t}\int dv \int_{-\infty}^{+\infty}\int_{-\infty}^{+\infty} \frac{e^{i\mu\left(t-\frac{r}{c}-\tau\right)}}{r}[\mathbf{C}_1\mathbf{r}_1]\, d\tau\, d\mu$$
$$+ \frac{1}{2\pi c}\int dv \int_{-\infty}^{+\infty}\int_{-\infty}^{+\infty} \frac{e^{i\mu\left(t-\frac{r}{c}-\tau\right)}}{r^2}[\mathbf{C}_1\mathbf{r}_1]\, d\tau\, d\mu,$$

wherein $\mathbf{r}_1$ denotes the unit vector in the line joining the typical element of integration to the field-point where the functions are calculated so that
$$\nabla r = \mathbf{r}_1.$$

223. If the total charge distribution in any field is of density ρ and the current flux is due solely to the convection of this density with velocity $\mathbf{v}$, then the above scalar and vector potentials can be written in the form in which

$$4\pi\phi = \frac{1}{2\pi}\int_{-\infty}^{+\infty}\int_{-\infty}^{+\infty}\int e^{i\mu\left(t-\frac{r}{c}-\tau\right)}\frac{dQ\, d\tau\, d\mu}{r},$$

$$4\pi\mathbf{A} = \frac{1}{2\pi c}\int_{-\infty}^{+\infty}\int_{-\infty}^{+\infty}\int e^{i\mu\left(t-\frac{r}{c}-\tau\right)}\frac{\mathbf{v}\, dQ\, d\tau\, d\mu}{r},$$

where $dQ = \rho\, dv$ is the charge element in the small volume dv at the time τ. When the charge is of amount e and concentrated in a small volume element round the point (x_e, y_e, z_e) these expressions reduce to

$$4\pi\phi = \frac{e}{2\pi}\int_{-\infty}^{+\infty}\int_{-\infty}^{+\infty} e^{i\mu\left(t-\frac{r}{c}-\tau\right)}\frac{d\tau\,d\mu}{r},$$

$$4\pi\mathbf{A} = \frac{e}{2\pi c}\int_{-\infty}^{+\infty}\int_{-\infty}^{+\infty} e^{i\mu\left(t-\frac{r}{c}-\tau\right)}\frac{\mathbf{v}\,d\tau\,d\mu}{r}.$$

Now change in these integrals the variable τ to τ' where

$$\tau' = \tau + \frac{r}{c}$$

so that $$d\tau' = d\tau\left(1 + \frac{1}{c}\frac{dr}{d\tau}\right) = d\tau\left(1 - \frac{\mathbf{v}_r}{c}\right),$$

then we get, for instance,

$$4\pi\phi = \frac{e}{2\pi}\int_{-\infty}^{+\infty}\int_{-\infty}^{+\infty} e^{i\mu(t-\tau')}\frac{d\tau'\,d\mu}{\left[r\left(1-\frac{\mathbf{v}_r}{c}\right)\right]_{t=\tau'-\frac{r}{c}}}$$

since the limits for τ' are the same as those for τ when $(\mathbf{v}) < c$. But the double integral in this last case is a proper Fourier integral whose value is

$$\frac{2\pi}{\left[r\left(1-\frac{\mathbf{v}_r}{c}\right)\right]_{\tau'=t}}$$

so that $$4\pi\phi = \frac{e}{\left[r\left(1-\frac{\mathbf{v}_r}{c}\right)\right]}.$$

Similarly we find that

$$4\pi\mathbf{A} = \frac{e\,[\mathbf{v}]}{\left[cr\left(1-\frac{\mathbf{v}_r}{c}\right)\right]},$$

where in the last two expressions square brackets serve to indicate the values of the functions affected at the time $\left(t - \frac{r}{c}\right)$.

The whole circumstances in the field of the moving charge element can now be readily deduced and the expressions for the electric and magnetic force vectors at any point obtained by simple differentiation of these potentials. It is however easier to deduce them from the general results. In fact if we put in these expressions $\mathbf{C}_1 = \rho\mathbf{v}$

and carry out the volume integration and then transform the integrals as we have just done for the potentials we find that

$$4\pi\mathbf{E} = \frac{e\mathbf{r}_1}{\left[r^2\left(1-\frac{\mathbf{v}_r}{c}\right)\right]} + \frac{d}{dt}\left[\frac{e\left(\mathbf{r}_1 - \frac{\mathbf{v}}{c}\right)}{cr\left(1-\frac{\mathbf{v}_r}{c}\right)}\right],$$

whilst

$$4\pi\mathbf{B} = \left[\frac{e\,[\mathbf{v}\mathbf{r}_1]}{cr^2\left(1-\frac{\mathbf{v}_r}{c}\right)}\right] + \frac{d}{dt}\left[\frac{e\,[\mathbf{v}\mathbf{r}_1]}{c^2 r\left(1-\frac{\mathbf{v}_r}{c}\right)}\right],$$

wherein $\mathbf{r}_1$ denotes the unit vector in the direction of the radius r.

We must now bear in mind that the quantities inside the large square brackets are functions of $\tau = t - \frac{r}{c}$ as well as (x, y, z) explicitly; but

$$\frac{d\tau}{dt} = \frac{1}{\left[1 - \frac{\mathbf{v}_r}{c}\right]}$$

and hence

$$\frac{1}{c}\frac{d}{dt}\left[r\left(1-\frac{\mathbf{v}_r}{c}\right)\right] = 1 - \left[\frac{(\dot{\mathbf{v}}\mathbf{r}) + c^2 - \mathbf{v}^2}{c^2\left(1-\frac{\mathbf{v}_r}{c}\right)}\right]$$

and

$$\frac{1}{c}\frac{d\mathbf{r}_1}{dt} = \left[\frac{\frac{\mathbf{r}_1\mathbf{v}_r}{c} - \frac{\mathbf{v}}{c}}{r\left(1-\frac{\mathbf{v}_r}{c}\right)}\right].$$

Hence finally we have

$$\mathbf{E} = \frac{e}{4\pi c^2}\left[-\frac{\dot{\mathbf{v}}}{r}\left(\frac{d\tau}{dt}\right)^2 + \frac{\left(\mathbf{r}_1 - \frac{\mathbf{v}}{c}\right)(c^2 - \mathbf{v}^2 + (\dot{\mathbf{v}}\mathbf{r}))}{r^2}\left(\frac{d\tau}{dt}\right)^3\right],$$

$$\mathbf{B} = \frac{e}{4\pi c^2}\left[\frac{[\dot{\mathbf{v}}\mathbf{r}_1]}{r}\left(\frac{d\tau}{dt}\right)^2 + \frac{[\mathbf{v}\mathbf{r}_1](c^2 - \mathbf{v}^2 + (\dot{\mathbf{v}}\mathbf{r}))}{cr^2}\left(\frac{d\tau}{dt}\right)^3\right].$$

224. We first notice that

$$\mathbf{B} = [\mathbf{r}_1\mathbf{E}],$$

so that the magnetic force is everywhere perpendicular to the electric force and to the radius from the field-point to the effective position of the charge element at the instant. On the other hand

$$4\pi\,(\mathbf{r}_1\mathbf{E}) = \left[\frac{e\,(c^2 - \mathbf{v}^2)}{c^2 r^2\left(1-\frac{\mathbf{v}_r}{c}\right)}\right],$$

so that the electric force is not transverse to the radius vector unless $c = |\mathbf{v}|$; but the deviation from perpendicularity becomes smaller and smaller as the distance from the charge increases.

Again since $1 - \frac{\mathbf{v}_r}{c} = 1 - \frac{(\mathbf{v}\mathbf{r}_1)}{c} = \left(\mathbf{r}_1, \mathbf{r}_1 - \frac{\mathbf{v}}{c}\right)$

it is easy to verify that

$$\left(\mathbf{r}_1 - \frac{\mathbf{v}}{c}\right)(\dot{\mathbf{v}}\mathbf{r}_1) - \dot{\mathbf{v}}\left(1 - \frac{\mathbf{v}_r}{c}\right) = \left[\mathbf{r}_1, \left[\mathbf{r}_1 - \frac{\mathbf{v}}{c}, \dot{\mathbf{v}}\right]\right]$$

so that

$$\mathbf{E} = \frac{e}{4\pi rc^2}\left(\frac{d\tau}{dt}\right)^3 \left[\frac{1}{r}\left(\mathbf{r}_1 - \frac{\mathbf{v}}{c}\right)(c^2 - \mathbf{v}^2) + \left[\mathbf{r}_1\left[\mathbf{r}_1 - \frac{\mathbf{v}}{c}, \dot{\mathbf{v}}\right]\right]\right].$$

Also since $$\mathbf{B} = [\mathbf{r}_1\mathbf{E}]$$

we have $$[\mathbf{B}\mathbf{r}_1] = -[\mathbf{r}_1[\mathbf{r}_1\mathbf{E}]] = \mathbf{E} - \mathbf{r}_1(\mathbf{r}_1\mathbf{E})$$

and thus also $\mathbf{E} = [\mathbf{B}\mathbf{r}_1] + \frac{e\mathbf{r}_1}{4\pi c^2 r^2}(c^2 - \mathbf{v})\left(\frac{d\tau}{dt}\right)^2.$

The part of the electric force not depending on the acceleration and the predominant part in the field near the electron is

$$\mathbf{E}_s = \frac{e}{4\pi c^2}\left[\left(\mathbf{r}_1 - \frac{\mathbf{v}}{c}\right)(c^2 - \mathbf{v}^2)\left(\frac{d\tau}{dt}\right)^3\right]$$

whilst the corresponding part of the magnetic force is

$$\mathbf{B}_s = \frac{1}{c}[\mathbf{v}\mathbf{E}_s].$$

Now the vector $\left[\mathbf{r}_1 - \frac{\mathbf{v}}{c}\right]$ is parallel to the direction of the radius from the field-point to what would be the instantaneous position of the moving charge (as distinct from its effective position) if it be assumed that it has moved from its effective position with the constant velocity [$\mathbf{v}$] that it then had.

If the motion of the particle is with constant velocity in a straight line this is the whole field, but for it the formulae can be expressed even more explicitly in terms of the position of the particle at the instant under consideration. If the motion is with velocity $\mathbf{v}$ in the straight line OO' and if the time taken from O to O' is τ then the length of OO' is $v\tau$. Further if the potentials are calculated at the point P distant $r = c\tau$

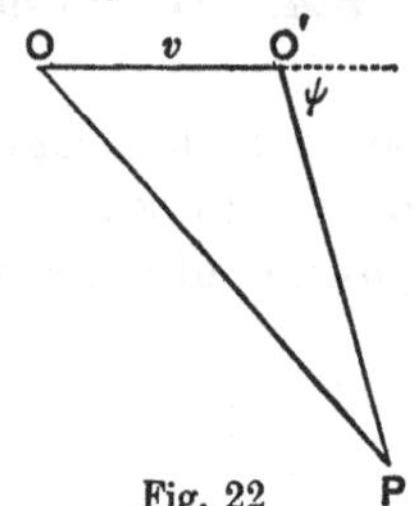

Fig. 22

from O and r' from O', then O is the position of the particle which gives rise to the field at P when the particle itself is at O'. Thus in the above formulae we have

$$\left[r\left(1-\frac{\mathrm{v}_r}{c}\right)\right] = OP\left(1-\frac{v}{c}\cos \widehat{O'OP}\right)$$
$$= OP - v\tau \cos \widehat{O'OP}$$
$$= OP - OO' \cos \widehat{O'OP}.$$

Now $\quad r'^2 = O'P^2 = OP^2 + OO'^2 - 2OO' \,.\, OP \cos \widehat{O'OP}$
$$= (OP - OO' \cos \widehat{O'OP})^2 + OO'^2 \sin^2 \widehat{O'OP}.$$

But from the figure it is clear that

$$OP \sin \widehat{O'OP} = O'P \sin \widehat{OO'P}.$$

Thus $\quad OO' \sin \widehat{O'OP} = \dfrac{OO'}{OP} \,.\, O'P \sin \widehat{OO'P} = \dfrac{vr'}{c} \sin \widehat{OO'P}$

and therefore

$$r'^2 = (OP - OO' \cos \widehat{O'OP})^2 + \frac{v^2 r'^2}{c^2} \sin^2 \widehat{OO'P},$$

or

$$r'^2\left(1-\frac{\mathbf{v}^2}{c^2}+\frac{\mathrm{v}_{r'}^2}{c^2}\right) = \left[r\left(1-\frac{\mathrm{v}_r}{c}\right)\right]^2.$$

Thus in terms of the position and velocity at P' we have

$$\phi = \frac{e}{4\pi r'} \frac{1}{\left(1-\dfrac{\mathbf{v}^2}{c^2}+\dfrac{\mathrm{v}_{r'}^2}{c^2}\right)^{\frac{1}{2}}},$$

$$\mathbf{A} = \frac{e\mathbf{v}}{4\pi c r'} \frac{1}{\left(1-\dfrac{\mathbf{v}^2}{c^2}+\dfrac{\mathrm{v}_{r'}^2}{c^2}\right)^{\frac{1}{2}}},$$

formulae which will be derived in another manner on a later occasion.

If the motion of the particle is accelerated the part of the field depending on the acceleration which predominates at large distances from the particle and is insignificant in the rest of the field is determined by

$$\mathbf{E} = \frac{e}{4\pi r c^2}\left(\frac{d\tau}{dt}\right)^3 \left[\mathbf{r}_1, \left[\mathbf{r}_1 - \frac{\mathbf{v}}{c}, \dot{\mathbf{v}}\right]\right]$$

and $\quad \mathbf{E} = [\mathbf{B}\mathbf{r}_1].$

Thus in this part of the field the electric and magnetic forces are both perpendicular to the radius vector from the effective position of the particle to the field-point under review; they are also perpendicular to one another and of equal magnitude.

If the velocity of the motion is small we have practically

$$\mathbf{E} = \frac{e\dot{\mathbf{v}}}{4\pi rc^2}$$

and

$$\mathbf{E} = [\mathbf{B}\mathbf{r}_1]$$

so that the electric force is tangential to the meridian circles and the magnetic force to the parallels of latitude, the polar axis being parallel to the effective acceleration. The full significance of these results will appear later.

225. The energy in the electromagnetic field. Since we became convinced of the impossibility of perpetual motion there has always been connected with our conceptions of natural phenomena the idea of that something which we call energy. The kinetic and potential energies of matter were the kinds of energy first recognised and for this reason it is customary to try to associate any new form of energy which turns up with something which in its properties is akin to matter. Thus arose for example the idea of the material aether which formed the basis of the older wave theory of light and which ascribed to this aether elastic and inertia properties and then regarded the energy of the light waves as composed of the kinetic and potential energy of the medium. The main object of Maxwell's electric theory is to adopt the idea of an elastic aether to explain the electrodynamical actions of electromagnetic systems, although care has been taken not to attribute to this aether any nature analogous to that of ordinary matter and the theory is in fact quite independent of any definite assumptions or hypotheses we may make as to its constitution. Thus far therefore we merely use the word 'aether' as a convenient means of describing those properties of space which are concerned in electromagnetic phenomena, those properties being mathematically expressed by the general equations of Maxwell's theory, which contain in themselves not only the totality of the older laws of electrostatic and electrodynamic phenomena, but also the laws for the propagation of light and electric waves in space.

In reality the universe consists of matter and the electromagnetic field in the aether. If the electromagnetic field were not present our eyes would not indicate to us the presence of the matter. In fact natural phenomena in general appear to us as a result of the interaction between matter and the aether. A simple materialistic conception of things regards the interaction of the material constituents as the essence of the affair and regards the electromagnetic theory merely as an auxiliary means of formulating the laws of these interactions; matter is the only actuality. We may however with equal justification regard the matter from the other point of view. We can regard the electromagnetic aether as the only actuality and matter as a special manifestation of a 'condition' in this aether. Such a view is of course one-sided but it is helpful in preventing us from going to the other extreme. There is in fact one point in its favour; our knowledge of electromagnetic phenomena is much more precise and extensive than our knowledge of matter.

For the present however we shall not definitely make any special hypothesis either one way or the other, beyond assuming that there is something to which we can attach the energy which is associated with any electromagnetic field. Whatever views we may take as to the aether we shall always find it at least expedient to retain the concept of energy with the associated idea that it is something definite which can be distributed and transferred through the space occupied by matter. Recent physical speculation is even tending to raise this energy to an even more important place in science: it dispenses entirely with aether and matter as independent entities and regards energy as the one fundamental quantity with which physical science deals; matter then is only manifest through effects the ultimate result of which is a change in its energy either as regards its total or distribution.

There are however certain difficulties in this new view, mainly concerned with the definition and specification of the energy in any body, and an even more general procedure has been suggested but not yet sufficiently developed for us to take much cognisance of it at present.

226. The energy of the electromagnetic field must be continuously distributed throughout the field and our present object is the investigation of its distribution and transfer in that field. For most dynamical purposes it is necessary to know not only the total amount of the energy in the field but also how it is distributed in the field. For example in a dynamical theory of an elastic solid it is always necessary to know the potential energy w per unit volume at any point (a quadratic function of the strains in the element dv at the place) as well as the total potential energy

$$W = \int w dv,$$

before using the results in any mechanical discussion. The important point is that we must be sure that this represents the actual distribution and not only the total amount. We might, for example, in the process of obtaining W, have integrated by parts and so obtained

$$W = \int_f w_n' df + \int w' dv,$$

where the surface integral is taken over a surface f bounding the system. If this surface is indefinitely extended we can neglect this part and thus

$$W = \int w' dv,$$

and $w' \neq w$. But by integrating by parts we always mix up the energy from different parts of the system to get that at the typical volume element; the energy in any element would then depend on all the distant elements and we should not then have a proper local distribution. In the case of media like the aether this is one of the complexities to be met with and we have to find out as best we can the proper distribution of energy; but in any case we are never absolutely certain that our simply obtained energy distributions have not after all been obtained by some such process as integration by parts and do not therefore represent the true distribution required.

227. With these preliminary remarks let us now consider the conservation of the total energy in any electrodynamic field*. In this case we interpret the general principle in the form that the diminution of energy inside any closed surface in the field is equal

* The present treatment follows the lines sketched by Larmor in *Phil. Trans.* A, 190 (1897), p. 285, and developed in further detail in his lectures.

to the flow of energy outwards across the surface. In other words if E is the total energy inside, $-dE/dt$ is equal to the flux of energy outwards over the surface; and we ought to be able to express this flux as a surface integral in the form

$$\int_f \mathbf{S}_n df,$$

$\mathbf{S}_n$ denoting the outward normal component of the vector determining the energy flux, which will of course have magnitude and direction like the flux of anything else.

The total energy E of course represents the available electrodynamic energy in the field. If W represents the potential energy and T the kinetic energy then

$$\frac{dE}{dt} = \frac{dW}{dt} + \frac{dT}{dt} + F = -\int_f \mathbf{S}_n df,$$

wherein F represents the electromagnetic energy dissipated in the space considered per unit time, either directly into heat or in the performance of mechanical work on the masses with free charges or polarisations.

The potential energy is very generally expressed by

$$W = \int_v dv \int_0^P (\mathbf{E} \, . \, \delta\mathbf{P}) + \tfrac{1}{2}\int_v \mathbf{E}^2 dv,$$

where $\mathbf{P}$ is the polarisation intensity of the molecules of the medium produced by the electric force $\mathbf{E}$. The first term represents the work done against the material reactions to the setting up of the polarisation and is stored up as internal potential energy in the polarisation of the medium, being simply the organised part of the energy of elastic stress. The second part represents the purely electrical part of the potential energy associated with the polarisations in the molecules representing energy of strain in the aethereal medium due to the presence and configuration of the electrical charges. We have of course excluded the existence of hysteretic effects in the polarisation of the medium so that it is generally reversible. The polarisation is presumed to be an elastic affair involving no dissipation.

Again we know that the energy dissipated per second is

$$\int (\mathbf{E}\mathbf{C}_1) \, dv,$$

$\mathbf{C}_1$ representing the part of the total current depending on the motion of the free electrons. This current consists of two essentially distinct parts concerned respectively with the true conduction electrons and with those giving rise to the convection currents. In both cases the electric force acts on the free electrons and increases their velocities, but in the former case this increase is dissipated by collision into irregular heat motion, whereas in the latter it is converted into mechanical energy of effectively non-electric nature. In the most general case the convection current will arise in the motion of charged bodies and polarised media.

We now have

$$\frac{dW}{dt} = \int_v (\mathbf{E} \,.\, \dot{\mathbf{E}})\, dv + \int_v (\dot{\mathbf{P}} \,.\, \mathbf{E})\, dv$$

or

$$= \int_v (\mathbf{E} \,.\, \dot{\mathbf{D}})\, dv,$$

and

$$F = \int (\mathbf{E} \,.\, \mathbf{C}_1)\, dv$$

$$= \int (\mathbf{E} \,.\, \mathbf{C} - \dot{\mathbf{D}})\, dv$$

$$= \int (\mathbf{E} \,.\, \mathbf{C})\, dv - \int (\mathbf{E} \,.\, \dot{\mathbf{D}})\, dv,$$

so that

$$\frac{dW}{dt} + F = \int (\mathbf{E} \,.\, \mathbf{C})\, dv.$$

We have therefore $\dfrac{dT}{dt} = \dfrac{dE}{dt} - \displaystyle\int (\mathbf{E} \,.\, \mathbf{C})\, dv.$

If Maxwell's ideas on the nature of the electromagnetic actions are tenable we must now be able to express $\dfrac{dE}{dt}$ as a surface integral so that

$$\frac{dT}{dt} = -\int_f \mathbf{S}_n df - \int (\mathbf{E} \,.\, \mathbf{C})\, dv.$$

Various possibilities are now open to us. This is all we can learn about the distribution of the kinetic energy from the energy principle alone. We can however make further simple hypotheses and thereby gain additional insight into the matter.

228. The most natural hypothesis is obtained by transforming the last integral by the substitution

$$\frac{1}{c}\,\mathbf{C} = \operatorname{curl}\mathbf{H}.$$

We get then

$$\int(\mathbf{E}\,.\operatorname{curl}\mathbf{H})\,dv = -\int[\mathbf{EH}]_n\,df + \int(\mathbf{H}\,.\operatorname{curl}\mathbf{E})\,dv$$
$$= -\int[\mathbf{EH}]_n\,df - \frac{1}{c}\int\left(\mathbf{H}\,.\frac{d\mathbf{B}}{dt}\right)dv,$$

so that $$\frac{dT}{dt} = -\int_f \mathbf{S}_n\,df + c\int[\mathbf{EH}]_n\,df + \int\left(\mathbf{H}\,.\frac{d\mathbf{B}}{dt}\right)dv.$$

We might now take $$\mathbf{S} = c\,[\mathbf{EH}],$$

and then we should have

$$\frac{dT}{dt} = \int_v\left(\mathbf{H}\,.\frac{d\mathbf{B}}{dt}\right)dv,$$

or $$T = \int dv\int_0^B(\mathbf{H}\,d\mathbf{B}),$$

provided of course that the integrand is a perfect differential. This means that this is a possible form for T if the induced magnetism is reversible.

But $$\int_0^B(\mathbf{H}\,d\mathbf{B}) = \tfrac{1}{2}\mathbf{B}^2 - \int_0^B(\mathbf{I}\,d\mathbf{B}),$$

so that in the absence of the magnetic media the kinetic energy would be distributed throughout the field with the density

$$\tfrac{1}{2}\mathbf{B}^2,$$

which is the form assumed by Maxwell. If we regard all magnetic energy as kinetic energy, then this special form of the density would suggest that the magnetic induction is a kind of velocity in the aether.

On this hypothesis we have

$$-\frac{dE}{dt} = c\int_f[\mathbf{EH}]_n\,df,$$

for this is what remains when we identify $\frac{dT}{dt}$ with the other part of the complete expression. This means that the vector

$$\mathbf{S} = c\,[\mathbf{EH}]$$

represents the flux of energy; the integral of the flux of this vector across any surface represents the rate of change of energy inside the surface. The resultant flux of energy at any point is

$$c\,(H\,.\,E\sin\widehat{\mathbf{HE}}),$$

and is directed perpendicular to both **H** and **E**; the energy flows perpendicular to both forces in the field.

This is Poynting's result and this vector **S** is usually called after him. It is however necessary to emphasise the fact that it represents the flux of energy in the field only on the hypothesis that the kinetic energy is distributed in the medium with a density $\int_0^B (\mathbf{H}\,d\mathbf{B})$ per unit volume; and even then it is uncertain to an additive vector quantity which integrates out when taken all over the surface f. However, following a usual practice in physics, it is best to adhere to the simplest hypothesis. The actual phenomena strongly suggest that the flux of energy is correctly represented by this vector and the addition of anything else is merely a gratuitous complication which is not, after all, necessary.

229. There is however no definite and precise reason why we should take the matter this way; we might have adopted some other scheme. The only other one of any importance is obtained by performing the first integration by parts in some other way. We found that

$$\frac{dT}{dt} = \frac{dE}{dt} - \int (\mathbf{E}\,.\,\mathbf{C})\,dv,$$

and we integrated by the substitution $\frac{1}{c}\,\mathbf{C} = \operatorname{curl}\mathbf{H}$. We might however follow another course and introduce the vector potential **A** by the substitution

$$\mathbf{E} = -\frac{1}{c}\frac{d\mathbf{A}}{dt} - \operatorname{grad}\phi,$$

and then we have

$$\int (\mathbf{E}\,.\,\mathbf{C})\,dv = -\frac{1}{c}\int \left(\frac{d\mathbf{A}}{dt}\,.\,\mathbf{C}\right) dv - \int (\mathbf{C}\,.\operatorname{grad}\phi)\,dv$$

$$= -\frac{1}{c}\int \left(\frac{d\mathbf{A}}{dt}\,.\,\mathbf{C}\right) dv + \int \phi\,.\operatorname{div}\mathbf{C}\,dv - \int_f \phi\mathbf{C}_n\,df,$$

and since

$$\operatorname{div}\mathbf{C} = 0,$$

we see that

$$\int (\mathbf{E} \cdot \mathbf{C})\, dv = -\frac{1}{c}\int \left(\mathbf{C}\,\frac{d\mathbf{A}}{dt}\right) dv - \int_f \phi \mathbf{C}_n\, df,$$

so that now we have

$$\frac{dT}{dt} = \frac{dE}{dt} + \int_f \phi \mathbf{C}_n\, df + \frac{1}{c}\int \left(\mathbf{C}\,\frac{d\mathbf{A}}{dt}\right) dv.$$

We might now take

$$-\frac{dE}{dt} = \int_f \mathbf{S}_n\, df = \int_f \phi \mathbf{C}_n\, df,$$

and then we should have

$$T = \frac{1}{c}\int dv \int_0^A (\mathbf{C}\, d\mathbf{A}).$$

This is the general form of a result which has received very influential support from some quarters* and there is something to be said for it. Consider the case of a conduction current flowing in complete circuits. If we regard the electrons as the energy carriers, then the work done is measured by the energy expended by the electric force in pulling them about. This is to a certain extent a reasonable hypothesis even if it entirely neglects the aether. On the aether theory the energy of an electric particle is really distributed in the field around it; the electrons themselves are thus simply keys or singular points which bind or lock up their own portion of the energy.

There are however difficulties of a far more fundamental kind involved in this new type: we cannot for instance assume in general that

$$\int_v dv \int^A (\mathbf{C}\, d\mathbf{A}) = \int_v dv \int^C (\mathbf{A}\, d\mathbf{C}),$$

so that, even if there is no magnetism, the intrinsic energy of the field of kinetic type is not all available energy. In other words we should have to assume something like an hysteretic quality for the free aether, so that the magnetic energy in any field will be a function of the history of the generation of the field and may not vanish when the field is again reduced to zero. We shall examine presently another aspect of this conclusion.

Again the new expression for the kinetic energy, involving as it does the vector potential, is correspondingly incomplete in its

* Cf. Macdonald, *Electric Waves*, chs. IV, V, VIII.

mathematical definition. Both the scalar and vector potentials are insufficiently defined and cannot therefore without further arbitrary restrictions represent definite physical entities, and a theory interpreted in terms of them necessarily remains indefinite from the physical point of view. Of course it is not *a priori* impossible that one of the sets of potentials does in fact represent some physical quantity so that this difficulty may not prove to be so serious as it seems at first sight.

In any case we cannot definitely say that either form is wrong, and the particular form of theory adopted is entirely a matter of preference and not proof. The chief point to be noticed is that we get different distributions of magnetic energy according to the assumptions we make; the differences are, it is true, unimportant in the ordinary statical and dynamical aspects of the theory so far examined, but cases will be examined where the two distributions are of fundamentally different types. In some types of fields, for example, the densities of the magnetic energy on the two theories are equal in magnitude but opposite in sign.

There are of course other forms of the transformation leading to formulae for the energy distribution and transfer which are in a way intermediate between the two forms here given, but we have given sufficient details to illustrate the uncertainty in the matter.

230. General electrodynamic theory*. The tendency of the physical investigations outlined in the previous chapters has been towards the construction of a dynamical theory which shall give a consistent account of electrodynamic phenomena, i.e. to answer the question as to the possibility of obtaining a complete parallel to the processes in any electromagnetic field from those observed in some imaginary system of masses moving according to the ordinary laws of mechanics. To reply completely to such a question it is not necessary to make any definite assumptions as to the mechanism underlying the phenomena; all we have to do is to show that they can be described by means of the general equations of mechanics.

The most general dynamical principle which determines the

* Cf. Larmor, *Aether and Matter*, ch. VI; Lorentz, *La théorie électromagnétique*, §§ 55–61; Helmholtz, *Ann. Phys. Chem.* 47 (1892), p. 1; Sommerfeld, *Ann. d. Phys.* 46 (1892), p. 139; Reiff, *Elastizität und Elektrizität* (Leipzig, 1893).

motion of every material system is the law of Least Action, expressible in the usual form

$$\delta \int (T - W)\, dt = 0,$$

wherein T denotes the kinetic energy and W the potential energy of the system in any configuration and formulated in terms of any coordinates that are sufficient to specify the configuration and motion in accordance with its known properties and connections; and where the variation refers to a fixed time of passage of the system from the initial to the final configuration considered. The power of this formula lies in the fact that once the energy function is obtained in terms of any measurements of the system that are convenient and sufficient for the purposes in view, the remainder of the investigation involves only the exact processes of mathematical analysis.

Now we have in the previous section succeeded in obtaining expressions for the potential and kinetic energies associated with any electromagnetic field and it thus only remains to interpret these functions in terms of suitable coordinates, before applying the general laws of dynamics to determine the sequence of events in any such system. But whatever view we may take as to the constitution of the aether and the electrons it is quite obvious from the whole of the preceding discussion that all electrical effects must be explicable on the hypothesis of the aether with the electrons or discrete atomic charges moving about in it freely or grouped into material atoms: thus as far as we are at present concerned the only difference between the aether and any material medium must simply be due to the presence of convection currents of electrons; that is, wherever there is matter there are these convection currents. Thus in a mechanical theory the electrodynamic state of any system will be completely known if we can specify the positions and motions of all the electrons in it together with the displacement, in Maxwell's sense, in the aether, and herein we have sufficient data for our present dynamical analysis. Of course for the purposes of electrodynamic phenomena of material which we can only test by observation and experiment on matter in bulk a complete atomic analysis of this kind is almost useless; for we are unable to take cognisance of the single electrons to which this analysis has regard. The development of the theory which is to be in line with experience ought instead to concern itself with an effective differential element of volume, containing a crowd of molecules numerous enough to be expressible continuously as regards their average relations, as a volume density of matter. As regards the actual distribution in the element of volume of the really discrete electrons, all that we can usually take cognisance of is an excess of one kind, positive or negative, which constitutes a volume density of electrification, or else an average polarisation in the arrangement of the groups of electrons in the molecules which must be specified as a vector by its intensity per unit volume: while the movements of the electrons, free and paired, in such elements of volume must be combined into statistical aggregates of translational fluxes and molecular whirls of electrification. With anything else than mean aggregates of the various types that can be thus separated out, each extended over the effective element of volume, mechanical science, which has for its object matter in bulk as it presents itself to our observation and experiment, is not directly concerned. Nevertheless it is convenient on account of simplicity to formulate the problem in terms of the separate electrons and to reserve the details of the process of averaging, which must in reality be implied throughout, until the mechanical relations of the system have been formulated in full.

231. Let us therefore proceed directly to the formulation of the mechanical relations of a system of discrete electrons in a field of aether, the potential or electrostatic energy of this system being expressed by the integral

$$W = \tfrac{1}{2}\int \mathbf{E}^2\, dv,$$

extended throughout the entire volume of the electrodynamic field, whilst the kinetic energy* is expressed by

$$T = \tfrac{1}{2}\int \mathbf{B}^2\, dv.$$

* The principle of Least Action was employed in the manner here adopted by Prof. Larmor but with the kinetic energy expressed in terms of the vector potential. The present deduction was given by the author, *Phil. Mag.* 32 (1916), p. 195. The most general formulation without the simplifying restrictions adopted above is given in *Phil. Trans.* A, 220 (1920), pp. 207–245.

The complete Lagrangian function for the system is therefore

$$L = L_0 + \tfrac{1}{2}\int (\mathbf{B}^2 - \mathbf{E}^2)\,dv,$$

L_0 being that part which does not depend on the aethereal configuration as specified in the displacement $\mathbf{E}$, but which does depend essentially on the size, constitution and motion of the electronic nucleus, as well as on the forces, not of electric origin, exerted on it from the material atoms, or otherwise, if such are presumed to exist.

We could now conduct the variation directly were it not for the circumstance that our variables are not wholly independent: in fact the variations of $\mathbf{E}$ and $\mathbf{B}$ are subject to the conditions as

$$\int \operatorname{div}\mathbf{E}\,dv - \Sigma q = 0,$$

and
$$\int\left(\operatorname{curl}\mathbf{B} - \frac{1}{c}\frac{d\mathbf{E}}{dt}\right)dv - \frac{1}{c}\Sigma q\dot{\mathbf{r}} = 0.$$

In these expressions Σ denotes a sum relative to all the electrons in the volume considered, each with its proper charge q and velocity $\dot{\mathbf{r}}$, $\mathbf{r}$ being the position vector of the typical electron. The second relation is a vector one and is therefore equivalent to three independent equations.

Hence we must now introduce into the variational equation four Lagrangian undetermined functions of position ϕ, $\mathbf{A}_x$, $\mathbf{A}_y$, $\mathbf{A}_z$, the last three of which may be regarded as the rectangular components of a vector $\mathbf{A}$. It is thus the variation of

$$\int L\,dt + \int dt\left[\int dv\left\{\phi\operatorname{div}\mathbf{E} - \left(\mathbf{A}\,.\operatorname{curl}\mathbf{B} - \frac{1}{c}\frac{d\mathbf{E}}{dt}\right)\right\} - \Sigma q\phi + \Sigma\frac{q}{c}(\mathbf{A}\dot{\mathbf{r}})\right]$$

that is to be made zero; afterwards determining the form of ϕ and $\mathbf{A}$ to satisfy the restrictions which necessitated their introduction. In conducting the variation we can now treat the electric force, magnetic induction and the position coordinates of the electrons as all independent.

As regards the electrons q the variation gives

$$\int dt\left\{\delta L_0 - \Sigma q\,(\delta\mathbf{r}\nabla)\,\phi + \Sigma\frac{q}{c}\,(\dot{\mathbf{r}},\,(\delta\mathbf{r}\nabla)\,\mathbf{A}) + \Sigma\frac{q}{c}\,(\delta\dot{\mathbf{r}}\mathbf{A})\right\},$$

where $\mathbf{A}$ must now be regarded as assuming the succession of values it takes as the point whose position is defined by the vector $\mathbf{r}$ moves through the aether, not the succession of values it takes at a fixed point. Integrating by parts we get that this part of the variation is equal to terms at the time limits together with

$$\int dt \left\{\delta L_0 - \Sigma q\,(\delta\mathbf{r}\nabla)\,\phi + \Sigma\frac{q}{c}\,(\dot{\mathbf{r}},\,(\delta\mathbf{r}\nabla)\,\mathbf{A}) - \Sigma\frac{q}{c}\left(\delta\mathbf{r}\,\frac{\delta\mathbf{A}}{dt}\right)\right\},$$

where the symbol $\dfrac{\delta\mathbf{A}}{dt}$ is used to denote the time rate of variation of $\mathbf{A}$ relative to the moving electron: thus

$$\frac{\delta\mathbf{A}}{dt} = \frac{\partial\mathbf{A}}{\partial t} + (\dot{\mathbf{r}}\nabla)\,\mathbf{A}.$$

Now $\quad (\dot{\mathbf{r}},\,(\delta\mathbf{r}\nabla)\,\mathbf{A}) - (\delta\mathbf{r},\,(\dot{\mathbf{r}}\nabla)\,\mathbf{A}) = ([\dot{\mathbf{r}},\,\text{curl}\,\mathbf{A}]\,\delta\mathbf{r})$

so that the main part of the variation of the integral due to the electrons is

$$\int dt \left\{\delta L_0 + \Sigma q\left(\delta\mathbf{r},\,-\nabla\phi - \frac{1}{c}\frac{\partial\mathbf{A}}{\partial t} + \frac{1}{c}\,[\dot{\mathbf{r}},\,\text{curl}\,\mathbf{A}]\right)\right\}.$$

As regards the variation of the state of the free aether defined by the vectors $\mathbf{E}$ and $\mathbf{B}$ we have the terms

$$\int dt\int dv\left\{(\mathbf{B}\,\delta\mathbf{B}) - (\mathbf{E}\,\delta\mathbf{E}) + \phi\,\text{div}\,\delta\mathbf{E} - (\mathbf{A}\,\text{curl}\,\delta\mathbf{B}) + \left(\mathbf{A}\,\frac{d\delta\mathbf{E}}{dt}\right)\right\}.$$

On integrating the last three terms by parts we get that this part of the variation is equal to terms at the time limits together with

$$\int dt\int dv\left[(\mathbf{B} - \text{curl}\,\mathbf{A},\,\delta\mathbf{B}) - \left(\mathbf{E} + \nabla\phi + \frac{1}{c}\frac{\partial\mathbf{A}}{\partial t},\,\delta\mathbf{E}\right)\right]$$
$$-\int dt\int df\,\{\phi\,\delta\mathbf{E}_n - [\mathbf{A}\,\delta\mathbf{B}]_n\},$$

wherein the last integral is taken over the infinite bounding surface of the field.

As usual in such problems we are not concerned with the terms at the time limits because they may be as a rule suitably chosen. Also if we impose the natural condition on ϕ that it should be continuous everywhere and vanish at the infinitely distant boundary, the surface integrals introduced also vanish and we are left with the complete variation of our generalised Lagrangian function in the form

$$\int \delta L_0 - \Sigma\int_{t_0}^{t} dt\,q\left(\frac{1}{c}\frac{\partial\mathbf{A}}{\partial t} - \frac{1}{c}\,[\dot{\mathbf{r}},\,\text{curl}\,\mathbf{A}] + \text{grad}\,\phi,\,\delta\mathbf{r}\right)$$
$$-\int_{t_0}^{t} dt\int\left(\frac{1}{c}\,\dot{\mathbf{A}} + \frac{1}{c}\,\mathbf{E} + \text{grad}\,\phi,\,\delta\mathbf{E}\right)dv + \int_{t_1}^{t_2} dt\int dv\,(\mathbf{B} - \text{curl}\,\mathbf{A},\,\delta\mathbf{B}).$$

232. Now the variation $\delta\mathbf{r}$ which determines the virtual displacement of the electron q and the variations $\delta\mathbf{E}$ and $\delta\mathbf{B}$ which specify the electric displacement of a point in the free aether, can now be considered as independent and arbitrary: hence the coefficient of each must vanish separately in the dynamical variational equation. We thus obtain three sets of equations of types

$$\mathbf{E} + \frac{1}{c}\frac{\partial \mathbf{A}}{\partial t} + \operatorname{grad}\phi = 0,$$

$$\mathbf{B} - \operatorname{curl}\mathbf{A} = 0,$$

and $$\frac{d}{dt}\left(\frac{\partial L_0}{\partial \dot{\mathbf{r}}}\right) - \frac{\partial L_0}{\partial \mathbf{r}} + \frac{1}{c}\dot{\mathbf{A}} + \operatorname{grad}\phi - \frac{1}{c}[\dot{\mathbf{r}}, \operatorname{curl}\mathbf{A}] = 0,$$

where for simplicity L_0 has been assumed to depend only on the coordinates and velocities of the electronic charges. These equations are the same as

$$\mathbf{E} = -\frac{1}{c}\dot{\mathbf{A}} - \operatorname{grad}\phi,$$

$$\mathbf{B} = \operatorname{curl}\mathbf{A},$$

$$\frac{d}{dt}\left(\frac{\partial L_0}{\partial \dot{\mathbf{r}}}\right) - \frac{\partial L_0}{\partial \mathbf{r}} = -\frac{1}{c}\dot{\mathbf{A}} - \operatorname{grad}\phi + \frac{1}{c}[\dot{\mathbf{r}}\mathbf{B}].$$

These are the differential equations which determine the sequence of events in the system. The first two show that the functions ϕ and $\mathbf{A}$, introduced as undetermined multipliers, are respectively the scalar and vector potentials of the theory. The third, expressed in the ordinary language of electrodynamics which avails itself of the conception of force, shows that

$$-\frac{1}{c}\frac{d\mathbf{A}}{dt} - \operatorname{grad}\phi + \frac{1}{c}[\dot{\mathbf{r}}\mathbf{B}] = \mathbf{E} + \frac{1}{c}[\dot{\mathbf{r}}\mathbf{B}]$$

is the electric force which tends to accelerate the motion of the electrons q, each electron being presumed to have a constitution which enables it to offer a kinetic reaction of an electric nature to the action of this force. We here speak of the electric force acting on the single electron which in strictness is really more than our analysis gives us. The equation thus interpreted should really have a Σ, a sign of summation, in front of it to show that it is an aggregate equation for all the electrons in the volume element, with which we are in reality dealing. Certain considerations will however be offered which point to the conclusion that the result is correct if interpreted for the single electron separately, and we shall therefore often make use of the result in this form.

233. Thus the whole mechanics of the electromagnetic system is summed up in terms of these forces of ordinary type acting in the aggregate on the individual electrons. Therefore for a complete specification of such a system it is merely necessary to know, in addition to the ordinary dynamical relations of the masses moving in it, also the aggregate of the 'applied' forces acting on the individual electrons which they contain; the force of electrodynamic origin acting on the matter in bulk is the aggregate of the forces acting on its electrons, and it is only in these impressed forces that the electrical conditions manifest themselves. The total force of electrodynamic origin on any body is thus*

$$\Sigma q\left(\mathbf{E} + \frac{1}{c}[\mathbf{uB}]\right),$$

which makes up in all a static part $\Sigma q\mathbf{E}$ and a kinetic part

$$\Sigma q \frac{1}{c}[\mathbf{uB}].$$

We shall return to a more detailed examination of these forces in a later paragraph, but it is perhaps worth while considering at the present stage the results of an experiment made by H. A. Wilson to distinguish between the electromotive force acting on the electrons and the electric force in the aether, by examining the effect on a dielectric body of motion through the aether.

Wilson rotated a hollow dielectric cylinder in a uniform magnetic field parallel to its axis, thus bringing in a force of electrodynamic origin which for all the electrons in the dielectric acts radially

* The occurrence of the magnetic induction instead of the magnetic force in this expression is important and must be emphasised. It points once again to the conclusion that the induction is the fundamental vector of the theory, as in fact is obvious from our previous discussions of the energy relations of the magnetic field; in fact from one point of view the only essential point where our treatment of these relations differs from that usually given lies in the choice of the magnetic induction instead of the more usual magnetic force, as the 'aethereal vector.' Again the conclusion that the induction is the true magnetomotive force would appear to invalidate the argument of Kempken (*Ann. d. Phys.* 20 (1906)) whereby he derives the fact that I/μ is constant in permanent magnets, instead of I, which is usually regarded as remaining constant in these cases. Cf. also Gans, *Ann. d. Phys.* [4], 16 (1905), p. 178.

outwards from the axis of the cylinder, i.e. perpendicular to the direction of their motion and to the lines of force of the magnetic field and of amount at any point equal to

$$\frac{1}{c}(\mathbf{Bv}),$$

$\mathbf{v}$ being the velocity. This force gives rise to an electric displacement across the shell of dielectric from the inner to the outer surface and in consequence there will be a difference of potential established between these surfaces; by coating them metallically and connecting by sliding contacts with the quadrants of a galvanometer a measurement of the potential difference was easily made.

The force producing the displacement being merely of kinetic origin, there will be no aethereal part in the displacement which will therefore be of intensity per unit volume equal to

$$\mathbf{D} = (\epsilon - 1)\,\mathbf{Bv},$$

ϵ being the dielectric constant. Wilson verified that the potential difference between the surfaces was proportional to

$$(\epsilon - 1)\,\mathbf{Bv},$$

with sufficient exactness to justify the basis of the explanation here offered. The result of this experiment also has another important bearing which will be mentioned later.

234. The transmission of force in the electromagnetic field. We now turn to the discussion of the mode of transmission of force in the electromagnetic field. We know from experience that forces are exerted by one body on another as a result of the existence of an electromagnetic field in the space surrounding them, or even as the result of an interchange of radiation, and we want to get a theory of the matter. According to the ideas which we have developed the mechanical actions on the parts of the material system resulting from the established electromagnetic field are to be regarded merely as the terminal aspects of a state of stress in the medium (the aether) between the bodies, the action of one body on another being transmitted through and by this medium. We should therefore be able to represent these forces as an imposed geometrical stress system applied in the medium between the bodies. The forces acting on the part of the system enclosed in any surface drawn in the field would then be expressed as statically equivalent to a system of tractions over the surface (statically equivalent meaning that the resultants are the same as if we imagined the forces to be applied to rigid systems).

The problem of determining this stress admits of an infinite number of solutions owing to our indefinite knowledge of the actual properties of the aether, which is the ultimate seat of the strain condition. A rough mechanical analogy would be obtained by the consideration of two oppositely electrified bodies set in an insulating jelly; they will tend to come together and will thus create a state of stress in the jelly around them, and this stress will balance their attractions. This balancing stress reversed would thus completely represent the actions between the bodies. But different kinds of jelly would give different stress-representations, and the problem of determining this representation is indefinite until the jelly is specified. This indefiniteness does not however trouble us much at the present stage. We only want a physical solution of the problem which can be expressed in general terms independent of the particular nature of the problem, and this can be obtained with certain limitations. The present discussion however leads to one of the points where electric theory has not yet been probed right to the bottom. The solution obtained is useful and suggestive but it cannot yet be linked with our general physical theories.

We must here emphasise that it is the mechanical forces on the material bodies that we are going to deal with. If there is no matter in the small volume, any system of tractions over its surface must balance among themselves.

235. We examine quite generally the forcive on the portion of any electrical system enclosed by an arbitrarily chosen surface f, assuming that it is ultimately the same as the resultant of the forces on the separate electrons associated with the matter of the system and constituting in their average relations its free charge and dielectric and magnetic polarisation. In estimating these forces account must however be taken of *all* the electrons properly associated with the matter, even if they are displaced across to the outside of the surface f, but not of those electrons temporarily inside f and really belonging to the matter outside. The force exerted by the field on any 'bound' electron is in reality applied to the matter at the point of it to which the electron is bound and not at the point where the electron may be found. For the purposes of the calculation we may therefore conveniently regard the portion of the material system under consideration as abruptly terminated by the surface f and therefore isolated from any portion outside this surface. In all other respects it will be assumed to be perfectly continuous throughout.

If $\mathbf{v}$ denote the vector velocity of the typical electron with charge e moving in a field at a point where the electric force and magnetic induction vectors are $\mathbf{E}$, $\mathbf{B}$ respectively, the force on it is

$$e\left\{\mathbf{E}+\frac{1}{c}[\mathbf{vB}]\right\}.$$

The force on the element δv of the matter inside the surface f is therefore

$$\left\{\rho'\mathbf{E}+\frac{1}{c}[\mathbf{C}'\mathbf{B}]\right\}\delta v,$$

where ρ' denotes the average charge density in the element and $\mathbf{C}'$ the average current density of the electric flux.

The average charge on a small element δv inside the surface is

$$(\rho - \operatorname{div}\mathbf{P})\,\delta v$$

if ρ is the density of the free charge and $\mathbf{P}$ the intensity of the polarisation at the point. In addition to this distribution there is however a surface charge of density $\mathbf{P}_n$ at the abrupt outer boundary of the portion of the system under consideration, that is the surface f itself.

Again the average current density in the interior of the medium is

$$\mathbf{C}_1+\rho\mathbf{u}+\frac{d\mathbf{P}}{dt}+c\operatorname{curl}\mathbf{I}_1,$$

where $\mathbf{C}_1$ is the true current of conduction; $\mathbf{u}$ the average velocity of the matter at the point, and

$$\mathbf{I}_1=\mathbf{I}+\frac{1}{c}[\mathbf{Pu}]$$

is the magnetic polarisation intensity, including the quasi-magnetism arising from the convection of electrically polarised molecules.

This current distribution has also to be adjusted at the boundary of f by the inclusion of a current sheet on that surface of density

$$c\,[\mathbf{I}_1\mathbf{n}_1],$$

where $\mathbf{n}_1$ is the unit normal vector at the point.

The electric part of the total forcive on the enclosed system will therefore be given as regards its linear component by

$$\int (\rho - \operatorname{div}\mathbf{P})\,\mathbf{E}\,dv + \int \mathbf{P}_n\mathbf{E}\,df,$$

the volume integral being taken throughout the space inside f and the surface integral over this surface itself. A reduction of the latter integral by Green's lemma shows that this forcive is the same as

$$\int \{\rho\mathbf{E} + (\mathbf{P}\nabla)\,\mathbf{E}\}\,dv.$$

Remembering now that the surface f was arbitrarily chosen we may interpret this as implying that there is a forcive per unit volume on the system of intensity

$$\mathbf{F}_e = \rho\mathbf{E} + (\mathbf{P}\nabla)\,\mathbf{E}.$$

This is the whole of the average linear electric force on the medium. In addition there is a torque per unit volume of intensity

$$\mathbf{C}_e = [\mathbf{PE}],$$

as is easily seen on analogy with the statical case.

236. The force on the portion of the medium under review due to the magnetic field is similarly

$$\frac{1}{c}\int \left[\mathbf{C}_1 + \rho\mathbf{u} + \frac{d\mathbf{P}}{dt} + c\operatorname{curl}\mathbf{I}_1,\,\mathbf{B}\right] dv + \int [[\mathbf{I}_1\mathbf{n}_1]\,\mathbf{B}]\,df.$$

The second integral transforms similarly by Green's lemma to the volume integral of

$$-\,[\operatorname{curl}\mathbf{I}_1,\,\mathbf{B}] + \operatorname{grad}(\mathbf{I}_1\mathbf{B}) - \mathbf{I}\operatorname{div}\mathbf{B},$$

where the differential operator in the second term affects only the $\mathbf{B}$ function. Thus since

$$\operatorname{div}\mathbf{B} = 0$$

we may take this part of the forcive as distributed throughout the medium with intensity

$$\mathbf{F}_m = \frac{1}{c}\left[\mathbf{C}_1 + \rho\mathbf{u} + \frac{d\mathbf{P}}{dt},\,\mathbf{B}\right] + \operatorname{grad}(\mathbf{I}_1\mathbf{B})$$

per unit volume at any place. We shall henceforth use $\mathbf{C}_1$ to include the convection and polarisation currents as well as the conduction currents, so that the expression for this part of the forcive reduces to

$$\mathbf{F}_m = \frac{1}{c}[\mathbf{C}_1\mathbf{B}] + \text{grad}\,(\mathbf{I}_1\mathbf{B}),$$

where the restriction is still implied in the operator in the last term.

This is the complete expression for the magnetic part of the total forcive per unit volume on the medium.

237. Following our previous theory we now try and express these forces on the electrical system or any part of it by surface integrals over the boundary of the volume containing it. To do this we must first express them in such a form that

$$\mathbf{F}_x = \frac{\partial \mathbf{T}_{xx}}{\partial x} + \frac{\partial \mathbf{T}_{xy}}{\partial y} + \frac{\partial \mathbf{T}_{xz}}{\partial z},$$

for then $\int_v \mathbf{F}_x dv$ can be transformed to

$$\int_f \mathbf{T}_{xn}\, df,$$

taken over the surface of the volume. This determines the $\mathbf{T}$'s as the components of the stress system in the usual way.

We consider the electric and magnetic forces separately. As regards the electric part we follow Maxwell's hint developed in our previous discussions* and try as before

$$\mathbf{F}_{e_x} = \frac{\partial}{\partial x}(\mathbf{E}_x\mathbf{D}_x - \tfrac{1}{2}\mathbf{E}^2) + \frac{\partial}{\partial y}(\mathbf{E}_x\mathbf{D}_y) + \frac{\partial}{\partial z}(\mathbf{E}_x\mathbf{D}_z).$$

In this case however the relationship does not hold: there is an outstanding term which cannot be included in the differentials. To see this easily we notice on differentiating this expression out that it is

$$(\mathbf{D}\nabla)\,\mathbf{E}_x - \frac{1}{2}\frac{\partial}{\partial x}(\mathbf{E}^2) + \mathbf{E}_x\rho,$$

which differs from the above value by

$$\mathbf{D}_y\frac{\partial \mathbf{E}_x}{\partial y} + \mathbf{D}_z\frac{\partial \mathbf{E}_x}{\partial z} - \left(\mathbf{E}_y\frac{\partial \mathbf{E}_y}{\partial x} + \mathbf{E}_z\frac{\partial \mathbf{E}_z}{\partial z}\right) - \mathbf{P}_y\frac{\partial \mathbf{E}_x}{\partial y} - \mathbf{P}_z\frac{\partial \mathbf{E}_x}{\partial z};$$

which is $$\mathbf{E}_y\left(\frac{\partial \mathbf{E}_x}{\partial y} - \frac{\partial \mathbf{E}_y}{\partial x}\right) - \mathbf{E}_z\left(\frac{\partial \mathbf{E}_z}{\partial x} - \frac{\partial \mathbf{E}_x}{\partial z}\right),$$

* See page 90.

or in vector notation $\quad -[\mathbf{E}\,\mathrm{curl}\,\mathbf{E}]_x;$

but $$\mathrm{curl}\,\mathbf{E} = -\frac{1}{c}\frac{d\mathbf{B}}{dt},$$

so that this is $$\frac{1}{c}\left[\mathbf{E}\,\frac{d\mathbf{B}}{dt}\right]_x.$$

The difference between this case and that discussed in chapter II is obvious. In the statical case we were able to put

$$\frac{\partial \mathbf{E}_x}{\partial y} = \frac{\partial \mathbf{E}_y}{\partial x}, \quad \frac{\partial \mathbf{E}_z}{\partial x} = \frac{\partial \mathbf{E}_x}{\partial z},$$

so that there was no discrepancy. These equalities are however no longer satisfied because they imply the existence of an electric potential.

Similar results apply of course to the other components of the stress system. Thus if we leave out for the present this outstanding part of the forcive, we see that the main part of the electric force acting on the matter is expressible as a stress system which can be specified by the matrix

$$\left|\begin{matrix} \mathbf{E}_x\mathbf{D}_x - \frac{1}{2}\mathbf{E}^2, & \mathbf{E}_x\mathbf{D}_y, & \mathbf{E}_x\mathbf{D}_z \\ \mathbf{E}_y\mathbf{D}_x, & \mathbf{E}_y\mathbf{D}_y - \frac{1}{2}\mathbf{E}^2, & \mathbf{E}_y\mathbf{D}_z \\ \mathbf{E}_z\mathbf{D}_x, & \mathbf{E}_z\mathbf{D}_y, & \mathbf{E}_z\mathbf{D}_z - \frac{1}{2}\mathbf{E}^2 \end{matrix}\right|.$$

This is identical with that obtained in the statical theory and is therefore amenable to the simpler specification there developed. It can in fact be dissected into parts.

(i) A simple hydrostatic pressure $\frac{1}{2}\mathbf{E}^2$ throughout the medium. The negative sign shows that it is a pressure.

(ii) A tension along the internal bisector of the angle between $\mathbf{D}$ and $\mathbf{E}$ equal to

$$\mathbf{E}\,.\,\mathbf{D}\cos^2\widehat{\mathbf{ED}}.$$

(iii) A pressure along the external bisector equal to

$$\mathbf{E}\,.\,\mathbf{D}\sin^2\widehat{\mathbf{ED}}.$$

(iv) And the torque per unit volume

$$[\mathbf{E}\,.\,\mathbf{D}].$$

This couple or torque is represented as part of the stress system and so the specification is complete, there is no discrepancy.

This is the general result; if the medium is isotropic it reduces to a pull along the lines of force equal to (**ED**) with a hydrostatic pressure $\frac{1}{2}\mathbf{E}^2$, and this is Maxwell's system.

238. A similar argument applies to the magnetic part of the forcive on the element. The x-component of this force is given by

$$\mathbf{F}_{mx} = \left(\mathbf{I}\frac{\partial \mathbf{B}}{\partial x}\right) + \frac{1}{c}[\mathbf{C}_1 \,.\, \mathbf{B}]_x = \left(\mathbf{I}\frac{\partial \mathbf{H}}{\partial x}\right) + 2\pi\frac{\partial \mathbf{I}^2}{\partial x} + \frac{1}{c}[\mathbf{C}_1\mathbf{B}]_x.$$

Now the usual analysis shows that

$$\left(\mathbf{I}\frac{\partial \mathbf{H}}{\partial x}\right) = \frac{\partial}{\partial x}\left[\mathbf{B}_x\mathbf{H}_x - \frac{1}{2}\mathbf{H}^2\right] + \frac{\partial}{\partial y}(\mathbf{B}_y\mathbf{H}_x) + \frac{\partial}{\partial z}(\mathbf{B}_z\mathbf{H}_x) + [\mathbf{B}, \operatorname{curl}\mathbf{H}]_x,$$

and since
$$\operatorname{curl}\mathbf{H} = \frac{4\pi\mathbf{C}}{c} = \frac{1}{c}\left(\mathbf{C}_1 + \frac{d\mathbf{E}}{dt}\right)$$

the last term in this expression is equal to

$$[\mathbf{B} \,.\, \operatorname{curl}\mathbf{H}] = \frac{1}{c}[\mathbf{BC}] = -\frac{1}{c}[\mathbf{CB}] = -\frac{1}{c}[\mathbf{C}_1\mathbf{B}] - \frac{1}{c}\left[\frac{d\mathbf{E}}{dt} \,.\, \mathbf{B}\right],$$

whence

$$\mathbf{F}_{mx} = \frac{\partial}{\partial x}\left[\mathbf{B}_x\mathbf{H}_x - \frac{1}{2}\mathbf{H}^2 + \frac{1}{2}\mathbf{I}^2\right] + \frac{\partial}{\partial y}(\mathbf{B}_y\mathbf{H}_x) + \frac{\partial}{\partial z}(\mathbf{B}_z\mathbf{H}_x) - \frac{1}{c}\left[\frac{d\mathbf{E}}{dt} \,.\, \mathbf{B}\right]_x,$$

and similar results apply to the other components. The main part of the forcive can thus be specified as a stress system whose components are given in the matrix

$$\begin{vmatrix} \mathbf{H}_x\mathbf{B}_x - \frac{1}{2}\mathbf{H}^2 + \frac{1}{2}\mathbf{I}^2, & \mathbf{B}_y\mathbf{H}_x, & \mathbf{B}_z\mathbf{H}_x \\ \mathbf{B}_x\mathbf{H}_y, & \mathbf{H}_y\mathbf{B}_y - \frac{1}{2}\mathbf{H}^2 + \frac{1}{2}\mathbf{I}^2, & \mathbf{B}_z\mathbf{H}_y \\ \mathbf{B}_x\mathbf{H}_z, & \mathbf{B}_y\mathbf{H}_z, & \mathbf{H}_z\mathbf{B}_z - \frac{1}{2}\mathbf{H}^2 + \frac{1}{2}\mathbf{I}^2 \end{vmatrix},$$

but this leaves out a part

$$-\frac{1}{c}\left[\frac{d\mathbf{E}}{dt} \,.\, \mathbf{B}\right],$$

which cannot be expressed by a surface integral.

Thus of the whole electrodynamic forcive per unit volume on the medium there is a total outstanding part

$$-\frac{1}{c}\left[\mathbf{E} \,.\, \frac{d\mathbf{B}}{dt}\right] - \frac{1}{c}\left[\frac{d\mathbf{E}}{dt} \,.\, \mathbf{B}\right]$$

which cannot be included in the stress specification.

239. Electromagnetic momentum. The result of the discussion of the previous paragraph is that the resultant electrodynamic force acting on the matter inside any given closed surface drawn in an electromagnetic field may in the main be expressed as the resultant of certain attractions across the surface, but that in addition there is a forcive per unit volume of amount

$$-\frac{1}{c}\frac{d}{dt}[\mathbf{EB}]$$

which cannot be so included. The fact that this outstanding force appears as a complete differential with respect to the time suggests that it may be regarded as representing the reaction to a rate of change of some kind of momentum. This leads directly to the idea of electromagnetic momentum which thus appears as being distributed throughout the field with a density at each point equal to

$$\frac{1}{c}[\mathbf{EB}].$$

This hypothesis provides a convenient representation for many purposes, and it brings the phenomena into line with dynamical theory. In fact in certain simple cases the quantity here defined as momentum behaves exactly as if it were an addition to the ordinary mechanical momentum of the system. But there are difficulties in other cases because this added momentum is of a very different kind to that involved in dynamical theory. In spite of these difficulties it has become usual and it is convenient to retain the term momentum in the sense of providing a simple and effective mode of expressing certain mathematical results. The actual force distribution would then be expressible partly as a static stress distribution on the surface and partly as the kinetic reaction to a rate of change of momentum in the interior.

240. To illustrate the meaning of the new quantity defined as momentum we consider the comparatively simple case where the whole field arises from a number of point-charges (electrons) moving in any manner under each other's influence. We restrict the discussion to the case of small and slowly varying motions so that the vectors in the surrounding field are functions of the velocities only, and we shall neglect all powers above the second of the ratios of these velocities to that of radiation (c). We take the typical particle with charge q_s situated at the instant t at the point $\mathbf{r}_s$ referred to a convenient framework of axes and moving with velocity v. The scalar and vector potentials at the point $\mathbf{r}$ of the surrounding field are then effectively determined by the formulae

$$\phi = \sum_s \phi_s = \frac{1}{4\pi} \sum_s \frac{q_s}{r_s},$$

and
$$\mathbf{A} = \sum_s \mathbf{A}_s = \frac{1}{4\pi} \sum_s \frac{q_s \mathbf{v}_s}{c r_s} = \sum \frac{\mathbf{v}_s \phi_s}{c}.$$

By definition the total linear momentum associated with the system of particles is

$$\mathbf{M} = \frac{1}{c} \int [\mathbf{EB}]\, dv$$

integrated throughout the infinite field. This is equal to

$$\frac{1}{c} \int [\mathbf{E} \,.\, \text{curl}\, \mathbf{A}]\, dv = \frac{1}{c} \sum_s \int [\mathbf{E}_s \,.\, \text{curl}\, \mathbf{A}_s]\, dv$$
$$+ \frac{1}{c} \sum_{s', s} \int \{[\mathbf{E}_s \,.\, \text{curl}\, \mathbf{A}_{s'}] + [\mathbf{E}_{s'} \,.\, \text{curl}\, \mathbf{A}_s]\}\, dv.$$

The terms in the first sum, corresponding each to a separate particle, cannot be evaluated without further detailed assumptions as to the intrinsic structure of these particles. It is easy to see however that each one must be of the form $m_s \mathbf{v}_s$, where m_s is a constant depending solely on the electrical constitution of the sth particle, and this will be verified by the calculation in greater detail in a later chapter of the value of m_s for certain types of particle. These terms therefore represent an addition to the momentum of each particle of additional momentum of ordinary type, just in fact as if its mass were increased by the amount m_s. To this extent therefore the concept of momentum of electromagnetic type is a reasonable one. The difficulties however present themselves when we pursue our calculations further.

As regards the other terms we have

$$\int [\mathbf{E}_s \,.\, \text{curl}\, \mathbf{A}_{s'}]\, dv = -\int [\nabla \phi_s \,.\, [\nabla \mathbf{A}_{s'}]]\, dv$$
$$= -\frac{1}{c} \int [\nabla_s \,.\, [\nabla_s \mathbf{v}_{s'}]]\, \phi_s \phi_{s'}\, dv$$
$$= -\frac{1}{c} [\nabla_s \,.\, [\nabla_{s'} \mathbf{v}_{s'}]] \int \phi_s \phi_{s'}\, dv.$$

Now approximately when the particles are small compared with their distance apart

$$\int \phi_s \phi_{s'} \, dv = \frac{q_s q_{s'}}{16\pi^2} \int \frac{dv}{r_s r_{s'}} = \frac{q_s q_{s'}}{8\pi} r_{ss'},$$

where $r_{ss'}$ is the radial distance between them. Thus

$$\int [\mathbf{E}_s \,.\, \text{curl}\, \mathbf{A}_{s'}]\, dv = -\frac{q_s q_{s'}}{8\pi c} [\nabla_s \,.\, [\nabla_{s'} \mathbf{v}_{s'}]]\, r_{ss'}$$

$$= -\frac{q_s q_{s'}}{8\pi c} \{\nabla_{s'} (\mathbf{v}_{s'} \nabla_s)\, r_{ss'} - \mathbf{v}_{s'} (\nabla_s \,.\, \nabla_{s'})\, r_{ss'}\}$$

$$= -\frac{q_s q_{s'}}{8\pi c r_{ss'}} \{(\mathbf{v}_{s'} \mathbf{u}_{ss'})\, \mathbf{u}_{ss'} + \mathbf{v}_{s'}\},$$

wherein $\mathbf{u}_{ss'}$ is the unit vector along the radius joining the positions of the sth and s'th particles. A similar result holds for the other integral, and thus finally by addition we have

$$M = \sum_s m_s \mathbf{v}_s + \sum_{s,\, s'} \frac{q_s q_{s'}}{8\pi c^2 r_{ss'}} [(\mathbf{v}_s + \mathbf{v}_{s'}) + (\mathbf{v}_s \mathbf{u}_{ss'})\, \mathbf{u}_{ss'} + (\mathbf{v}_{s'} \mathbf{u}_{s's})\, \mathbf{u}_{s's}].$$

We see at once from the form of this expression that it is not merely the sum of terms arising simply and solely from each separate electron. The momentum of each one separately must involve terms depending on the positions and motions of all the others, and so it is a very different quantity to the momentum of ordinary dynamics.

241. The above analysis does not in reality determine the momentum of each particle separately. This can however be done by a more indirect method depending on a use of principles established in an earlier paragraph. We have seen that the Lagrangian function from which the motion of the particles can be determined is quite generally of the form

$$L = L_0 + \tfrac{1}{2} \int (\mathbf{B}^2 - \mathbf{E}^2)\, dv.$$

By writing $\mathbf{B} = \text{curl}\, \mathbf{A}$, integrating by parts and using the relations

$$\int \text{curl}\, \mathbf{B}\, dv = \frac{1}{c} \Sigma q_s \mathbf{v}_s + \frac{1}{c} \int \frac{d\mathbf{E}}{dt}\, dv$$

and

$$\mathbf{E} = -\frac{1}{c} \frac{d\mathbf{A}}{dt} - \nabla\phi,$$

it is easy to prove that

$$L = L_0 + \tfrac{1}{2} \sum_s q_s \left[\frac{1}{c} (\mathbf{v}_s \mathbf{A}) - \phi \right] - \frac{1}{2c} \frac{d}{dt} \int (\mathbf{AE})\, dv.$$

The evaluation of the last integral requires a little care, but if we notice that to our order of approximation

$$\frac{d\mathbf{E}_s}{dt} = (\mathbf{v}_s \nabla_s)\, \mathbf{E}_s$$

it can be carried through on exactly the same lines as before and it proves to be

$$\sum_{s, s'} \frac{q_s q_{s'}}{16\pi c^2 r_{ss'}} [\mathbf{v}_s^2 + \mathbf{v}_{s'}^2 - 2 (\mathbf{v}_s \mathbf{v}_{s'}) - (\mathbf{v}_s \mathbf{u}_{ss'})^2 - (\mathbf{v}_{s'} \mathbf{u}_{s's})^2 + 2 (\mathbf{v}_s \mathbf{u}_{ss'}) (\mathbf{v}_{s'} \mathbf{u}_{s's})],$$

together with a series of intrinsic terms, one for each particle, depending solely on the charge and motion of the particle, but not in any way on those of the others, and proportional to the square of its velocity.

In the evaluation of the first terms in the expression for L we must use more accurate forms for the potentials than those already given. For approximately uniform motions we can use the formulae for uniform motion

$$\phi_s = \frac{q_s}{4\pi r_s} \left[1 - \frac{\mathbf{v}_s^2}{c^2} + \frac{(\mathbf{v}_s \mathbf{u}_s)^2}{c^2} \right]^{\frac{1}{2}},$$

and

$$\mathbf{A}_s = \frac{q_s \mathbf{v}_s}{4\pi c r_s} \left[1 - \frac{\mathbf{v}_s^2}{c^2} + \frac{(\mathbf{v}_s \mathbf{u}_s)^2}{c^2} \right]^{-\frac{1}{2}},$$

which to our order of approximation are simply

$$\phi_s = \frac{q_s}{4\pi r_s} \left[1 + \frac{\mathbf{v}_s^2}{2c^2} - \frac{(\mathbf{v}_s \mathbf{u}_s)^2}{2c^2} \right],$$

$$\mathbf{A}_s = \frac{q_s \mathbf{v}_s}{4\pi c r_s} \left[1 + \frac{\mathbf{v}_s^2}{2c^2} - \frac{(\mathbf{v}_s \mathbf{u}_s)^2}{2c^2} \right].$$

Inserting these values we find that

$$L = L_0 + \tfrac{1}{2} \sum_s (m_s \mathbf{v}_s^2 - m_{s'}) + \frac{1}{8\pi c^2} \sum_{s, s'} \frac{q_s q_{s'}}{r_{ss'}} [- 2c^2 + (\mathbf{v}_s \mathbf{u}_{ss'})^2 + (\mathbf{v}_{s'} \mathbf{u}_{s's})^2 - \mathbf{v}_s^2 - \mathbf{v}_{s'}^2 - (\mathbf{v}_s \mathbf{u}_{ss'}) (\mathbf{v}_{s'} \mathbf{u}_{s's}) + 3 (\mathbf{v}_s \mathbf{v}_{s'})].$$

The part L_0 of this function is inserted for complete generality: it is the part—if such exist—which does not depend on the electrical conditions of the particles.

Adopting now the general definition of momentum in analytical dynamics we can say that the linear momentum of the sth particle has an electromagnetic component equal to

$$\frac{\partial (L - L_0)}{\partial \mathbf{v}_s} = m_s \mathbf{v}_s + \Sigma \frac{q_s q_{s'}}{4\pi c^2 r_{ss'}} [(\mathbf{v}_s \mathbf{u}_{ss'}) \mathbf{u}_{ss'} - \tfrac{1}{2} (\mathbf{v}_{s'} \mathbf{u}_{ss'}) \mathbf{u}_{ss'} - \mathbf{v}_s + \tfrac{3}{2} \mathbf{v}_{s'}].$$

This is of course a new definition of electromagnetic momentum which is more specific than that given above: but it is in reality consistent with the previous definition because the sum of these separate momenta is in fact, as is easily seen, equal to the total momentum of the system determined above from the first definition. The previous method unfortunately does not give precise details for the separate particles, but in so far as the dynamical method leads to results consistent with it, we may say that in a general sense the idea of electromagnetic momentum is not so far removed from the dynamical one as suggested at the outset.

In any case the linear momentum of a particle in addition to an intrinsic part always contains terms depending on its own charge and on the position and motion and charge of all the other particles of the system. The more important of these extra terms is that which is proportional to the sum

$$\Sigma_{s'} \frac{q_{s'} \mathbf{v}_{s'}}{c r_{ss'}}$$

which is precisely the vector potential, in the field of the charges, at the position of that particular one under investigation; and this part will exist even if the particle itself is at rest. We thus derive the result that an electrified particle *at rest* in the neighbourhood of other and moving particles possesses in general a certain amount of momentum owing mainly to the asymmetric disposition of its electric field in the magnetic field of the moving particles.

In the simple case where all but one of the particles constitute by their motion a linear electric current unaccompanied by an electric field we have

$$\Sigma_{s'} \frac{q_{s'}}{r_{ss'}} = 0,$$

and also
$$\sum_{s'} \frac{q_{s'}(\mathbf{v}_{s'}\mathbf{u}_{ss'})^2}{r_{ss'}} = \sum_{s'} \frac{q_{s'}\mathbf{v}_{s'}^2}{r_{ss'}} = 0.$$

Further since the motions $\mathbf{v}_{s'}$ form a closed current we have
$$\Sigma\, q_{s'}\,[(\mathbf{v}_s\mathbf{u}_{ss'})\,(\mathbf{v}_{s'}\mathbf{u}_{ss'}) + (\mathbf{v}_s\mathbf{v}_{s'})] = 0,$$
as in Ampère's analysis. Thus now we have practically
$$L = \frac{1}{2}m\mathbf{v}^2 + \frac{q}{c}(\mathbf{A}\mathbf{v}) + \frac{1}{2c}aJ^2,$$
where J is the current in the conductor, a the coefficient of self-induction of the conductor and $\mathbf{A}$ the vector potential of the field of J. In this case the momentum of the single particle is simply
$$m\mathbf{v} + \frac{q}{c}\mathbf{A},$$
so that the term in the vector potential is all there is besides the intrinsic part.

242. We shall return to further aspects of this question at a later stage, but before we leave the matter now it might be as well to add one further warning note. We have obtained the expression for the momentum originally by trying to reduce the volume integral of the electromagnetic forces to a surface integral. Since we were only partially successful in this attempt—the outstanding terms suggesting the momentum idea—it follows that the expressions we have obtained both for the stresses and the momentum must be arbitrary, but of course in a mutual manner. We cannot be quite certain that we have obtained the correct stress, and a part of what we have obtained may really be simply a part of the reaction to momentum which we have succeeded in expressing as a surface integral. For example, we have
$$\begin{aligned}\frac{1}{c}[\mathbf{E}\mathbf{B}] &= -\frac{1}{c^2}\left[\frac{d\mathbf{A}}{dt}\,.\,\mathbf{B}\right] - \frac{1}{c}[\nabla\phi\,.\,\mathbf{B}] \\ &= -\frac{1}{c^2}\left[\frac{d\mathbf{A}}{dt}\,.\,\mathbf{B}\right] - \frac{1}{c}\{[\nabla\phi\,.\,\mathbf{B}] - [\phi\nabla\,.\,\mathbf{B}]\} + \frac{1}{c^2}\phi\mathbf{C}.\end{aligned}$$
Now the two terms in the middle bracket when differentiated with respect to the time can be included in the stress specification because they are complete spatial derivatives. This would leave a new expression for the momentum density, viz.
$$\frac{1}{c^2}\left(\phi\mathbf{C} + \left[\mathbf{B}\frac{d\mathbf{A}}{dt}\right]\right),$$

which differs fundamentally from the previous form under certain circumstances. Of course in cases of quasi-stationary motion of the type investigated in detail above there is no real distinction between any of the forms as the independent dynamical derivation requires, but in less simple cases, particularly those involving radiation, there will be very vital differences.

The most general form of the relation between the momentum and stress vectors of the field will be obtained in another connexion on a future occasion*.

* Cf. page 386.

CHAPTER VII

ELECTROMAGNETIC OSCILLATIONS AND WAVES

243. The general problem with electromagnetic waves. We have already had cause to investigate in a previous chapter a possible source of a very rapidly oscillating field (viz. that associated with a condenser discharging through an induction) in which we can no longer neglect the time taken to smooth out the field, and as this case has an important theoretical as well as practical bearing, we shall examine it more closely by more general methods.

The general problem is the investigation of the conditions in any electromagnetic field consequent on a rapid alteration of the conditions in any one part of it; perhaps by discharging one conductor in the field by connecting it through an induction to another conductor in the same field. Complete generality will be obtained by the disposition of dielectric and other conducting bodies throughout the field, the whole being then included in one scheme.

The generalised scheme adopted by Maxwell for the treatment of these cases has already been set out in detail, the underlying idea being that the electromagnetic phenomena are the result of some action, mechanical or otherwise, transmitted from one body to the other by means of a supposed medium, the aether, occupying the space between them, the mode of action of this medium being completely specified by the two fundamental circuital relations of the theory.

The essential point of this scheme involves the assumption of a quasi-current in the aether of density

$$\frac{d\mathbf{E}}{dt},$$

which is equal to the time rate of change of the aethereal displacement. This current in addition to the real displacement current in the dielectrics* being sufficient to secure that all currents flow in complete cycles. We can thus adopt the two circuital relations of

* Throughout this chapter where not otherwise specified we shall assume that linear isotropic relations hold between the electric and magnetic forces and the induced polarisations respectively.

electrodynamics as descriptive of the general state of affairs. In their differential form they are

$$\frac{\mathbf{C}}{c} = \operatorname{curl} \mathbf{H},$$

$$-\frac{1}{c}\frac{d\mathbf{B}}{dt} = \operatorname{curl} \mathbf{E}.$$

These equations represent the simplest conception of a general electromagnetic theory of these things. They could not of course be right unless in addition

$$\operatorname{div} \mathbf{C} = 0,$$

and also

$$\operatorname{div} \mathbf{B} = 0,$$

the first being secured by Maxwell's hypothesis and the second indicating that it is **B** and not **H** that must be used to count the flux of force.

These two dynamical equations expressing exact physical principles are independent of the constitution of the substances in which the action takes place. As however they involve four vectors they are not sufficient for a complete scheme and we must again introduce the constitutive relations depending on the nature of the media occupying the field. The first one expresses the total current as a function of the electric force in the form

$$\mathbf{C} = \sigma\mathbf{E} + \frac{d}{dt}(\epsilon\mathbf{E}).$$

The first term, representing the conduction current, expresses an exact relation as far as experiment can follow it, but the second expresses the best we can do in our theory; it represents however a fairly good approximation to the facts. The second relation between the magnetic induction and magnetic force also assumes the form

$$\mathbf{H} = \mu\mathbf{B},$$

and in the simplest cases μ is constant. This relation is however not so exact as the above.

244. These four relations represent Maxwell's complete scheme: adopting the latter we can write the first two in the form

$$\operatorname{curl} \mu\mathbf{B} = \frac{\sigma\mathbf{E}}{c} + \frac{\epsilon}{c}\frac{d\mathbf{E}}{dt},$$

$$\operatorname{curl} \mathbf{E} = -\frac{1}{c}\frac{d\mathbf{B}}{dt}.$$

We shall now limit the complete generality of our scheme by the assumption that there are no magnetic bodies present so that we can take $\mu = 1$ everywhere.

We deduce at once that

$$\frac{\sigma}{c}\frac{d\mathbf{E}}{dt} + \frac{\epsilon}{c}\frac{d^2\mathbf{E}}{dt^2} = \text{curl}\,\frac{d\mathbf{B}}{dt}$$

$$= -\,c\,\text{curl curl}\,\mathbf{E}$$

$$= c\nabla^2\mathbf{E} - c\,\text{grad div}\,\mathbf{E}.$$

But we have $$\rho = \text{div}\,\epsilon\mathbf{E},$$

and yet $$0 = \text{div}\,(\epsilon\mathbf{E} + \mathbf{C}_1)$$

$$= \frac{d\rho}{dt} + \sigma\,\text{div}\,\mathbf{E},$$

or $$\frac{d\rho}{dt} + \frac{\sigma}{\epsilon}\,\text{div}\,(\epsilon\mathbf{E}) = 0,$$

or again $$\frac{d\rho}{dt} + \frac{\sigma\rho}{\epsilon} = 0,$$

which means that $$\rho = \rho_0 e^{-\frac{\sigma t}{\epsilon}}.$$

This equation shows that the changes in ρ are independent of the external electromagnetic influence, so that even if there is an initial electrical volume charge distribution it will decrease very rapidly except in the very improbable case when ϵ/σ is a very large quantity. We may thus consider that $\rho = 0$ always, for we may take the origin of time when $\rho = 0$. The equations then become

$$\frac{d\mathbf{E}}{dt} + \epsilon\,\frac{d^2\mathbf{E}}{dt^2} = c^2\,\nabla^2\mathbf{E},$$

since now $$\text{div}\,\epsilon\mathbf{E} = 0.$$

In general we can neglect the displacement current in the metallic conductors in comparison with the conduction currents. Our equation thus reduces to the form

$$\nabla^2\mathbf{E} = \frac{\sigma}{c^2}\cdot\frac{d\mathbf{E}}{dt},$$

which exhibits the propagation of the electromagnetic disturbances into the conducting substances as a simple process of diffusion.

In the dielectric parts of the field $\sigma = 0$ and the equation becomes

$$\nabla^2 \mathbf{E} = \frac{\epsilon}{c^2} \frac{d^2 \mathbf{E}}{dt^2},$$

which shows that the electric field in the dielectric can be propagated as a simple wave motion in the medium with a velocity

$$\frac{c}{\sqrt{\epsilon}}.$$

A similar discussion easily shows that the magnetic force is propagated in an exactly similar manner, the equations satisfied by it in the separate media being

(i) in the conductors $$\nabla^2 \mathbf{B} = \frac{\sigma}{c^2} \frac{d\mathbf{B}}{dt},$$

(ii) in the dielectric $$\nabla^2 \mathbf{B} = \frac{\epsilon}{c^2} \frac{d^2 \mathbf{B}}{dt^2}.$$

These equations represent the characteristic differential equations of the theory. In attacking any problem where there are different regions (conducting or dielectric) we have to solve the different equations for each region and the corresponding solutions have then to be connected by the appropriate continuity conditions at the boundary.

We notice that any discontinuities in crossing the surface at any point clearly arise from the distribution over a surface element δf surrounding that point, for the disturbances propagated from the more distant parts are virtually the same at points on the two sides of the surface whose distance apart is infinitesimal compared with the linear dimensions of δf. Moreover if this is the case the discontinuities will be the same as in the corresponding static or stationary condition of that part of the boundary δf, because the field from it produces almost instantaneously its effect at an infinitely near point. We may therefore conclude at once that, in the general case,

(i) the tangential electric force must be continuous unless a double sheet distribution exists on the surface. This case is excluded.

(ii) the normal magnetic induction is also continuous: this follows also as a consequence of the general circuital property of that vector.

(iii) the total normal electric current component is also continuous.

The first and second of these conditions are however not independent for we know that

$$-\frac{1}{c}\frac{d\mathbf{B}}{dt} = \text{curl}\,\mathbf{E},$$

so that if the tangential components of **E** are continuous the normal component of **B** must also be continuous.

Also since
$$\frac{\mathbf{C}}{c} = \text{curl}\,\mathbf{H},$$

we see from the third relation that unless there are surface current sheets at the surface the tangential components of the magnetic force must also be continuous.

There are thus in all two independent boundary conditions which have to be satisfied.

The previous general equations with these boundary conditions provide us with a complete scheme of equations for all electromagnetic wave problems.

245. The skin-effect and perfect conductors. Before proceeding to the consideration of particular problems we shall apply these results to the general case discussed above with the additional assumption that all the conductors in the field are perfect.

In this case we have $\sigma = 0$ for all the conductors and therefore inside them

$$\nabla^2\mathbf{E} = 0,$$

and also
$$\nabla^2\mathbf{H} = 0,$$

which combined with the general results

$$\text{div}\,\mathbf{E} = 0 \text{ and } \text{div}\,\mathbf{H} = 0$$

show that inside the conductors

$$\mathbf{E} = \mathbf{H} = 0;$$

there is no field inside the conductors. In external space on the other hand we have still

$$\nabla^2\mathbf{E} = \frac{\epsilon}{c^2}\frac{d^2\mathbf{E}}{dt^2},$$

and also
$$\nabla^2\mathbf{H} = \frac{\epsilon}{c^2}\frac{d^2\mathbf{H}}{dt^2},$$

but if the alternations of the field are not too fast the time variations on the right-hand side containing the very small factor $\frac{\epsilon}{c^2}$ is negligible and the equations can then be written in the simpler form

$$\nabla^2 \mathbf{E} = 0,$$

and
$$\nabla^2 \mathbf{H} = 0.$$

We know moreover from the boundary conditions that the electric force just outside the conductors is normal to their surface but the magnetic force is tangential even in the general case.

The physical explanation of these solutions is now obvious. The electromagnetic field exists only in the dielectric between the conductors. The lines of electric force go across from one conductor to another and the magnetic ones are round about. The positive and negative charges on the surfaces of the conductors are the terminations of the tubes of electric force in the intermediate field. The real propagation of the effects thus takes place in the dielectric medium, a given field being propagated through that medium with a velocity depending only on the medium. Wherever the electric force arrives at a conductor it pulls the electric charges on the conductors (the electrons) about until the statical force due to their rearranged distribution counterbalances the electric force in the field on any one of them, i.e. until there is no resultant electric force inside the conductors. The conductors are full of charges (positive and negative) more or less free which slightly adjust themselves, concentrating on the surface so as to get the necessary field in the interior of the conductors which cancels that of the oncoming wave. If the conduction is perfect the redistribution of charge at each instant takes place instantaneously and thus the field in the conductor right up to its surface is annulled; the charge on the conductor creating the induced electric force is entirely on its surface. If the conduction were not so good, the electrons would not be so free and the electric field would at each instant penetrate into the conductors a little way before being annulled by the reaction of the field due to the electrons which it pulls about.

This is the general idea of the phenomena. Before electrons were discovered the matter could not be put so definitely. As early as 1884 however Poynting, Hertz and Heaviside emphasised the point that where a current is used to transmit the power the energy travels in the dielectric round about, which is the real elastic thing, the conductors only acting as guides to prevent the disturbance spreading.

There is a rough mechanical analogy in the propagation of waves in an elastic medium with holes in it. In this case the transverse waves or waves of shear travel along through the elastic material adjusting itself by material deformation of the surfaces of the holes so that there is no elasticity inside the holes. In the electrical case the dielectric is the elastic medium through which the field is propagated by wave motion; this field (or the elasticity in it) is annulled at the surfaces of the conductors (the holes) by the pulling about of the mobile electrons on their surface. The conductors thus appear as places where there is no elasticity, where the electrical elasticity of the aether is annulled by the mobile electrons.

The property of perfect conductors thus appears to be merely a negative one, viz. that of cancelling the elasticity of the aether. An imperfect conductor is one in which this damping action is only partially successful.

246. We are thus led to the general idea that the whole affair of the electric currents is actually in the field outside the conductors, the current merely providing a convenient mode of describing the changes taking place. On this view all the energy of the current is to be found in the dielectric medium surrounding the conductors; and the heat developed in these conductors is the energy which has soaked in as it were from the store in the surrounding field. Let us consider for example the case of a long straight cylindrical conductor carrying an alternating current of maximum intensity J. The conductor is surrounded by an electromagnetic field in which the lines of magnetic force are circles round the wire. As regards the electric force we know that just inside the conductor it is directed along the axis and is just sufficient to drive the current, this being secured by a slight initial accumulation of charge on the conductor. But this force cannot change in going across the surface of the conductor so that just outside it is also tangential and equal to the internal value. Thus at a place just outside the conductor the flux of energy is into the conductor normally and is equal per unit area to

$$cH \,.\, E.$$

Thus the total energy flowing into unit length of the conductor from the external field is the integral

$$c\int_s H \,.\, E\,ds,$$

taken round a ring on the conductor. By symmetry this is equal to

$$cE\int H\,ds,$$

and
$$\int H\,ds = \frac{J}{c},$$

so that the energy flowing in per unit length is

$$EJ,$$

which is the energy dissipated into heat in the unit length of the conductor, by Joule's law. We thus see that when a current is flowing along a wire energy flows in sideways from the external field in just sufficient quantities to account for the energy converted into heat. From this point of view the energy appearing in the form of heat is supplied from the aether.

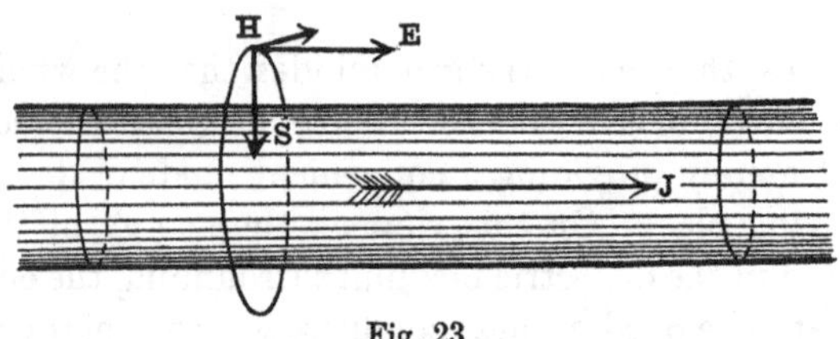

Fig. 23

Let us now assume that the wire is a long straight cylindrical one with a circular cross-section of radius a. Then if the current flow is uniform all along the wire the magnetic force will be in circles round its axis and the electric force directed along the axis. To obtain a closer insight into the field thus specified we shall find it convenient to refer the field to a system of cylindrical polar coordinates (r, θ, z) with the axis along the axis of the cylinder. The only components of the field vectors at any point which are not zero are $\mathbf{E}_z$ and $\mathbf{H}_\theta$, and these are symmetrical round the axes of the field. In the conductor these two components are connected by the relations

$$\frac{4\pi\sigma}{c}\mathbf{E}_z = \frac{1}{r}\frac{d}{dr}(r\mathbf{H}_\theta) - \frac{1}{c}\frac{d\mathbf{H}_\theta}{dt} = \frac{\partial \mathbf{E}_z}{\partial r},$$

to which the fundamental field equations of Ampère and Faraday reduce under the special circumstances of the present problem. It is of course assumed that it is possible to neglect the displacement current in comparison with conduction current; this is justified in most cases of any real importance.

It follows then that

$$\frac{4\pi\sigma}{c^2}\frac{d\mathbf{E}_z}{dt} = \frac{1}{r}\frac{d}{dr}\left(r\frac{d\mathbf{E}_z}{dr}\right).$$

A particular solution of this partial differential equation is*

$$\mathbf{E}_z = Ce^{ipt} J_0(x),$$

where we have written

$$x = \nu r, \quad \nu^2 = -\frac{4\pi i\nu\sigma}{c^2},$$

and where $J_0(x)$ satisfied Bessel's equation

$$J_0''(x) + \frac{1}{x} J_0'(x) + J_0 = 0.$$

Since the field cannot become infinite inside the cylinder J_0 must be taken as the Bessel function of the first kind, viz.

$$J_0(x) = \frac{1}{\pi}\int_0^\pi e^{ix\cos\alpha}\, d\alpha.$$

The density of the current flux at any point in the interior of the cylinder is given by

$$\mathbf{C}_z = \sigma\mathbf{E}_z = \sigma Ce^{ipt} J_0(x),$$

and its distribution over the cross-section is thus determined. Remembering the particular approximate forms of the function J_0 when its argument is small and large we notice that for very slowly alternating currents the distribution is practically uniform across the section, but for very rapid oscillations it is confined to a very thin layer at the surface.

To excite this field in the interior of the conductor we must apply along its outer surface a field of strength

$$\mathbf{E}_{0_z} = Ce^{ipt} J_0(\nu' a),$$

a being the radius.

* Cf. Rayleigh, *Phil. Mag.* [5] 21 (1886), p. 381; O. Heaviside, *Electrical Papers*, II. pp. 39, 168; J. Stefan, *Wien. Ber.* 95 (1887), p. 917; *Ann. d. Phys.* 41 (1890), p. 400; Abraham, *Encykl. d. Math. Wissensch.* v. 18, p. 514.

The total current flowing through a section of the cylinder is

$$J = 2\pi \int_0^a \sigma \mathbf{E}_z r dr$$

$$= \frac{c^2}{2ip}\left\{r\frac{\partial \mathbf{E}_z}{\partial r}\right\}_{r=a},$$

or

$$J = \frac{c^2 C e^{ipt}}{2ip}\{xJ_0'(x)\}_{x=\nu a}.$$

We may thus write the relation between the complex expressions for J and $\mathbf{E}_{0_z}$ in the form

$$\mathbf{E}_{0_z} = RJ + \frac{1}{c^2}L\frac{dJ}{dt}$$

$$= J\left(R + \frac{ip}{c^2}L\right),$$

whence it follows that

$$R + \frac{ip}{c^2}L = \frac{2ip}{c^2}\left\{\frac{J_0(x)}{xJ_0'(x)}\right\}_{x=\nu a}.$$

From this relation we can get the two real quantities R and L.

The physical meaning of these coefficients R and L can now be directly deduced by an application of Poynting's theorem as above explained. If we now denote by $\mathbf{E}_{0_z}$, $\mathbf{H}_\theta$, J the real parts of their respective complex representations as above, we shall find for the energy which enters per unit time and length into the conductor from the external field

$$-2\pi(r\mathbf{S}_r)_{r=a} = \frac{c}{2}(\mathbf{E}_z \,.\, r\mathbf{H}_\theta)_{r=a}$$

$$= \mathbf{E}_{0_z}J.$$

But from the above this is

$$\mathbf{E}_{0_z}J = RJ^2 + \frac{d}{dt}\left(\frac{1}{2c^2}LJ^2\right).$$

Suppose now we integrate this equation over a complete oscillation; the energy entering the conductor must then just be equal to the heat developed in the circuit as Joule's heat: this is given by the first term on the right of the above equation. The second term on the right of this equation which disappears on integration over the whole oscillation must then give the increase of the magnetic energy of the field of the current inside the wire. According to this explanation it is usual to call R the *effective resistance* and L the *effective self-inductance* (per unit length) of the conductor for the particular period of the oscillating current.

If the period of oscillation is long or the radius of the wire small νa will be small, and we may use the approximate form for J_0 which expresses it as an ascending power series; it is then found that

$$R = R_0\left(1 + \frac{(\theta a^2)^2}{12} - \frac{(\theta a^2)^4}{180} + \dots\right),$$

$$L = \frac{1}{8\pi}\left(1 - \frac{(\theta a^2)^2}{24} + \frac{13\,(\theta a^2)^4}{4320} - \dots\right),$$

where we have used $\theta = \dfrac{p\sigma}{c^2}$; $R_0 = \dfrac{1}{\sigma\pi a^2}$.

The effective resistance only departs slightly from the value for a uniform current.

247. The fundamental equation of wave propagation. We have so far tacitly assumed that any quantity which is determined mathematically by certain scalar or vector component quantities which satisfy an equation of the type

$$\nabla^2\phi = \frac{1}{c^2}\frac{d^2\phi}{dt^2} + \frac{2\lambda}{c}\frac{d\phi}{dt}$$

is essentially propagated by a wave motion throughout the field. That this is so follows from our knowledge of such phenomena as, for example, accompany the propagation of sound through any elastic medium; but it may be inferred directly as a mathematical consequence of the implied condition involved in the characteristic equation.

The general problem in the present aspect of the theory is to determine how any electromagnetic disturbance is propagated across space filled with dielectric and conducting masses in any specified configuration, and to see how the conditions at any one point of this field are affected by those occurring at any other. The complexity of the conditions involved naturally excludes the determination of a simple solution for the general problem, and we must therefore be content with the examination of simpler problems with restricted circumstances.

We first examine the general case of the propagation of effects from a specified type of disturbance located in a finite region of an infinite homogeneous isotropic dielectric with zero conductivity. The disturbance will for simplicity be assumed to be of a continuous character and to have been in operation for an indefinite period previous to the instant at which the field is examined*.

Suppose we enclose the origin of the disturbance by any closed surface f: then at all points in the region outside this surface the field vectors will satisfy an equation of the type

$$\nabla^2\phi = \frac{1}{c^2}\frac{\partial^2\phi}{\partial t^2}.$$

The function ϕ satisfying this equation must necessarily be regular at all points outside f even at an infinite distance, and it therefore follows from the general problem analysed in the introduction (§§ 25–29) that the appropriate form for the function at time t at the typical field-point outside f is

$$\phi = \frac{1}{4\pi}\int_f \frac{df}{r}\left[\frac{\partial\phi}{\partial n}\right] + \int_f df \frac{d}{dn}\left(\frac{[\phi]}{r}\right),$$

where each integral is taken over the surface f bounding the origin of the disturbance; r is the distance of the element of this surface from the point of the external field where the conditions are examined, and square brackets as usual indicate that the functions affected are to be taken for the time

$$\left(t - \frac{r}{c}\right).$$

Now let us see what this formula means. The conditions at any point in the dielectric medium outside the surface f depend only on the conditions of the field on the surface itself, so that any alteration of condition in the disturbing system inside f affects the external field only through the medium of the field on the arbitrary separating surface. Moreover the conditions existing on any element df of this surface at a given instant are not effective at any external point distant r from it until after the time r/c. This suggests the view that the conditions originated at any point in the field travel out from that point into the surrounding field, traversing each part of the intervening field in turn and proceeding from point to point with the velocity c.

* More general cases are examined by Love, *Proc. L. M. S.* [2] 1 (1903), p. 37.

This is the essence of a radiation theory and is exactly analogous to the phenomenon with which we are familiar in the theory of sound, and although it will appear that the type of radiation is essentially different from that met with in all such material phenomena, it is convenient to talk of electromagnetic waves and radiation in the same sense as we talk of waves of sound.

248. The analytical formula under review has an important physical significance which it is worth while examining in detail. The potential propagated from a point source variable with the time and of strength $f(t)$ is with the same characteristic equation

$$\frac{1}{4\pi r} f\left(t - \frac{r}{c}\right).$$

The potential propagated from a doublet consisting of simple sources $f(t)$ and $-f(t)$ separated by an interval δn is consequently

$$\frac{1}{4\pi} \delta n \frac{d}{dn} \left\{\frac{1}{r} f\left(t - \frac{r}{c}\right)\right\}.$$

Thus for a doublet of strength $F(t)$, the equivalent of $f(t)\,\delta n$, it is

$$\frac{1}{4\pi} \frac{d}{dn} \left\{\frac{F\left(t - \frac{r}{c}\right)}{r}\right\},$$

in which the function F comes under the differentiation.

The formula quoted above for ϕ thus implies that each differential element of the surface f acts, as regards the point P inside it, as a complex radiating element consisting of a simple source of strength

$$\frac{\partial \phi}{\partial n} df,$$

and a normal doublet of strength

$$\phi\, df.$$

The wave disturbance originated by these elements travels out into the space inside f as a simple spherical wave propagation. The disturbance at P is thus just the same as if the surface itself acted as a sort of secondary radiator, and this is the essence of Huygens' well-known principle in physical optics. The above mode of deduction of the formula, not free from analytical difficulties, can hardly however be said to throw much light on the character of the simple principle which is thus demonstrated. In this connection however the following discussion due to Prof. Larmor* is of special interest as indicating the exact amount of precision in the specification of the secondary disturbance thereby introduced. Reference may also be made back to the discussions of Green's theorem and the equivalent stratum.

249. Consider a potential specified throughout all space as follows. It is a function ϕ, single valued and continuous as to itself and its first gradient, and satisfying $\nabla^2\phi = 0$, in all the space outside a boundary f, and as a consequence diminishing towards infinity according to the law r^{-1} or higher inverse power: it is zero everywhere inside the boundary. What is the distribution of attracting mass to which this belongs? This distribution is as usual in Green's manner determined by the singularities and discontinuities of the potential function. It consists of a surface density σ over f and a double sheet τ over f also; where

$$\sigma = -\frac{1}{4\pi}\frac{\partial\phi}{\partial n}, \quad \tau = \frac{1}{4\pi}\phi,$$

δn is an element of the outward normal. For it follows by the usual procedure that if ϕ' is the potential of this distribution, then $(\phi - \phi')$ is a potential function which has no singularities, or discontinuities throughout all space, and is therefore identically null. Expressed analytically the potential at a point in space is

$$-\frac{1}{4\pi}\int_f \frac{1}{r}\frac{\partial\phi}{\partial n}\,df + \frac{1}{4\pi}\int \phi\frac{d}{dn}\left(\frac{1}{r}\right)df,$$

in other words the formula gives the value of a potential function ϕ in the free space outside the surface in terms of the values which it and its gradient assume on the surface, it constitutes in fact the analytical continuation of the function outward from the surface, while inside the surface the value of the expression is everywhere null. In the case of a closed surface as well as that of an open sheet either side may be called the outside for the present purpose. This continuation of the function is necessarily unique and determinate, but the form of the integral expressing it is far from being so. We may in fact generalise the formula immediately in Green's manner. Consider a function ϕ which is the potential throughout space of

* *Proc. L. M. S.* [2] 1 (1903), p. 1.

any assigned distribution of mass. Draw any surface dividing space into two regions A and B each of which contains part of the mass, these parts being represented by M_A and M_B. What distribution of masses and of surface densities and normal doublets on the surface f is required to produce a potential equal to ϕ in the region A and equal to zero in the region B? Clearly M_A together with

$$\sigma = -\frac{1}{4\pi}\frac{\partial\phi}{\partial n}, \quad \tau = \frac{1}{4\pi}\phi.$$

What distribution is required to make the potential zero in the region A and ϕ in the region B? Clearly M_B with the same distribution on the surface but with the sign changed if δn is measured in the same way. Thus a distribution of surface density and normal doublets is found which exactly cancels the effect of M_A on the other side of the dividing sheet f; moreover an infinite number of such distributions can be found, for in determining it M_B is entirely arbitrary.

The same procedure can now be extended to a scalar potential propagated in time, i.e. which satisfies a characteristic equation involving the time as a variable.

Consider first the simplest case of a velocity potential ϕ satisfying

$$\nabla^2\phi = \frac{1}{c^2}\frac{\partial^2\phi}{\partial t^2}.$$

It is necessary to ascertain what distribution of sources on a surface f will create given discontinuity in the values of ϕ, and of its normal gradient, in crossing the surface, it being clear that such discontinuities in ϕ and $\partial\phi/\partial n$ constitute the most general type, involving only first differential coefficients of ϕ, that can exist.

We have already seen that the velocity potential propagated from a doublet consisting of simple sources $f(t)$ and $-f(t)$ separated by an interval δn is

$$\delta n \frac{d}{dn}\left\{\frac{1}{r} f\left(t - \frac{r}{c}\right)\right\};$$

and that for a doublet of strength $F(t)$, the equivalent of $f(t)\,\delta n$, it is

$$\frac{d}{dn}\left[\frac{F\left(t - \frac{r}{c}\right)}{r}\right],$$

in which the function F comes under the differentiation. Within a region of such small extent that the functions $f(t)$ and $F(t)$ do not sensibly change in the time required for the disturbance to pass across it, these potentials are of types $\frac{f(t)}{r}$ and $F(t)\frac{d}{dn}\left(\frac{1}{r}\right)$, so far as they relate to sources inside the region of which $f(t)$ and $F(t)$ are the strengths at this interval of time; for this modification only neglects lower inverse powers of r than those retained. Thus for such a region enclosing an element δf of the surface f, and at times for which the functions $f(t)$ and $F(t)$ do not change there abruptly, the potentials, subject to exceptions to be presently encountered in the case of double sheets, take, throughout any time of the order above specified, the form of simple gravitational potentials, the circumstances of propagation not sensibly inferring.

We are therefore invited to follow the procedure of Coulomb and Laplace for the ordinary potential and investigate the discontinuities arising from a surface distribution of simple sources and one of doublets orientated normally to the surface. The discontinuities, on crossing the surface at any point, clearly arise from the distribution over a surface element δf surrounding that point; for the disturbances propagated from the more distant sources are virtually the same at points on the two sides of the surface, whose distance apart is infinitesimal compared with the linear dimensions of δf, so that, as regards their effect, no discontinuities can arise.

Taking first, then, the case of a simple surface density $\sigma(t)$ spread over δf, which we may take to be uniform all over it at each instant, its effect is to transmit towards both sides a train of plane waves with fronts parallel to δf, which remain plane until the distance n to which they have travelled becomes comparable with the linear dimensions of δf. For them the value of $d\phi/dn$ at a distance n at time t is $2\pi\sigma\left(t-\frac{n}{c}\right)$, but with different sign on the two sides; such a surface distribution $\sigma(t)$ of simple sources thus accounts for a discontinuity in $d\phi/dn$ of amount $4\pi\sigma(t)$, but introduces no discontinuity in ϕ itself.

This result now assists us to analyse the circumstances of a sheet of normal doublets of strength $\tau(t)$ per unit area, for we can replace it by two simple parallel sheets of densities $\sigma_1(t)$ and $-\sigma_1(t)$ at an infinitesimal distance δn apart, such that $\sigma_1(t)\,\delta n = \tau_1(t)$. At a point at a distance n from the sheet of density $+\sigma_1(t)$ the values of $\frac{d\phi}{dn}$ at time t arising from these two sheets are, as above, $\pm 2\pi\sigma_1\left(t_1 - \frac{n}{c}\right)$ and $\mp 2\pi\sigma_1\left(t - \frac{n+\delta n}{c}\right)$, in which signs are to be determined by the sides of the respective component sheets on which this point lies. If the point is not between the sheets, the signs are opposite and the sum for both is $-2\pi\frac{\delta n}{c}\frac{d}{dt}\sigma_1\left(t - \frac{n}{c}\right)$, which is the value of $\frac{d\phi}{dn}$ due to the element $\tau\delta f$; but it has the same sign on both sides of the double sheet: so that in crossing the double sheet there is no discontinuity in the value of $\frac{\partial\phi}{\partial n}$, though the element $\tau\delta f$ of the double sheet contributes $\frac{4\pi}{c}\frac{d\tau}{dt}\delta f$ to that quantity on each side. But between the sheets the value of $\frac{d\phi}{dn}$ arising from them is of a higher order of magnitude, being a sum instead of a difference, and is $4\pi\sigma_1\left(t - \frac{n}{c}\right)$, or simply $4\pi\sigma_1(t)$, when σ_1 is not discontinuous in the time; and this value integrated across the interval δn gives a discontinuity in ϕ itself, on crossing the double sheet, of amount $4\pi\delta n\sigma_1(t)$; that is $4\pi\tau(t)$.

Collecting these results we see that a discontinuity in ϕ over a surface f of amount $\chi(x, y, z, t)$ and a discontinuity in $\frac{\partial\phi}{\partial n}$ equal to $\psi(x, y, z, t)$ over the same surface are accounted for respectively by a double sheet on the surface of strength τ equal to $\frac{1}{4\pi}\chi$ and a single sheet of density σ equal to $\frac{1}{4\pi}\psi$.

We are thus in a position to proceed exactly as in the first instance. Consider any system of sources, and let ϕ, a function of (x, y, z, t), be their potential function in infinite free space. Assign any surface f_1 dividing space into two regions A and B and let m_A and m_B stand for the sources as divided between the two regions. What distribution of sources would give rise to a potential equal to ϕ in region A and equal to zero in region B? Clearly the sources m_A, together with a distribution $(\sigma_1\tau)$ over f given by

$$4\pi\sigma = -\frac{\partial\phi}{\partial n}, \quad 4\pi\tau = \phi;$$

for if ϕ' is the potential arising from this distribution and Φ is a function equal to ϕ in region A and to zero in region B, then $\Phi - \phi'$ will be a potential having no singularities or discontinuities throughout infinite space, and must therefore be null by simple physical intuition, or analytically by the usual type of theorem of determinacy based on the energy of the relative disturbance being of necessity essentially positive. As the total effect within the region B is zero, we can say thus that ϕ_A the part of it arising from the sources m_A outside the region is given at time t by

$$-4\pi\phi_A = -\int_f \frac{df}{r}\left[\frac{\partial\phi}{\partial n}\right] + \int_f df\,\frac{d}{dn}\left[\frac{[\phi]}{r}\right],$$

where $[\phi]$ and $\left[\frac{\partial\phi}{\partial n}\right]$ are the values of these quantities for the element δf at time $t - \frac{r}{c}$ and in forming $\frac{d}{dn}\frac{[\phi]}{r}$, the variation of $[\phi]$ with regard to the coordinates is calculated only in as far as it involves them implicitly as a function of $\left(t - \frac{r}{c}\right)$*. This formula expresses the vibration potential due to sources m_A within the surface f, throughout the region B outside, as determined by the values which it and its gradient assume on that surface. It is so to speak an analytical continuation beyond the surface of a function satisfying the aforesaid characteristic differential equation. Such a continuation must be unique, and it is determined by the value assumed by ϕ alone on the surface: as therefore $\frac{d\phi}{dn}$ is determined by a knowledge of ϕ over the surface, the data for the formula here given are redundant; if arbitrarily assigned they will usually be self-contradictory and the formula thus nugatory. Moreover the formula determines ϕ_A in terms of the surface distribution of ϕ, equal to $\phi_A + \phi_B$, where ϕ_B is due to an entirely arbitrary distribution of sources within the region B to which the formula relates. Thus the quantities integrated in it are very widely arbitrary and the

* This point was overlooked by Prof. Larmor in the original paper.

element of the integral corresponding to δf in no sense represents any influence actually propagated from that part of the surface. The formula is purely analytical and in no degree a mathematical formulation of the principle of Huygens, relating to propagation of actual disturbance. In fact if m_B vanishes the formula represents a distribution of surface disturbances, which does not radiate at all into the region A.

250. In the more general case when the uniform medium possesses conducting qualities the above analysis is no longer applicable. The general character of the solution in this case can however be demonstrated by another method which is also suited to the simpler problem.

In this case the characteristic potential equation assumes the form

$$c^2 \nabla^2 \phi = \frac{\partial^2 \phi}{\partial t^2} + 2\lambda c \frac{\partial \phi}{\partial t}.$$

If we transform this equation to a spherical polar coordinate system, then multiply it by $r^2 d\omega$, where $d\omega$ is the element of solid angle at the polar origin, and finally integrate over the unit sphere, it reduces immediately to the form

$$c^2 \frac{\partial^2 \Phi}{\partial r^2} = \frac{\partial^2 \Phi}{\partial t^2} + 2\lambda c \frac{\partial \Phi}{\partial t},$$

where

$$\Phi = \frac{r}{4\pi} \int \phi \, d\omega.$$

Let us now examine the propagation of conditions from an initial disturbance specified by

$$\phi = f(x, y, z), \quad \frac{1}{c} \frac{\partial \phi}{\partial t} = g(x, y, z)$$

at the time $t = 0$; the propagation is assumed to take place in a uniform isotropic medium possessing dielectric and conducting properties corresponding to the equation chosen.

We first transform the functions f and g to the same spherical polar coordinates and then use

$$F(r) = \frac{r}{4\pi} \int f \, d\omega, \quad G(r) = \frac{r}{4\pi} \int g \, d\omega,$$

so that $F(r)$ and $G(r)$ are respectively the initial values of Φ and $\dfrac{\partial \Phi}{\partial t}$.

We are of course concerned only with the positive values of r, so we can choose the functions F and G for negative values as we please. We choose them so that

$$F(-r) = -F(r), \quad G(r) = -G(-r).$$

We then have, by applying a familiar method in this theory, the general solution for Φ in the form

$$2\Phi e^{\lambda ct} = F(r+ct) + F(r-ct) + \int_{r-ct}^{r+ct} \left[G(s) + \lambda F(s) + \frac{\lambda ct}{z} F(s) \frac{d}{dz}\right] I_0(z)\, ds$$

where again I_0 is the Bessel function of zero order and imaginary argument and

$$z = \lambda\sqrt{c^2t^2 - (s-r)^2}.$$

We conclude that the average conditions propagated from the disturbance at any point in the field travel outwards radially from that point in exactly the same way as a signal travels along a telegraph cable. In other words if there is no friction the propagation is like that of a simple undamped wave form with the velocity c, but if there is appreciable conductivity rapid distortion and dissipation occur to destroy these simple propagation effects.

The value of the more general function ϕ can be easily obtained from the value found above for Φ by determining the limiting value of the ratio Φ/r as r tends to zero. The result is that*

$$\phi = e^{-\lambda ct}\left[F'(ct) + G(ct) + \frac{\lambda ct}{2} F(ct) + \lambda \int_0^{ct} \left\{G(s) + \lambda ct\, F(s) \frac{1}{z}\frac{d}{dz}\right\} \frac{1}{z}\frac{dI_0}{dz}\, s\, ds\right],$$

where now

$$z = \lambda\sqrt{c^2t^2 - s^2}.$$

This formula determines completely the way in which the conditions at any one point in the field depend on those at the other points, and is in complete accord with our physical conception of these things.

251. Plane waves. We turn now to the consideration of certain cases of wave motion where the circumstances are much simpler than those just analysed. If the radiating system is at a very great

* Cf. Riemann-Weber, *Die partielle Differentialgleichungen, etc.*, II. pp. 299–312 (4th Ed. 1901).

distance from the part of the field under investigation the expanding wave front surfaces have become so large that in the part of the field where they are investigated they may be treated as practically plane surfaces. We then realise the idea of plane waves and the consequent rectilinear propagation of electromagnetic disturbances, and the analytical problems become much simplified.

Analytically the conception is obtained by a simple type of solution of the fundamental equations of propagation, which directly suggests itself. Let us choose our rectangular axes so that the z-axis is along the direction of propagation and the other two conveniently in the perpendicular plane which is parallel to the front of the wave. The general solutions for this case are then of the form

$$\left.\begin{matrix}\mathbf{E}\\ \mathbf{B}\end{matrix}\right\} = Ae^{int-(a+ib)\,z+i\theta}$$

which represents plane waves of period $\frac{2\pi}{n}$ advancing in the medium with a velocity $c_1 = \frac{n}{b}$ along the positive direction of the z-axis. The solution for any vector is of course represented by the real part of the general solution thus obtained for it.

If we assume, for the isotropic medium, that both the two simple constitutive relations

$$\mathbf{C} = \sigma\mathbf{E} + \frac{\epsilon}{4}\frac{d\mathbf{E}}{dt}$$

and

$$\mathbf{H} = \mathbf{B}$$

are valid even for the general case of electric waves under consideration, then both vectors $\mathbf{E}$ and $\mathbf{B}$ satisfy the equation of the previous paragraph,

$$\nabla^2\mathbf{E} = \frac{\sigma}{c^2}\frac{d\mathbf{E}}{dt} + \frac{\epsilon}{c^2}\frac{d^2\mathbf{E}}{dt^2},$$

and thus the above-mentioned forms for $\mathbf{E}$ and $\mathbf{H}$ are valid if

$$-\epsilon n^2 + in\sigma = c^2(a+ib)^2,$$

or

$$-\epsilon n^2 = c^2(a^2 - b^2),$$

$$n\sigma = 2c^2ab.$$

Thus if σ is different from zero, a is so also, and thus the amplitude A of each vector contains the factor e^{-az}, which means that as the waves are propagated along the positive direction of the z-axis, the amplitudes of the two vectors gradually decrease as the wave proceeds. Conductivity in the medium implies a damping of the waves, the energy being absorbed by the medium and converted into heat.

We can now show that these electric waves are transverse. To do this we merely prove that in the most general aeolotropic medium any vector which has the property of the ordinary stream vector of hydrodynamic theory has its direction parallel to the wave front; for if $\mathbf{V}$ is any such vector

$$\operatorname{div} \mathbf{V} = \frac{\partial \mathbf{V}_x}{\partial x} + \frac{\partial \mathbf{V}_y}{\partial y} + \frac{\partial \mathbf{V}_z}{\partial z} = 0,$$

and since all quantities in a plane wave specified as above only depend on the z-coordinate we must have

$$\frac{\partial \mathbf{V}_x}{\partial x} = \frac{\partial \mathbf{V}_y}{\partial y} = 0,$$

and therefore also

$$\frac{\partial \mathbf{V}_z}{\partial z} = 0,$$

which implies that $\mathbf{V}_z = 0$.

This is a general property of a train of plane waves whatever the constitution of the medium, whether it be crystalline or not. It means in any case that the total current and magnetic force vectors are in the plane of the wave front. Thus in this sense all such electromagnetic waves are transverse, the fluxes associated with them being both transverse to the direction of propagation. The electric force and magnetic induction vectors are of course in the general case not in the wave front; this is true only when the medium is isotropic as already assumed.

252. We now assume that the magnetic and electric forces are in the wave front and thus are expressed by their components

$$(\mathbf{B}_x, \mathbf{B}_y, \mathbf{B}_z) = (A_m \cos \phi_m, \quad A_m \sin \phi_m, 0)\, e^{int-(a+ib)z+i\theta_m},$$

$$(\mathbf{E}_x, \mathbf{E}_y, \mathbf{E}_z) = (A_e \cos \phi_e, \quad A_e \sin \phi_e, \quad 0)\, e^{int-(a+ib)z+i\theta_e},$$

and these must satisfy the fundamental equations of the field. Faraday's relation implies that

$$-\frac{1}{c}\frac{d}{dt}(\mathbf{B}_x, \mathbf{B}_y) = \frac{\partial}{\partial z}(\mathbf{E}_y, -\mathbf{E}_x),$$

so that
$$-in\frac{A_m}{c}\cos\phi_m = A_e\sin\phi_e\,(a+ib)\,e^{i(\theta_e-\theta_m)},$$
$$-in\frac{A_m}{c}\sin\phi_m = -A_e\cos\phi_e\,(a+ib)\,e^{i(\theta_e-\theta_m)}.$$

Whence we deduce directly the two important conclusions:

(i) By division we see that
$$\tan\phi_e = -\cot\phi_m,$$
$$\phi_m = \phi_e + \frac{\pi}{2}.$$
Thus the electric and magnetic force vectors in the wave front are perpendicular to one another.

(ii) Using the fact that $\phi_m = \phi_e + \frac{\pi}{2}$ we see also that
$$-inA_m = c\,(a+ib)\,A_e\,e^{i(\theta_e-\theta_m)}.$$
Now from Ampère's circuital relation we can deduce similarly that
$$A_m c\,(a+ib)\,e^{i(\theta_m-\theta_e)} = A_e\,(\sigma + i\epsilon n),$$
so that since A_e and A_m are real we must have
$$\tan(\theta_m-\theta_e) = \frac{a}{b} = \frac{1}{2}\frac{\sigma}{n}\frac{c_1^2}{c^2},$$
where $c_1 = \frac{n}{b}$ is the velocity of propagation in the medium.

Also
$$A_m = \frac{cA_e}{n}\sqrt{a^2+b^2}.$$
Thus there is in the general case always a phase difference $(\theta_m-\theta_e)$ between the electric and magnetic force vibrations, this phase difference vanishing only for the case when $a = 0$, i.e. $\sigma = 0$, or for a perfect non-conductor. The amplitudes of the waves are also different in the general case of absorption.

The velocity of the wave $\frac{n}{b}$ is given by
$$c_1 = \frac{c}{\sqrt{\epsilon + \dfrac{a^2c^2}{n^2}}},$$
which reduces in the case of non-absorbing media to
$$\frac{c}{\sqrt{\epsilon}}.$$

If we now choose the real part of the general solutions and also write

$$\Theta = nt - bz + \theta_e,$$

then for the magnitudes of the respective vectors we can put

$$E = A_e \cos \Theta e^{-az},$$

$$H = A_m \cos (\Theta + \theta_e - \theta_m)\, e^{-az}.$$

The electric energy per unit volume at a place is

$$W = \frac{\epsilon}{2} E^2 = \frac{\epsilon}{2} A_e^2 \cos^2 \Theta e^{-2az},$$

and the mean value at the place taken over a whole oscillation is

$$\overline{W} = \frac{n}{2\pi}\int_0^{\frac{2\pi}{n}} W\,dt = \frac{\epsilon A_e^2}{4}\, e^{-2az},$$

similarly the mean kinetic or magnetic energy is

$$\overline{T} = \frac{n}{2\pi}\int_0^{\frac{2\pi}{n}} T\,dt = \frac{n}{4\pi}\int_0^{\frac{2\pi}{n}} H^2 dt = \frac{A_m^2}{4}\, e^{-2az}$$

$$= \frac{A_e^2}{4}\frac{c^2}{n^2}(a^2 + b^2)\, e^{-2az}$$

$$= \frac{\epsilon A_e^2}{4}\frac{b^2 + a^2}{b^2 - a^2}\, e^{-2az}.$$

Whence it follows that the mean magnetic energy is in general larger than the electric energy, equality occurring only in the non-conducting substances when $a = 0$. This result also exhibits clearly the way in which the energy in the wave is absorbed as it progresses, the mean total energy of the wave at any place being the sum of the electric and magnetic energies, viz.

$$\frac{\epsilon A_e^2}{8\pi}\frac{2b^2}{b^2 - a^2}\, e^{-2az}.$$

253. These results enable us to explain in greater detail the behaviour of metallic conductors in a radiating dielectric field. They show that the propagation of the waves in the dielectric takes place without any absorption at all. As soon, however, as a disturbance reaches a conducting surface and starts off through the conducting medium the damping factor

$$e^{-az}$$

at once enters into the expression for all the vectors. If the conductivity is considerable a is large and the wave is practically damped right out before it gets far into the metal. The larger the value of a the shorter the distance the waves penetrate into the conductor, and by sufficiently large values we may neglect the penetration altogether. We can also increase a by increasing n, so that for very rapid oscillations the conducting material will always act as if it were a perfect conductor. Let us take an example to illustrate the matter further. For copper $\sigma = \frac{c^2}{1600}$ and ϵ is negligible. Now consider the incidence of waves of length 100 cms.: then

$$n = \frac{2\pi \times 3 \times 10^{10}}{10^2} = 2 \times 10^9 \text{ approx.}$$

and since ϵ is negligible

$$i4\pi n\sigma = c^2 (a + ib)^2,$$

whence

$$\frac{1+i}{2}\sqrt{4\pi n\sigma} = c\,(a + ib),$$

or

$$2ca = 2cb = \sqrt{4\pi n\sigma},$$

$$a = \frac{1}{2}\frac{\sqrt{4\pi n\sigma}}{c} = \frac{1}{2}\sqrt{\frac{2 \times 10^9 \times 4\pi}{16 \times 10^2}}$$

$$= 2 \times 10^3 \text{ approx.}$$

Thus at a depth $5 \,.\, 10^{-4}$ cms. the amplitudes are reduced to $1/e$ times their initial value. The penetration is therefore very slight in a real case of this kind.

254. The generation of electric waves. Having now definitely established the compatibility of the existence of electric waves with Maxwell's scheme and having derived their main characteristics it remains to discuss briefly their mode of generation. The essence of simple wave motion is that it is oscillatory in character, so we should expect it to result from oscillation more or less rapid of certain electric or magnetic distributions. The simplest oscillation of this kind and the first to be discovered was that resulting from the discharge of a condenser through an induction coil in the circuit connecting its plates. This experiment was carried out even in Ampère's time and attempts were made to discover the direction of the flow by inserting an iron needle in the coil to see in which direction it was magnetised, it being not at first realised that the current was of an oscillatory kind. The magnetism induced was however sometimes in one direction and sometimes in the other and the whole matter remained a mystery until Helmholtz suggested that the discharge was really an oscillation, in which case it would be largely a matter of chance whether the first, second, third or any subsequent swing gave the preponderating magnetisation in the needle. Kelvin developed the idea more precisely a few years later and on the following lines.

If the current in the wire is supplied from a condenser in the circuit instead of from a battery then the current is

$$J = -\frac{dQ}{dt},$$

when Q is the charge on the condenser. Further the electromotive force driving the current is the difference in potential between the plates of the condenser and this is Q/b if b is the capacity of the condenser. The equation for Q or J thus becomes

$$\frac{Q}{b} + \frac{a}{c}\frac{d^2Q}{dt^2} = -k\frac{dQ}{dt},$$

or

$$\frac{d^2Q}{dt^2} + \frac{ck}{a}\frac{dQ}{dt} + \frac{c}{ab}Q = 0,$$

the solution of which determines the complete circumstances of the affair. This equation, first obtained by Kelvin, is easily solved, for if

$$Q = Q_0 e^{pt}$$

is a solution then

$$p^2 + \frac{kc}{a}p + \frac{c}{ab} = 0,$$

or

$$p = -\frac{kc}{2a} \pm \sqrt{-\left(\frac{c}{ab} - \frac{k^2c^2}{4a^2}\right)},$$

say p_1 and p_2, so that the complete solution of the equation is of a type

$$Q = Q_1 e^{p_1 t} + Q_2 e^{p_2 t}.$$

There are two distinct types of solution of an equation of this kind. If

$$q^2 = +\left(\frac{c}{ab} - \frac{k^2c^2}{4a^2}\right)$$

is positive, i.e. if $k < \sqrt{\frac{4a}{bc}}$, imaginary values are obtained for p

and the solution is an oscillatory one. In this case we can in fact write the general solution in the form

$$Q = Q_0 e^{-\frac{kct}{2a}} \sin [qt + \chi],$$

the integration constants Q_0 and χ being obtained from the initial conditions.

If the resistance k is practically negligible or the conduction nearly perfect, the solution reduces to

$$Q = Q_0 \sin \left(\frac{\sqrt{c}\, t}{\sqrt{ab}} + \chi\right)$$

and represents a permanent oscillation of period

$$2\pi \sqrt{\frac{ab}{c}}.$$

Even if k is sensible its effect on the period is practically always negligible; in fact in the general case

$$q = \sqrt{\frac{c}{ab} - \frac{k^2c^2}{4a^2}} = \frac{\sqrt{c}}{\sqrt{ab}}\left(1 - \frac{k^2cb}{8a}\right) \text{ approx.},$$

so that the period of the oscillation is increased in the ratio

$$1 + \frac{k^2cb}{8a} : 1,$$

the effect of k thus being of the second order and therefore negligible unless k is very big. This is a general result in dynamics, when the dissipation is comparatively small its effect on the period is of the second order of smallness; the main effect of resistance is in damping the amplitude, which gradually decreases to zero.

Sometimes however the resistance is so very considerable that it actually destroys the periodic nature of the motion altogether. This is the case when $k > \sqrt{\frac{2a}{bc}}$ when the solution is of the type

$$Q = Q_0 e^{-\frac{kct}{2a} - qt}.$$

255. This theoretical possibility of the oscillatory nature of the currents in a conductor discharging a condenser was subsequently verified in detail by Feddersen, who examined the spark of the discharge by a revolving mirror, and by Hertz who succeeded in reducing the period of the oscillations sufficiently to make the wave characteristics of the surrounding field appreciable. To reduce the period Hertz separated his condenser plates and thus reduced the elastic spring in the electrical arrangement. He used a form of condenser similar to that exhibited in the figure. A, B are two metallic plates which in Hertz's original experiments were of zinc and about 40 cms. square: to these were soldered brass rods C, D (30 cms. long) terminating in brass balls E, F. Such a condenser has very little capacity and induction. In order to excite the waves the balls E, F are charged to very different potentials by connecting C, D to the terminals of an induction coil. When a sufficient potential is attained sparks cross the air gap between E, F which then

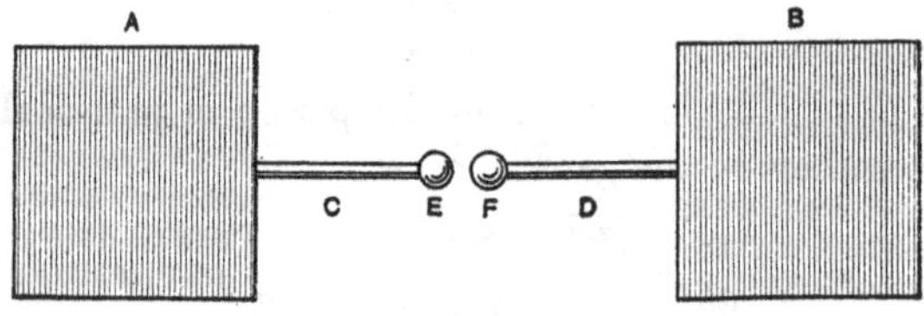

Fig. 24

becomes a conductor and the charges on the plates can then oscillate backwards and forwards like the charges on the coatings of a Leyden jar.

256. To obtain some idea of the radiation field in the dielectric surrounding a condenser of this type we may notice that the main and most vigorous part of the electrical motions occurs in the rods C, D and across the discharge gap EF, so that to all intents and purposes the field should be symmetrical round an axis along the rods and also about the origin mid-way between the balls E, F with the magnetic force in horizontal circles round this axis. To obtain a solution of the fundamental equations of this type we shall find it most convenient to refer the field to a system of spherical polar coordinates with the pole at the origin and axis along the axis of symmetry of the field: we shall then have

$$\mathbf{B}_r = \mathbf{B}_\theta = \mathbf{E}_\phi = 0,$$

and thus, assuming for the present that the dielectric medium surrounding the apparatus is a pure vacuum the remaining equations of Ampère give

$$\frac{1}{c}\frac{d\mathbf{E}_r}{dt} = \frac{1}{r^2 \sin\theta}\frac{\partial}{\partial\theta}(r\mathbf{B}_\phi \sin\theta),$$

$$\frac{1}{c}\frac{d\mathbf{E}_\theta}{dt} = -\frac{1}{r\sin\theta}\frac{\partial}{\partial r}(r\mathbf{B}_\phi \sin\theta),$$

whilst the third equation of the Faraday type becomes

$$-\frac{1}{c}\frac{d\mathbf{B}_\phi}{dt} = \frac{1}{r}\left(\frac{\partial}{\partial r}(r\mathbf{E}_\theta) - \frac{\partial \mathbf{E}_r}{\partial\theta}\right).$$

Thus if we write $\qquad \psi \equiv r\mathbf{B}_\phi \sin\theta,$

we have $\qquad \frac{1}{c}\frac{d\mathbf{E}_r}{dt} = \frac{1}{r^2\sin\theta}\frac{\partial\psi}{\partial\theta}, \quad \frac{1}{c}\frac{d\mathbf{E}_\theta}{dt} = -\frac{1}{r\sin\theta}\frac{\partial\psi}{\partial r},$

and on using these values in the last equation we find on putting $\mu \equiv \cos\theta$

$$\frac{1}{c^2}\frac{d^2\psi}{dt^2} = \frac{\partial^2\psi}{\partial r^2} + \frac{1-\mu^2}{r^2}\frac{\partial^2\psi}{\partial\mu^2}.$$

A simple solution of this equation is obtained by putting

$$\psi = R(1-\mu^2),$$

and regarding R as a function of r only: this function satisfies the equation

$$\frac{1}{c^2}\frac{d^2R}{dt^2} = \frac{d^2R}{dr^2} - \frac{2R}{r^2},$$

of which the general solution is easily verified to be

$$R = r\frac{d}{dr}\left(\frac{F_1(ct-r)}{r}\right) + r\frac{d}{dr}\left(\frac{F_2(ct+r)}{r}\right).$$

Of the two parts of this solution the first will represent an expanding wave whilst the second represents a condensing wave: it is with the first alone that we shall be concerned and we shall also find it more convenient to write

$$F(x) = \frac{\partial f(x)}{\partial x},$$

so that $\qquad \psi = \sin^2\theta\left\{f''(ct-r) + \frac{1}{r}f'(ct-r)\right\}.$

It is then easily verified that

$$\mathbf{E}_r = \frac{2\cos\theta}{r^2}\left\{f'(ct-r) + \frac{1}{r}f(ct-r)\right\},$$

$$\mathbf{E}_\theta = \frac{\sin\theta}{r}\left\{f''(ct-r) + \frac{1}{r}f'(ct-r) + \frac{1}{r^2}f(ct-r)\right\},$$

$$\mathbf{H}_\phi = \frac{\sin\theta}{r}\left\{f''(ct-r) + \frac{1}{r}f'(ct-r)\right\},$$

dashes being used to denote differentiation of the function f with respect to its argument.

Near the origin, that is near the oscillator itself, the electric field reduces to

$$\mathbf{E}_r = \frac{2f(ct)\cos\theta}{r^3}, \quad \mathbf{E}_\theta = \frac{f(ct)\sin\theta}{r^3},$$

the other terms being very small compared with these: the magnetic field reduces to

$$\mathbf{H}_\phi = \frac{f'(ct)\sin\theta}{r^2}.$$

Here then we have the type of solution obtained. In the immediate neighbourhood of the system the electromagnetic field is identical as regards its electric part with the electrostatic field of a simple doublet of strength at any time given by $f(ct)$, while as regards its magnetic part it is identical with the field of the current produced in the changing of this doublet. It is only so far as the actual electrical motions in the vibrator described above approximate to this simple specification that the solution obtained will represent the field actually investigated by Hertz; in any case however it indicates the type of solution to be expected.

At a great distance from the origin it is the other terms of the solution which become appreciable and if we assume that the function f and its differential coefficients are of the ordinary type of regular function then we may put

$$\mathbf{E}_r = 0, \quad \mathbf{E}_\theta = \frac{f''(ct-r)\sin\theta}{r},$$

whilst

$$\mathbf{H}_\phi = \frac{f''(ct-r)\sin\theta}{r},$$

and this is of course probably more representative of the conditions in the actual case than the field near the oscillator is likely to be.

257. Although it is hardly representative of the conditions realised in actual practice we may for the present assume that the dissipation of the energy in the system is so slight that the oscillations are maintained for a considerable time, and as the motion is oscillatory we may represent it in such a case by taking

$$f(x) = A \sin px,$$

so that the surrounding field is determined by the force vectors in it which at the point (r, θ, ϕ) are given by

$$\mathbf{E}_r = \frac{2A \cos \theta}{r^3} \{pr \cos p (ct - r) + \sin p (ct - r)\},$$

$$\mathbf{E}_\theta = \frac{A \sin \theta}{r^3} \{pr \cos p (ct - r) + (1 - p^2 r^2) \sin p (ct - r)\},$$

$$\mathbf{H}_\phi = \frac{Ap \sin \theta}{r^2} \{- pr \sin p (ct - r) + \cos p (ct - r)\},$$

all the others being zero.

Thus the intensity of the field at any point is oscillating in full accord with the vibrator with the period $\frac{2\pi}{p}$. The actual conditions are best exhibited by plotting the lines of force in it. The equations to these lines are easily obtained, being in fact in any meridian plane the lines along which ψ is constant. Several of these curves have been drawn by Hertz and are depicted below; they correspond to the instants $t = 0, \frac{\tau}{4}, \frac{\tau}{2}, \frac{3\tau}{4}$ when τ is half a complete period. From these figures we see that the lines of force running between the opposed parts of the vibrator gradually expand in all directions. This expansion goes on continuously but the ends on the vibrator itself slowly close up and finally coalesce; the line then breaks itself away and travels out into space as a closed line of force. This breaking away of the field and its travelling to a distance is the essence of the radiation emitted by a vibrator of this type.

The distant field in this case is specified by

$$\mathbf{E}_\theta = -\frac{p^2 A \sin \theta \sin p (ct - r)}{r},$$

$$\mathbf{H}_\phi = -\frac{p^2 A \sin \theta \sin p (ct - r)}{r},$$

so that the surfaces over which the phase of the vibration is constant are the spheres

$$r = \text{const.}$$

To use the ordinary phraseology we may call these the wave front

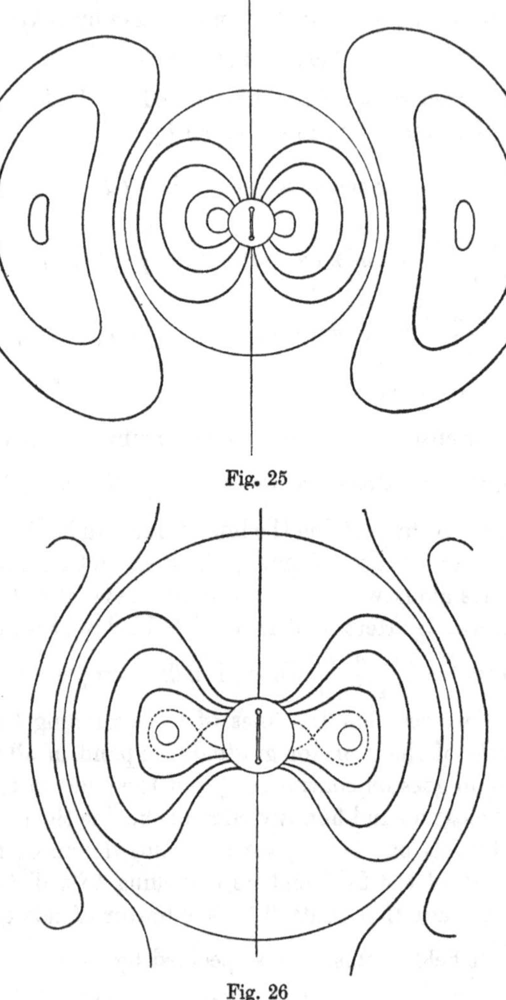

Fig. 25

Fig. 26

surfaces: each of them is advancing outwards with a velocity of linear expansion equal to c.

We see that in the present case also the directions of both the electric and magnetic vectors at any point in the field are tangential to the wave front, and they are in addition of equal magnitude and perpendicular to one another. The electromagnetic waves emitted

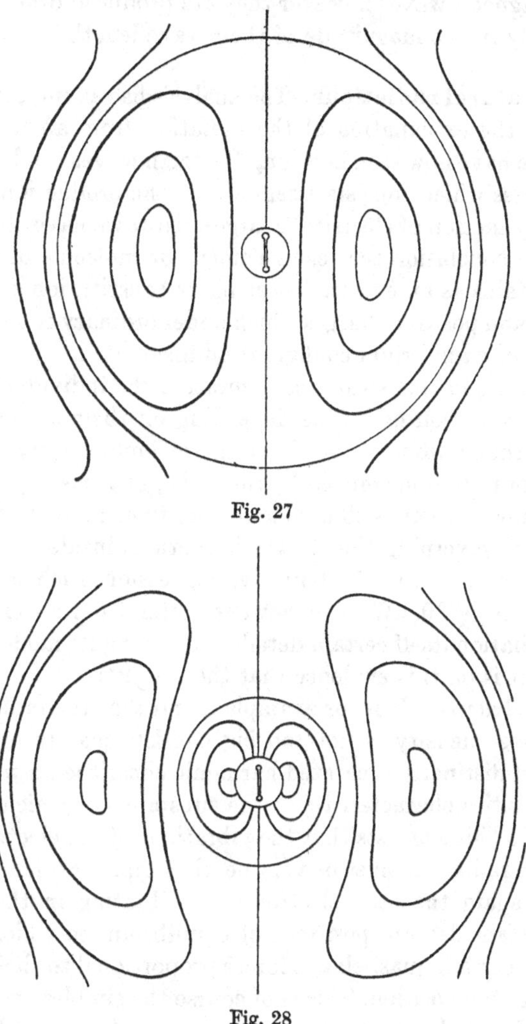

Fig. 27

Fig. 28

by a vibrator of this type are therefore transverse waves. We shall see presently that properties of the waves here illustrated by a special case are characteristic of the waves produced under all circumstances and, combined with the fact that the velocity of their propagation in a vacuum is c, a velocity identical in magnitude with the velocity of light, they point to the conclusion that electromagnetic waves however they are produced differ from light waves only in the magnitude of their wave length.

258. Natural radiations. The analysis has an important application to the explanation of the radiation from an incandescent body. We have now seen how long Hertzian waves can be produced by a process which consists essentially in the production of a rapid oscillatory motion of electrical charges. Now we have already been led to the conclusion that each element or molecule of a material body contains as an essential part in its constitution a number of electrons and positive charges which under ordinary conditions take up a sort of equilibrium configuration inside it: if we can produce a disturbance in this steady configuration the individual electrons will emit radiation of a type depending on their motion. Thus if we knew the motions of the electrons we could specify completely the type of radiation emitted by the body; but this is just what we do not know. We are still unable to specify completely the type of mechanism governing the electronic motions inside an atom, and we can therefore only offer tentative suggestions such as that given above on page 60. We can however infer from an examination of the radiation itself certain details concerning its mode of generation and it is on this evidence that the suggestive mechanisms are being constructed. It is for example found that the radiation from a gas whose density is not too big consists mainly of a limited number of distinctly separated harmonic constituents with periods and intensities characteristic of the substance of which the gas is composed. This suggests that the vibrations of the electrons giving rise to the radiation must be very nearly simple harmonic; and this suggests again that the electrons are vibrating in the molecule about certain definite positions of equilibrium to which they are bound by certain quasi-elastic forces proportional to their displacement from the position. This is of course the simplest possible idea and has already been introduced on a previous occasion, but further evidence seems to indicate the impossibility of its validity: it would for instance seem to imply that one electron cannot be responsible for more than three of the harmonic constituents of the radiation, and an almost incredibly large number of electrons would thus have to be assumed to exist in the molecules of certain substances. It appears however that no completely satisfactory explanation of these difficulties has yet been offered and we shall therefore content ourselves with this simple explanation.

In striking contrast with the radiation from a gas, the radiation from an incandescent solid or liquid presents as a general rule nothing of a periodic character, for it arises from the independent and irregular disturbances of countless molecules and electrons: it thus has the appearance of a formless mass of radiant disturbance advancing with the velocity of light: it is possible however even in these cases to have certain more or less predominant constituents of a definite period, but at high temperatures these are completely covered by the irregular radiation which is of a purely thermal character. It is a problem in these theories to determine how, if the thermal radiation of a substance is resolved by a prism into its harmonic spectrum, the energy of the total radiation is distributed among the harmonic constituents thus separated out: but no satisfactory solution is yet forthcoming and we can therefore give no more than this passing reference to it.

259. This discussion suggests that we have to deal in actual practice not with the single electrons but with whole groups of them more or less tightly bound to the elements of the ponderable matter or moving about freely in the interstices between these elements: and since the formulation of this more general problem brings out further aspects of the radiation from incandescent bodies, it seems desirable to give at least its barest outlines*. We first suppose that the motion of the electron under consideration is confined to a certain very small region v, one point of which is chosen for origin of coordinates. Referred to the axes of coordinates thus chosen let $\mathbf{r}_e$ be the position vector of the electron so that its velocity is $\dot{\mathbf{r}}_e$ and its acceleration $\ddot{\mathbf{r}}_e$. We shall regard all these quantities as so small that we may neglect any terms involving their squares and products. Next let $\mathbf{r}$ denote the coordinate vector of a point P at some distance from the origin of coordinates, outside the small surface considered, for which we want to determine the field. Now if Q is the effective position of the electron as

* Cf. Lorentz, *The Theory of Electrons*, p. 55.

regards the field at the point P at time t, the distance PQ will differ from r only by terms of the first order, and the effective time t_0 will differ from the time $t - \frac{r}{c}$ in the same way. The changes of position and motion of the electron in a very small time being infinitely small of the second order we may regard Q as the effective position at the instant $t - \frac{r}{c}$ and the velocity there as the velocity at this time. Moreover

$$\frac{1}{PQ} = \frac{1}{r} - ([\mathbf{r}_e]\,\nabla)\,\frac{1}{r},$$

because the difference between the distances PQ and PO is equal to the difference between the vectors $\mathbf{r}$ and $\mathbf{r}_e$ taken at P or Q. The square brackets of course serve to indicate the values of the quantities affected at the instant $t - \frac{r}{c}$. Thus if we use also

$$\frac{1}{1 - \frac{[\mathbf{v}_r]}{c}} = 1 + \frac{[\mathbf{v}_r]}{c},$$

we have

$$\phi = \frac{e}{4\pi}\left\{\frac{1}{r} - ([\mathbf{r}_e]\,\nabla)\,\frac{1}{r} + \frac{[\mathbf{v}_r]}{cr}\right\}.$$

Now as regards the last term in the expression we may write

$$\frac{[\mathbf{v}_r]}{c} = \frac{1}{cr}\,[(\mathbf{r}\,.\,\mathbf{v})]$$

$$= \frac{[(\mathbf{r}\,.\,\dot{\mathbf{r}}_e)]}{cr}$$

and this is

$$= -\operatorname{div}\,[\mathbf{r}_e],$$

since for example

$$\frac{\partial\,[x_e]}{\partial x} = \frac{\partial\,[x_e]}{\partial t_0}\cdot\frac{\partial t_0}{\partial r}\cdot\frac{\partial r}{\partial x}$$

$$= [\dot{x}_e]\left(-\frac{1}{c}\cdot\frac{x}{r}\right).$$

Thus we have finally for the scalar potential at the external point of the field

$$\phi = e\left\{\frac{1}{r} - \operatorname{div}\left(\frac{[\mathbf{r}_e]}{r}\right)\right\}.$$

The expression for the vector potential is similarly deduced and is

$$\mathbf{A} = \frac{e}{c}\left\{1 - (\nabla \mathbf{r}_e)\right\} \frac{[\mathbf{v}]^*}{r}.$$

The radiation field that predominates at large distances, and in which we find the flow of energy of which we have already spoken, is determined by the second term in ϕ and by the vector potential. At smaller distances it is superposed on the field represented by the first term of ϕ, which is the static potential of the electron at rest.

260. Now suppose that there are a number of electrons and elements of charge inside the small volume v under consideration. The field of each of these charge elements will then be exactly of the type thus specified and the total field of them all together will be simply obtained by superposition of their separate fields. Thus if Σ is used to denote a sum over all the elements of charge, we have in this total field

$$\phi = \frac{(\Sigma e)}{r} - \text{div}\left(\frac{\Sigma\,[e\mathbf{r}_1]}{r}\right)$$

and to the first order the vector potential is given by

$$\mathbf{A} = \frac{[\Sigma e\mathbf{v}]}{r} = \frac{[\Sigma e\dot{\mathbf{r}}_e]}{r}.$$

But if the volume element under consideration is very small and contains only the constitutional electrons and charges of the molecules of matter inside, so that the total charge is zero $\Sigma e = 0$, and then

$$\Sigma e\mathbf{r}_e$$

is the vector quantity which we have previously recognised as the polarisation of the volume element: we may denote this by

$$\mathbf{P}dv$$

and then we see that

$$\phi = -\,\text{div}\,\frac{[\mathbf{P}]}{r}\,dv,$$

$$\mathbf{A} = \frac{[\dot{\mathbf{P}}]}{cr}\,dv.$$

These relations would also hold in the case of a single uncharged molecule if the appropriate value of the vector **P** is implied. They

* In this formula the operator ∇ is presumed to affect all quantities following it.

show that the single molecule or small volume element of a material body will be a centre of radiation whenever the polarisation $\mathbf{P}$ is changing: the distant field will again be determined by

$$\mathbf{E}_\theta = \mathbf{H}_\phi = \frac{\ddot{\mathbf{P}} \sin \theta}{cr},$$

with the usual spherical polar frame of reference with the pole at the centre of the element and axis parallel to the direction of polarisation. Thus however rapidly the individual electrons may be rushing about inside the molecule or element of volume there will be no radiation if the acceleration of the polarisation or the vectorial sum $(\Sigma e\dot{\mathbf{v}})$ taken over them all is constantly zero*.

261. Before concluding this paragraph reference must be made again to the so-called Roentgen or X-rays. These rays were first observed by Roentgen in the neighbourhood of a discharge tube when the vivid green phosphorescence is exhibited on the walls of the tube, and they have been found also as an important constituent (the γ-rays) of the radiation emitted by radio-active substances. These rays exhibit a remarkable resemblance to light. Their rectilinear propagation, as evidenced by the sharp shadow thrown by bodies which intercept them, their power of affecting a photographic plate and their power of passing through solid bodies are obvious examples of this resemblance. But there are equally striking differences between Roentgen rays and rays of light. They are not refracted in their passage from one medium to another: they show some sort of reflexion, but the laws governing it are totally different from those of the corresponding phenomenon in light.

The generally accepted view of this radiation as originated from the discharge tube is that first proposed by Stokes: it is composed of thin spherical sheets of disturbance sent out into the aether by the sudden impacts of the rapidly moving electrons of the kathode stream against the walls of the tube: it may also in part be due to the shocks imparted to the molecules forming the walls of the tube. The similar radiation from radio-active substances would then have its origin in part in the sudden generation of the rapid motion in the electrons thrown off from those substances and again in part

* Cf. Larmor, *Aether and Matter*, p. 228.

in the readjustment of the remaining molecule to its new conditions after the electron has left.

In so far as these sheets of radiation are due to sudden but transient disturbance of the electrons in the molecules themselves, the magnetic force belonging to them alternates in direction in crossing each thin shell of pulse so that the average value taken across it is null. In so far as they are due to the sudden start or stoppage of the kathode particles or electrons, each of which is a single moving electron, this balanced alternation of magnetic force across the thickness of the sheet does not hold; the force may be in the same direction all the way across. As during the progress of the emission or impact the accelerations of the kathode particles arrested or emitted and of the disturbed electrons of the molecules will be presumably of the same order of magnitude, we would naturally conclude from the formula expressing the radiation in terms of the acceleration of the electron that these are both concerned in the emission of radiant energy, and the fact that the radiation is found to contain certain constituents characteristic of the substances on which the particles collide or from which they are thrown off supports this view.

In addition to the thin pulse arising from the sudden shock imparted to the molecules of the substance stopping or starting the electron, we would expect to find also more continuous radiation due to their state of vibration which would ensue: this would be represented by the phosphorescent light which accompanies the phenomenon.

As regards the X-radiation the present explanation has received remarkable confirmation in the last few years in a wonderful series of experiments originated by Laue and extensively developed by many workers, among the most prominent of whom are Prof. Bragg and his son*. In these experiments beams of carefully sifted homogeneous X-rays are reflected or refracted by crystalline media, and it is found that the regular arrangement of the molecules of such substances makes them behave towards the radiation more or less

* Cf. W. L. Bragg, *Proc. Camb. Phil. Soc.* 17 (1913), p. 43; *Proc. R. S. A.* 88 (1913), p. 428; M. Laue, *München. Ber.* (1912), p. 363; *Ann. d. Phys.* 41 (1913), p. 989; 42 (1913), p. 397; P. P. Ewald, *Phys. Zeitschr.* (1913), p. 465; L. S. Ornstein, *Amsterdam. Proc.* (1913); M. Born u. T. von Karman, *Phys. Zeitschr.* (1912), p. 297.

like a three-dimensional optical grating. The radiation passing across each molecule sets the electrons in that molecule in rapid vibration and these in their turn emit the secondary radiation which is observed as the transmitted or reflected beam, the regular arrangement of the molecules or vibrating centres giving rise to a measurable regularity in the radiation. It has thus been found possible not only to discover the arrangement of molecules in the crystals but also to establish definitely the periodic character of the X-radiation, even to the extent of obtaining an accurate estimate of its frequency. The wave length of the radiation is characteristic of the exciting substance but is much shorter than the radiation in the visible spectrum.

262. On the mechanism of the establishment of radiation fields*. We have so far confined our discussions only to the continuous propagation of electromagnetic disturbances in homogeneous fields of indefinite extent. It remains therefore to examine the mode of generation of such radiation fields. It is the essence of a propagation theory that the conditions of the field at any point P are affected by the conditions at any other point Q only after the time $\frac{PQ}{c}$ after the establishment of the conditions at Q. To illustrate the point in more detail let us consider the simple case of a Hertzian oscillator at the beginning of its oscillations. If we assume that the charge distribution on the oscillator before it collapses has been held there for an indefinite time previously, the field surrounding the oscillator will be identical with the simple electrostatic field appropriate to the charge distribution involved. Now suppose that at the time $t = 0$ the discharge takes place and the consequent series of oscillations started. The radiation field which now begins to be generated does not however instantaneously cover the whole field because the conditions at any point in the field at a distance r from the oscillator will remain unaffected by the changes produced by the discharge for the time $\frac{r}{c}$ after the instant $t = 0$ when the discharge takes place, in other words the conditions which existed there at the time $t = 0$ remain unaltered until the disturbance in the radiation field in the surrounding aether which travels outwards

* A. E. H. Love, *Proc. L. M. S.* [2], 1 (1903), p. 37.

in all directions with the velocity c has reached the point. This means that the new radiation field is at the instant t confined within the sphere $r = ct$ surrounding the oscillator: outside this sphere which is a wave surface for the advancing waves the old electrostatic field remains undisturbed, although of course the extent of the field covered by it is gradually diminishing.

The first question that naturally arises is as to the manner in which the two essentially different fields thus involved are connected across the advancing wave front: the radiation field has resulted mainly from a collapse of the initial electrostatic field and must therefore be connected with it in some way or other; by some boundary conditions applicable at the surface of the advancing wave front. These conditions were first directly formulated by Prof. Love* by applying the fundamental equations of the theory to small circuits at the wave front.

Let us assume quite generally that the wave front in the neighbourhood of the point under investigation is practically a plane surface advancing through the field with the velocity c_1 in a direction parallel to the axis of x in a conveniently chosen coordinate system. Draw a small rectangle parallel to the plane Oxz and through

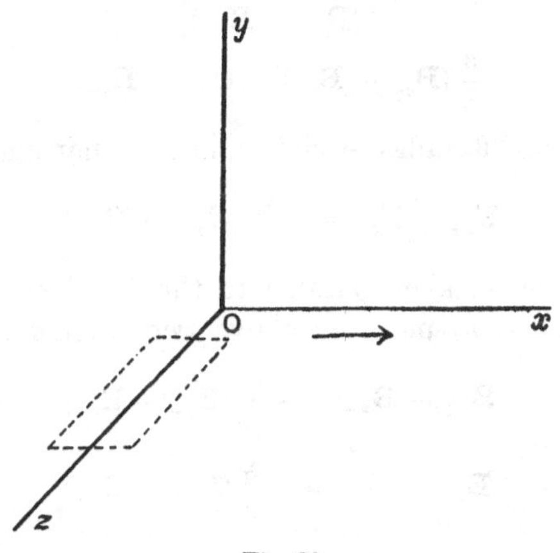

Fig. 29

* They were however previously known to Heaviside (*Electrical Papers*, II. p. 405) and Duhem (*Comptes Rendus*, t. 131 (1900), p. 1171) An elegant analytical proof is given by Bateman in *Electrical and Optical Wave Motion* (Cambridge, 1915), p. 20.

which the wave front cuts (dotted in the figure): the dimensions of this rectangle which are assumed to be small compared with the radius of curvature of the wave front surface and the wave length of the radiation are such that the sides parallel to Oy are of length l and those parallel to Ox of length l', which is extremely small compared with l and at the instant t is divided by the wave front into portions of lengths ξ and $l'-\xi$. On the positive side of the wave front the field is determined by the vectors $\mathbf{E}_+$, $\mathbf{B}_+$ and on the negative side by $\mathbf{E}_-$, $\mathbf{B}_-$. Now the displacement current through the small rectangle is the rate of change of the quantity

$$l\{\mathbf{E}_{y+}\xi+\mathbf{E}_{y-}(l'-\xi)\},$$

or since $\dot{\xi}=c_1$, this current is

$$l\{\dot{\mathbf{E}}_{y+}\xi+\dot{\mathbf{E}}_{y-}(l'-\xi)+c_1(\mathbf{E}_{y+}-\mathbf{E}_{y-})\},$$

which, on account of the extreme smallness of ξ and $l'-\xi$ compared with the wave length of the radiation, is practically equivalent to

$$lc_1(\mathbf{E}_{y+}-\mathbf{E}_{y-}).$$

But by Ampère's relation this will be proportional to the line integral of the magnetic force round the small circuit which on account of the smallness of l' is practically equal to

$$l(\mathbf{B}_{z+}-\mathbf{B}_{z-}),$$

and thus
$$\frac{c}{c_1}(\mathbf{B}_{z-}-\mathbf{B}_{z+})=\mathbf{E}_{y-}-\mathbf{E}_{y+}.$$

An application of Faraday's relation in a similar manner gives

$$\mathbf{E}_{z+}-\mathbf{E}_{z-}=-\frac{c_1}{c}(\mathbf{B}_{y+}-\mathbf{B}_{y-}).$$

If the rectangle is taken parallel to the (x, z) coordinate plane instead of the (x, y) plane as above two further conditions are obtained, viz.

$$\mathbf{B}_{y+}-\mathbf{B}_{y-}=-\frac{c_1}{c}(\mathbf{E}_{z+}-\mathbf{E}_{z-}),$$

$$\mathbf{E}_{y+}-\mathbf{E}_{y-}=\frac{c_1}{c}(\mathbf{B}_{z+}-\mathbf{B}_{z-}).$$

These four conditions break up into two pairs for they are equivalent to the condition that
$$c=c_1,$$

and also
$$(\mathbf{E}_y-\mathbf{B}_z)_+=(\mathbf{E}_y-\mathbf{B}_z)_-,$$
$$(\mathbf{E}_z+\mathbf{B}_y)_+=(\mathbf{E}_z+\mathbf{B}_y)_-.$$

The propagation of the wave front is thus verified to be with the velocity c, and the new field inside the front is connected with the old field outside by a boundary condition which expresses that the tangential component of the vector

$$\mathbf{F} = \mathbf{E} + \frac{1}{c}[\mathbf{cB}]$$

is continuous across the surface: **c** denotes a vector defining the direction and magnitude of the velocity of transmission of the conditions in the radiation field.

The condition expressed in this form which has been deduced on the assumption of an approximately plane wave front is not necessarily restricted by this assumption and it will apply to all sufficiently extended wave front surfaces of ordinary type without discontinuity.

263. The first case examined by Love is that of the oscillations on the perfectly conducting sphere (of radius a) of the charge distribution induced by a uniform field, when that field is suddenly removed. The initial state of the aethereal field outside the sphere is that expressed by the electrostatic vector components

$$(\mathbf{E}_r, \mathbf{E}_\theta, \mathbf{E}_\phi) = \left(\frac{2E\cos\theta}{r^3}, \frac{E\sin\theta}{r^3}, 0\right),$$

$$\mathbf{B} = 0.$$

At the instant $t = 0$ the cause which previously maintained the field thus expressed is supposed to cease to operate. It is required to determine the subsequent state of the field to agree with this initial field at time $t = 0$ and to be such that the tangential electromotive force is continuous at the wave front which at the time t will be the sphere $r = ct + a$ and the tangential electric force at the sphere is zero. The form of solution which suggests itself is naturally of the type previously obtained in which

$$\mathbf{E}_r = \frac{2\cos\theta}{r^3}\{f(ct - r + a) + rf'(ct - r + a)\},$$

$$\mathbf{E}_\theta = \frac{\sin\theta}{r^3}\{f(ct - r + a) + rf'(ct - r + a) + r^2 f''(ct - r + a)\},$$

$$\mathbf{B}_\phi = \frac{\sin\theta}{r^3}\{rf'(ct - r + a) + r^2 f''(ct - r + a)\}.$$

This solution for the field can only apply inside the sphere $r = ct + a$: outside it the old electrostatic conditions still prevail.

At the surface of the sphere the tangential electric force is zero always so that

$$f(ct) + af'(ct) + a^2 f''(ct) = 0,$$

which provides the differential equation for the arbitrary function f: it leads to the solution

$$f(x) = Ae^{-\frac{x}{2a}} \sin\left(\frac{\sqrt{3}x}{2a} + \alpha\right),$$

and this will apply for values of $x \equiv ct - r + a$ greater than zero. The conditions at the wave front imply that

$$\mathbf{E}_{r_1} = \mathbf{E}_{r_2},$$

$$\mathbf{E}_{\theta_1} - \frac{1}{c}\mathbf{B}_{\phi_1} = \mathbf{E}_{\theta_2},$$

giving on substitution

$$f(0) = E,$$

$$f'(0) = 0,$$

and thus we must have

$$A = \frac{E}{\sin\alpha},$$

where

$$\tan\alpha = \sqrt{3}, \quad \alpha = \frac{\pi}{3},$$

and the problem is completely determined. With this form of f it is easily verified that the expressions for the force vectors in the field can be put in the form

$$\mathbf{E}_r = \frac{4E}{\sqrt{3}} \frac{\cos\theta}{r^3} \sqrt{1 - \frac{r}{a} + \frac{r^2}{a^2}}\, e^{-\Theta} \sin(\sqrt{3}\Theta + \beta_r),$$

$$\mathbf{E}_\theta = \frac{2E}{\sqrt{3}} \frac{\sin\theta}{r^3} \left(1 - \frac{r}{a}\right) \sqrt{1 - \frac{r}{a} + \frac{r^2}{a^2}}\, e^{-\Theta} \sin(\sqrt{3}\Theta + \beta_\theta),$$

$$\mathbf{B}_\phi = -\frac{2E}{\sqrt{3}} \frac{\sin\theta}{ar^2} \sqrt{1 - \frac{r}{a} + \frac{r^2}{a^2}}\, e^{-\Theta} \sin(\sqrt{3}\Theta + \beta_\phi),$$

wherein

$$\Theta = \frac{ct - r + a}{2a},$$

$$\tan\beta_r = \frac{a\sqrt{3}}{a - 2r}, \quad \tan\beta_\theta = \frac{a + r}{a - r}\sqrt{3}, \quad \tan\beta_\phi = \frac{r\sqrt{3}}{2a - r},$$

results which differ from those given in a previous paragraph only by the phase of the motion.

It is important to notice that the field of the radiation is strongest close up near the wave front and just inside it; in fact the intensity of the field diminishes exponentially as the distance from the front is increased. In the next chapter we shall examine the energy in this field and will there show that by far the greatest proportion of the total energy radiated out is concentrated close behind the wave front.

It appears from this solution that the damped harmonic wave train can advance into a region in which the electric field is the statical one described above. It is also clear that it cannot advance into a region free from electric and magnetic forces.

264. Aided by the solution thus obtained for a simple mathematical case, Prof. Love* attempted to specify an appropriate solution for the more practical case, the Hertzian oscillator, taking into account the damping which is really existent. In this case also the original field is the electrostatic field of a doublet at the origin giving a field in which

$$\mathbf{E}_r = \frac{2E\cos\theta}{r^3}, \quad \mathbf{E}_\theta = \frac{E\sin\theta}{r^3}, \quad \mathbf{B}_\phi = 0,$$

the radiation field which originates on the collapse of the distribution giving this statical field (presumed to take place at the time $t = 0$) will be of the usual Hertzian type in which

$$\mathbf{E}_r = \frac{2\cos\theta}{r^3}(f + rf'),$$

$$\mathbf{E}_\theta = \frac{\sin\theta}{r^3}(f + rf' + r^2f''),$$

$$\mathbf{B}_\phi = \frac{\sin\theta}{r^3}(rf' + r^2f''),$$

a specification which will at the time t hold at all points inside the wave surface which to a first approximation may be treated as the sphere $r = ct$.

We then try a solution of the appropriate type in which

$$f(x) = Ae^{-\kappa x}\sin p(x + \alpha).$$

* *Proc. R. S.* 74 (1904), p. 73. I am greatly indebted to Prof. Love and the Royal Society for permission to reproduce some of the diagrams illustrating this paper: they are appended to this section.

We have to connect this with the external statical field by the boundary conditions deduced above which in this case are

$$\mathbf{E}_{r_1} = \mathbf{E}_{r_2}, \quad \mathbf{E}_{\theta_1} - \mathbf{B}_{\phi_1} = \mathbf{E}_{\theta_2},$$

and initially we must have $f = E$.

The boundary conditions give

$$f'(0) + \frac{1}{r} f(0) = \frac{E}{r},$$

so that $$f'(0) = 0, \quad f(0) = E:$$

this means that

$$E = A \sin pa, \quad 0 = p \cos pa - \kappa \sin pa,$$

whence $$\tan pa = \frac{p}{\kappa}.$$

Thus the arbitrary function starts in a definite phase which is equal to $\frac{\pi}{2}$ if there is no damping.

The field is thus in the general case defined by the vectors

$$\mathbf{E}_r = \frac{2 \cos \theta}{r^3} A e^{-\kappa(ct-r)} [(1 - \kappa r) \sin \Theta + pr \cos \Theta],$$

$$\mathbf{E}_\theta = \frac{\sin \theta}{r^3} A e^{-\kappa(ct-r)} [(1 - \kappa r + r^2 \overline{\kappa^2 - p^2}) \sin \Theta + pr (1 - 2\kappa r) \cos \Theta],$$

$$\mathbf{B}_\phi = \frac{\sin \theta}{r^2} A e^{-\kappa(ct-r)} [(\kappa - r \overline{\kappa^2 - p^2}) \sin \Theta - p (1 - 2\kappa r) \cos \Theta],$$

where we have used $\Theta = p(ct - r + a)$.

These formulae apply only for $r < ct$; for regions beyond the ordinary electrostatic field remains valid. The radial electric force is continuous at the front of the wave, i.e. at $r = ct$. The discontinuity of the transverse component of the electric force at the front of the wave is

$$\frac{\sin \theta}{r} \cdot A (\kappa^2 + p^2) \sin pa,$$

and this is equal as it should be to the magnetic force at the front of the wave.

The curves of electric force in this case are easily obtained for we know that

$$\frac{1}{c} \frac{d\mathbf{E}_r}{dt} = \frac{1}{r^2 \sin \theta} \frac{\partial \psi}{\partial \theta}, \quad \frac{1}{c} \frac{d\mathbf{E}_\theta}{dt} = -\frac{1}{r \sin \theta} \frac{\partial \psi}{\partial r},$$

where $$\psi = \frac{\mathbf{B}_\phi}{r \sin\theta}.$$

Thus if we write $$\psi = -\frac{1}{c}\frac{dQ}{dt},$$

then $$\mathbf{E}_r = -\frac{1}{r^2 \sin\theta}\frac{\partial Q}{\partial\theta}, \quad \mathbf{E}_\theta = \frac{1}{r\sin\theta}\frac{\partial Q}{\partial r},$$

and thus the curves of intersection of the surfaces

$$Q = \text{const.}$$

with planes through the axes of the doublet are the lines of electric force. Now it is easily verified that

$$Q = -\frac{\sin^2\theta}{r} A e^{-\kappa(ct-r)} [(1-\kappa r)\sin\Theta + pr\cos\Theta],$$

when $ct < r$ and $$Q = -\frac{\sin^2\theta}{r} A \sin p\alpha,$$

when $ct > r$.

Some of these curves have been drawn in a special case by Prof. Love and are depicted below. The case taken is that for which

$$\frac{\kappa}{p} = \cdot 25 \text{ approx.},$$

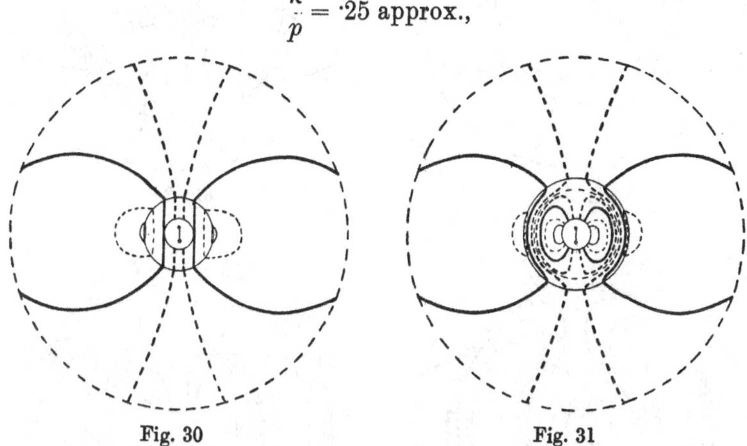

Fig. 30 Fig. 31

and if τ is the period the curves are plotted for the times

$$t = \cdot 26\tau,\ \cdot 385\tau,\ \cdot 51\tau,\ \cdot 635\tau,\ \cdot 76\tau,\ \cdot 885\tau,\ 1\cdot 01\tau, \text{ and } 1\cdot 135\tau,$$

after the initial instant of starting. In the figures the fine continuous circle represents the wave front at the time t. The discontinuity of the field there is shown by the change of direction

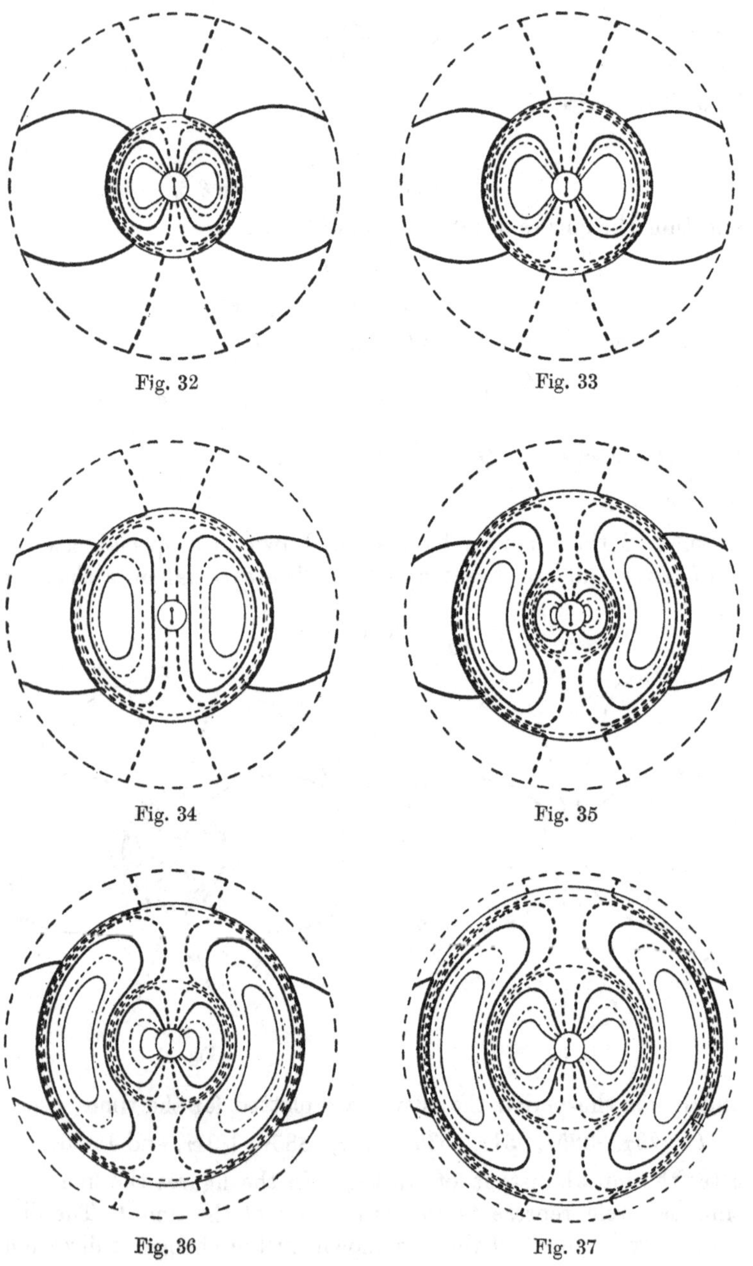

Fig. 32

Fig. 33

Fig. 34

Fig. 35

Fig. 36

Fig. 37

of the lines of force at this circle: the lines of force themselves are shown by the heavy dotted, heavy continuous, fine dotted and fine continuous lines respectively. The dotted circles that lie within the wave front are curves at which $Q = 0$ or the electric force has no radial component. It appears that no spherical surface of the set $Q = 0$ is the front of the advancing wave train but one of these surfaces tends to coincidence with this front as the wave train advances.

265. On the flux of energy in radiation fields. According to the usual conceptions of physical science, when energy travels by radiation the direction of the flux is along the ray, so that the flux vector gives not only the direction but also the intensity of the ray (the intensity of a ray being measured by the energy that passes along it per unit of time). In ordinary propagation in isotropic media the direction of the beam is perpendicular to the wave front, because the electric and magnetic vectors are both in this surface. The energy in this case travels along the beam normally to the wave surfaces. In crystalline media however it is the electric displacement vector that is in the wave front and the electric force is not coincident with the displacement so that the energy flux vector is no longer normal to the wave front. The direction of the ray, that is the path of the energy, is then oblique to the wave front surfaces, but in any case its direction at any point is the same as that of the energy flux vector at that point.

In entering into a more detailed analysis of these phenomena the first difficulty encountered is the ambiguity in the definition of the flux vector. The usual procedure is to base the whole discussion on Poynting's form of the theory, which appears to provide the simplest view of the phenomena, and to ignore the possibility of alternatives. We must not however forget that our view-point may be coloured by a long use of this particular form of the theory as the sole possibility so that its apparent suitability may be at least misleading. It is therefore essential that we bear in mind that Poynting's theory is not the only one which is consistent with the rest of the electromagnetic scheme and we shall therefore follow the usual discussion along the lines laid down by Poynting by a brief review of at least one simple alternative.

266. Let us first consider the circumstances involved in the propagation of a train of plane waves travelling parallel to the axis Ox in an absorbing medium, the waves being polarised so that the magnetic force is parallel to the axis Oz and the electric force parallel to the axis Oy in a system of rectangular coordinates. The equations of propagation are then, as before,

$$\frac{1}{c}\left(\sigma\mathbf{E}_y + \epsilon\frac{d\mathbf{E}_y}{dt}\right) = -\frac{\partial\mathbf{B}_z}{\partial x}, \quad -\frac{1}{c}\frac{d\mathbf{B}_z}{dt} = \frac{\partial\mathbf{E}_y}{\partial x}.$$

It follows that
$$\frac{\sigma}{c^2}\frac{d\mathbf{E}_y}{dt} + \frac{\epsilon}{c^2}\frac{d^2\mathbf{E}_y}{dt^2} = \frac{\partial^2\mathbf{E}_y}{\partial x^2},$$

which is the equation of propagation. Considering the case of radiation of period $\frac{2\pi}{p}$ so that

$$\mathbf{E}_y = E_0 e^{ipct-(a+ib)x},$$

this gives
$$\frac{ip\sigma}{c} - \epsilon p^2 = (a+ib)^2,$$

$$a + ib = +\,ip\sqrt{\epsilon}\left(1 - \frac{\sigma i}{pc}\right)^{\frac{1}{2}}.$$

On separating the real parts we have

$$\mathbf{E}_y = E_0 e^{-ax}\cos(pct - bx),$$

corresponding to

$$\mathbf{B}_z = \frac{E_0}{p}(a^2+b^2)^{\frac{1}{2}} e^{-ax}\sin(pct - bx + \theta),$$

where
$$\tan\theta = \frac{b}{a},$$

and
$$\mathbf{C}_y = \frac{E_0}{p}(a^2+b^2)^{\frac{1}{2}} e^{-ax}\sin(pct - bx + 2\theta).$$

Thus as we have already seen the magnetic flux is in a different phase from the electric force, involving a diminution in their vector product which determines the energy transmitted across any plane.

The energy per unit volume of the radiation at any part of the wave consists of an electric part $\frac{1}{2}\mathbf{E}_y\mathbf{D}_y$, or $\frac{1}{2}\epsilon\mathbf{E}_y^2$ and a magnetic part $\frac{1}{2}\mathbf{B}^2$: and the ratio of the time averages of these is exactly as before

$$\frac{\epsilon}{p^2(a^2+b^2)} \text{ or } \left(1 + \frac{\sigma^2}{p^2c^2\epsilon^2}\right)^{-\frac{1}{2}},$$

which is constant, but not unity except for transparent media.

The time rate of propagation of energy is, by Poynting's theorem,

$$\frac{dE}{dt} = \frac{[\mathbf{HE}]}{c} = \frac{E_0^2}{pc^2} e^{-2ax} (\tfrac{1}{2}b^2 + \text{periodic term}).$$

Across the plane $x = 0$ it is therefore on the average

$$\frac{E_0^2 b^2}{pc^2},$$

which corresponds to a density of energy equal to the mean square of electric force travelling with the speed $\frac{pc}{a}$ of the waves. This involves the result* that only the fraction

$$\frac{2}{\epsilon} \Big/ \left\{1 + \left(1 + \frac{\sigma^2}{p^2c^2\epsilon^2}\right)^{\frac{1}{2}}\right\}$$

of the total energy of the wave system can be considered as propagated; in the case of an undamped wave train this is only the purely aethereal part. The aethereal wave train, passing across the material medium, sets its molecules into sympathetic independent vibration: the energy of these vibrations constitutes a part of the total energy per unit volume, but that part is not propagated. This remark applies equally to all optical theories in which change of velocity of propagation is traced to the influence of sympathetic vibrations of the molecules, in fact it applies to all cases in which velocity depends upon the wave length.

267. We must however leave these considerations and return to the discussion of further aspects of the general flux of energy in radiation fields. We first consider the flux of energy in the field surrounding the ideally simple type of vibrator discussed in § 256. It was there shown that the field of a small vibrating electric doublet of moment $f(ct)$ at time t and placed at the origin and along the axis of a system of spherical polar coordinates reduces at a large distance from the vibrator to the simple radiation field in which the electric and magnetic forces are simply

$$\mathbf{E}_\theta = \mathbf{B}_\phi = \frac{\sin\theta f''(ct - r)}{r},$$

and the wave front surfaces are the spheres

$$r = \text{const.}$$

* Cf. Larmor, *Aether and Matter*, p. 135.

It follows therefore by direct application of Poynting's theorem that the flux of energy at the point (r, θ, ϕ) is radially outwards, i.e. in the direction of propagation, and of density per unit area

$$\frac{\sin^2\theta}{c}\frac{\{f''(ct-r)\}^2}{r^2}.$$

The total flux over the sphere of radius r is thus

$$\frac{\{f''(ct-r)\}^2}{c}\int_0^{2\pi}\int_0^{\pi}\sin^3\theta\,d\theta\,d\phi=\frac{8\pi}{3c}\{f''(ct-r)\}^2.$$

In the particular case when the vibrations are periodic so that we may take

$$f(ct-r)=A\sin p\,(ct-r),$$

it is on the time average $\quad -\dfrac{4\pi A^2p^4}{3c},$

or if λ is the wave length this is

$$\frac{64\pi^5A^2}{3c\lambda^4}.$$

This energy which is radiated outwards from the vibrator is of course lost to the system, and it must have been drawn from the store of the energy in the original vibrations. Thus unless the oscillations of such a system can be maintained by external agency, they will gradually decay owing to the dissipation of their energy by radiation. It is important to notice that the dissipation increases rapidly as the wave length is decreased and always represents an irreversible loss of energy to the vibrator*.

268. The whole of this discussion has been based on Poynting's theory of the processes involved. If we turn to the single alternative theory suggested in paragraph 229 where the radiation vector appears not as the vector product of the force vectors but as the product of the complete vector current by the scalar potential

$$\mathbf{S}=\phi\mathbf{C},$$

we shall find a remarkably different aspect of the whole of the processes.

We first examine the circumstances in a simple radiation field

* Cf. Lorentz, *The Theory of Electrons*, p. 140.

in which the propagation takes place by simple harmonic plane-polarised waves as in paragraph 251. In such a case

$$\mathbf{E}_y = E_0 e^{ipct-(a+ib)x}, \quad \mathbf{B}_z = \frac{E_0}{ip}(a+ib)\, e^{ipct-(a+ib)x},$$

so that the vector potential in the field is sufficiently defined by the relation

$$\mathbf{A}_y = ip\mathbf{E}_y$$

whilst the scalar potential is constant in both space and time.

If the constant scalar potential is zero then according to this new theory no transfer of energy is taking place in the field at all, for the vector $\phi\mathbf{C}$ vanishes everywhere. In any other case, when ϕ does not vanish, the transfer would merely take place parallel to the main component of the current, which is generally parallel to the electric force in isotropic media, and it would therefore be perpendicular to the direction of propagation of the radiation itself.

Of course in a theory where there is to be no transfer of the energy, the whole conception of the energy at a point must be different. That this is so in our present case is immediately obvious. According to the general discussion the appropriate formula for the kinetic energy density is

$$T = \frac{1}{c}\int^t \left(\mathbf{C}\frac{d\mathbf{A}}{dt}\right).$$

But in our case

$$\mathbf{C}_y = \epsilon\frac{d\mathbf{E}_y}{dt} + \sigma\mathbf{E}_y$$

$$= (ipc\epsilon + \sigma)\,\mathbf{E}_y.$$

Thus we have

$$T = (ipc\epsilon + \sigma)\,\mathbf{E}_y^2,$$

and the condition of propagation, viz.

$$(a+ib)^2 = -\epsilon p^2 + \frac{ip\sigma}{c},$$

shows that this is the same as

$$T = -\tfrac{1}{2}\mathbf{B}_z^2,$$

that is the kinetic energy now has the same value but the opposite sign to that usually employed in Poynting's theory, so that the total energy is on the modified theory simply the excess of the electric potential energy over the magnetic kinetic energy on the older interpretation. In the case of no absorption these are equal and the present theory does not associate energy at all with the radiation, so that no question of its transference arises. In the case of absorption it will be seen that the new theory identifies as the total energy in the field just that part of the energy which on Poynting's theory is not transferred.

269. Consider next the circumstances in the field surrounding a simple Hertzian vibrator. The complete circumstances in this field have been examined above and the force vectors are derived from scalar and vector potentials which are determined by

$$\mathbf{A}_r = \frac{\cos\theta}{r} f'(ct - r), \quad \mathbf{A}_\theta = \frac{\sin\theta}{r} f'(ct - r)$$

and

$$\phi = -\frac{\partial}{\partial r} \frac{\cos\theta\, f(ct - r)}{r}.$$

In this case the total current at the typical field-point has components

$$\mathbf{C}_r = \frac{c\cos\theta}{2\pi r^3}(rf'' + f'),$$

$$\mathbf{C}_\theta = \frac{c\sin\theta}{4\pi r^3}(r^2 f''' + rf'' + f'),$$

the aethereal constituent being the only one existing.

The radiation vector has therefore components

$$\mathbf{S}_r = \frac{c\cos^2\theta}{2\pi r^5}(f + rf')(rf'' + f'),$$

$$\mathbf{S}_\theta = \frac{c\sin\theta\cos\theta}{4\pi r^4}(r^2 f''' + rf'' + f')(f + rf').$$

The total outward flux over the sphere of radius r is therefore now equal to

$$\frac{2c}{3r^3}(r^2 f' f'' + rf'^2 + rff'' + ff')$$

and vanishes at a great distance. Thus whereas in Poynting's theory the transfer of energy in the distant field always exists and is directed outwards if the rate of change of the moment of the vibrator is accelerated, there is on the present theory no such thing as a radiation of energy away from the vibrator in the distant field: all that happens is a rearrangement taking place by flux in the spherical surfaces round the vibrator as centre and generally along the lines of electric force at all parts of the field.

Again of course the new interpretation places an entirely different value on the kinetic energy density; it is now equal to

$$\frac{1}{c}\int^t \left(\mathbf{C}\frac{d\mathbf{A}}{dt}\right) dt = c\int dt \left[\frac{2\cos^2\theta}{r^4}(rf''^2 + f'f'') + \frac{\sin^2\theta}{r^4}(r^2 f''f''' + rf''^2 + f'f'')\right]$$

which gives for the amount of energy between the spheres of radii r and $r + \delta r$ the total $T_r \delta r$, where

$$T_r = \frac{8\pi}{3r^2}\int^t (2rf''^2 + 2f'f'' + r^2 f''f''')\, dt$$

$$= \frac{4\pi}{3r^2} r^2 f''^2 + 2f'^2 + 4rc\int^t f''^2\, dt.$$

On Poynting's form of the theory this same quantity is

$$\frac{4\pi}{3r^2}(r^2 f''^2 + 2rf'f'' + f'^2).$$

The distribution of energy is now therefore essentially different. At a great distance from the vibrator, in the purely radiation portions of the field, both expressions agree in a form which corresponds to a value of T_r equal to

$$\tfrac{1}{3} f''^2.$$

Close to the radiator however the main part of the energy density on the new theory is just twice what it is on Poynting's theory. There is in addition also another part, more significant at a medium distance from the vibrator, that is in the transition field, which increases gradually with the time, even to the extent of becoming infinite if there is no damping. On such a theory, therefore, the process of increasing or reducing an electromagnetic one is an irreversible one, and in order to destroy a field by reducing the forces in it to zero, a certain amount of work is necessary, depending essentially on the process adopted to secure the vanishing of the field, and the equivalent energy remains stored in the space previously occupied by the field.

270. Thus whereas on Poynting's theory the energy supplied to the field at the vibrator is transferred outwards and radiated away, on the new form of the theory the energy, now however differently interpreted, is stored up in the field surrounding the vibrator and counted there in the kinetic energy. Modified forms of the theory can be given, which occupy a sort of intermediate position between these two extreme forms; some of them avoid the fundamental difficulty of the new form but then they involve radiation in a manner which is not essentially different to that in which it occurs in Poynting's theory. We need not however further extend these considerations: we have raised the point as to the possibility of the existence of perfectly consistent mathematical theories differing essentially from Poynting's merely to emphasise the one really uncertain aspect of electromagnetic theory, viz. that which concerns itself with the energy of the field. We have in reality not yet found an exclusive guiding principle to help us to a definite choice of functions to represent the kinetic and potential energies of the field in the general case, and until we can do this it is really of little use bothering about how the energy is transferred from one point of the field to another.

We know without ambiguity the difference of the energies $(T - V = L)$, the Lagrangian function, which is of necessity correct, as it leads to equations which have been proved by experience to represent the motions of the observable electrons. But beyond this the rest is pure conjecture, because we cannot possibly know to what extent the particular form of function we find has been modified by the ignoration of coordinates before we find it.

271. The mechanical pressure of radiation. An important application of our general dynamical principles is to the explanation of the pressure of radiation on absorbing and reflecting bodies.

Consider the case of a train of plane polarised waves advancing through an isotropic medium in the direction of the axis of x, so that all the quantities are functions of x only, the electric force being $(0, \mathbf{E}_y, 0)$ and the magnetic force $(0, 0, \mathbf{B}_z)$. We assume for simplicity that the permeability is unity. The equations of propagation

$$\frac{1}{c}\mathbf{C}_y = -\frac{d\mathbf{B}_z}{dx}, \quad \frac{d\mathbf{E}_y}{dx} = -\frac{1}{c}\frac{d\mathbf{B}_z}{dt},$$

wherein

$$\mathbf{C}_y = \sigma\mathbf{E}_y + \epsilon\frac{d\mathbf{E}_y}{dt},$$

are satisfied by

$$\mathbf{E}_y = A \,.\, e^{-ax}\cos(nt - bx),$$

$$\mathbf{B}_z = \frac{cA}{n}\sqrt{a^2 + b^2}\, e^{-ax}\cos(nt - bx + \theta),$$

wherein $$-\epsilon n^2 + in\sigma = c^2 (a + ib)^2,$$
or $$c^2 (a^2 - b^2) = -\epsilon n^2 \text{ and } n\sigma = 2c^2ab.$$

The mechanical forcive per unit volume is as before given by its single component*
$$\mathbf{F}_x = \frac{1}{c}\left(\mathbf{C}_y - \frac{1}{4\pi}\frac{d\mathbf{E}_y}{dt}\right)\mathbf{B}_z.$$

Now $$\int_{x_1}^{x_2} \mathbf{C}_y \mathbf{B}_z dx = \int_{x_1}^{x_2} - c\mathbf{B}_z \frac{d\mathbf{B}_z}{dx} dx = -\frac{c}{2}\Big|\mathbf{B}_z{}^2\Big|_{x_1}^{x_2},$$
so long as $\mathbf{B}_z$ is continuous between the limits of integration. Also
$$-\int_{x_1}^{x_2} \mathbf{B}_z \frac{d\mathbf{E}_y}{dt} dx = -\frac{c}{n^2}\int_{x_1}^{x_2} \frac{d\mathbf{E}_y}{dt} \cdot \frac{d^2\mathbf{E}_y}{dt\,dx} dx,$$
since for the harmonic oscillatory motion assumed
$$-n^2\mathbf{B}_z = \frac{d^2\mathbf{B}_z}{dt^2} = -c\frac{d^2\mathbf{E}_y}{dt\,dx},$$
and thus this integral is
$$-\frac{c}{2n^2}\Big|\left(\frac{d\mathbf{E}_y}{dt}\right)^2\Big|_{x_1}^{x_2},$$
provided $\frac{d\mathbf{E}_y}{dt}$ is continuous throughout the range of integration as is always the case, though ϵ may change gradually or abruptly. We have thus
$$\int_{x_1}^{x_2} \mathbf{F}_x dx = -\frac{1}{2}\Big|\mathbf{B}_z{}^2 + \frac{1}{n^2}\left(\frac{d\mathbf{E}_y}{dt}\right)^2\Big|_{x_1}^{x_2},$$
which gives the aggregate mechanical forcive on the stretch of the medium between x_1 and x_2 in the form of pressures on its ends. Thus for the simple forms of $\mathbf{E}_y$ and $\mathbf{B}_z$ assumed the time average of the pressure on either end is
$$\tfrac{1}{2}(A_e{}^2 + A_m{}^2),$$
A_e and A_m representing the amplitudes of the magnetic and electric vibrations. This is however the sum of the mean kinetic and potential energies per unit volume of the radiation, less that involved in the electric polarisation in the molecules; hence on any portion of the medium there is a mechanical force, directed along the waves equal per unit cross-section to the difference of these densities of energy at its ends.

* We have followed Larmor in deducing the expression for the forcive directly from first principles. The same result can however be readily obtained by an application of Maxwell's stress formulae, but the present procedure deduces it without resort to these formulae, the validity of which may be doubted.

In a transparent medium

$$\left(\frac{d\mathbf{E}_y}{dt}\right)^2 = \frac{c^2}{\epsilon}\left(\frac{d\mathbf{E}_y}{dx}\right)^2 = \frac{c^2}{\epsilon}\left(\frac{d\mathbf{B}_z}{dt}\right)^2,$$

so that the above internal pressure may be expressed in the form

$$\frac{1}{2}\left\{\mathbf{B}_z^{\,2} + \frac{1}{n^2\epsilon}\left(\frac{d\mathbf{B}_z}{dt}\right)^2\right\}.$$

If there is in the medium a directly incident wave whose vibration at the interface is $A_{m_i}\cos nt$ and also a reflected wave $A_{m_r}\cos(nt-\epsilon)$ and also a refracted wave, this result may be applied to a layer of the medium containing the interface; thus there will be a mechanical traction on the interface represented by a difference of pressure on its two sides, that on the incident side being

$$\frac{1}{2}\left[\{A_{m_i}\cos nt + A_{m_r}\cos\overline{nt-\epsilon}\}^2 + \frac{1}{\epsilon}\{A_{m_i}\sin nt + A_{m_r}\sin\overline{nt-\epsilon}\}^2\right].$$

In air or vacuum this is

$$\tfrac{1}{2}\left(A_{m_i}^{\,2} + A_{m_r}^{\,2} + 2A_{m_i}A_{m_r}\cos\epsilon\right),$$

or

$$\tfrac{1}{2}A_m^{\,2},$$

where A_m is the amplitude of the resultant magnetic vibration on that side.

When the radiation is directly incident on an opaque medium $\mathbf{B}_z$ and $\frac{d\mathbf{E}_y}{dt}$ are null in its interior; so that, when the surrounding medium is air or vacuum, its surface sustains in all a mechanical inward normal traction of intensity $\frac{1}{2}A_m^{\,2}$, that is, equal to the mean energy per unit volume of the radiation just outside it, in agreement with Maxwell's original statement*.

This is the pressure of radiation which has now been experimentally examined and the theoretical conclusions as to its intensity verified to within one per cent. by the measurements of Nicholls and Hull†. We can in a case of this kind thus say that radiation exerts a force just as if it carried momentum. This leaves open the question of the actual existence of the momentum spoken of.

* Treatise, II. § 792.

† *Phys. Rev.* 13 (1901), p. 293. Cf. also Lebedew, *Arch. des Sciences Phys. et Nat.* [4], 8 (1899), p. 184; Poynting, *Phil. Mag.* 9 (1905), pp. 169, 475.

272. To illustrate these matters* further and to bring out another aspect of the subject let us consider the opposite action, viz. the reaction or back pressure exerted by the radiation emitted by a perfectly black body into free space. Similar reasoning to the above will show that the back pressure is of a similar amount to that there calculated. To exhibit the argument in a simpler manner let us examine the component of the radiation emitted from a small patch of the surface as a plane beam travelling out into space. Surround the patch by a small closed surface cutting across the ray on the one side. Our previous general theory then shows that

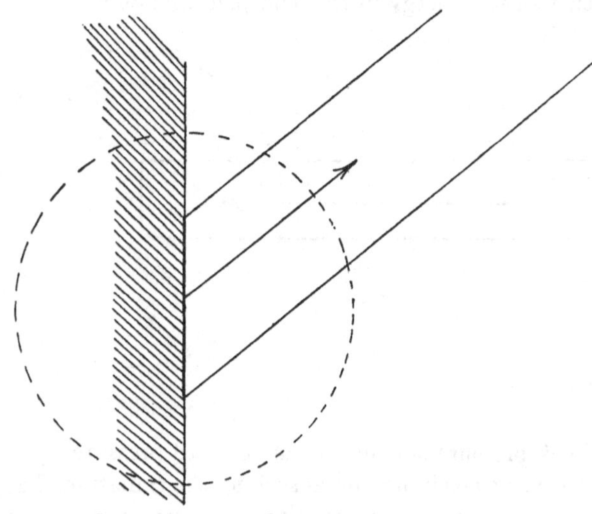

Fig. 38

the static resultant of the forces on everything inside this surface is represented by a stress system over the single patch of the surface where the ray cuts through it, this being the only part on the surface where the field in this ray is different from zero. If the ray cuts through this part of the surface normally it appears that the normal stress at that point is a normal pull of amount

$$\tfrac{1}{2}(\mathbf{E}^2 + \mathbf{B}^2)\, df_1,$$

* The treatment here followed is given by Larmor in his lectures: certain aspects of it are dealt with in his address on "The dynamics of radiation" to the Mathematical Congress in 1912. (Cf. *Proceedings*, 1, p. 206.)

and thus there is a total pull normally on the element of the emitting surface equal to

$$\frac{1}{2}(\mathbf{E}^2 + \mathbf{B}^2)\,\mathbf{n}_1 df_1,$$

$\mathbf{n}_1$ being the normal vector direction to the elementary patch of area df_1.

As it does not matter how big we draw this bounding sphere, this stress is the representative of a real force on the patch of the surface emitting the beam of radiation; and thus this beam exerts a back pressure on the body emitting it which is equal per unit area to the density of the energy in the field just outside it.

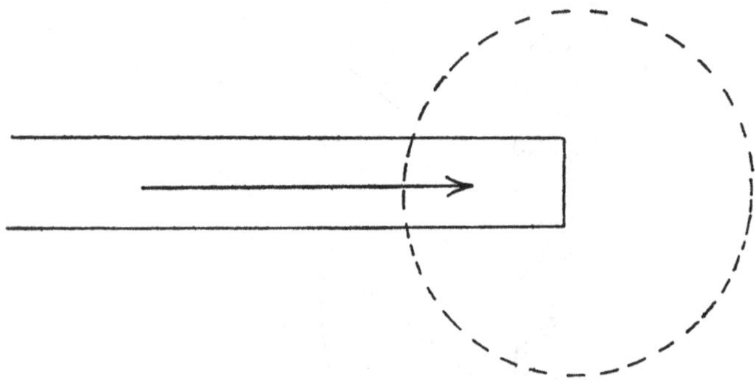

Fig. 39

This back pressure is equal to the forward pressure on the body at the other end receiving and absorbing the radiation. The action and reaction are equal and opposite at the two ends of the beam, so that on this view of things the ray behaves as if it were a carrier of momentum.

The question naturally arises as to what happens before the ray reaches the second body. Where is then the corresponding reaction to the back pressure on the body giving out the radiation? To answer this question completely we must return to our original scheme. According to the general theory the stresses acting over any geometrical surface drawn in the field which does not contain any matter are balanced by the kinetic reaction to the rate of change of the quasi-momentum in the aether inside the surface. Now consider a small parallel beam of light advancing into free space. The plane perpendicular front of the beam is advancing with the velocity c of radiation. Draw a surface in the field much as that shown in the figure. The stresses over the boundary of this surface are represented by the pressure of radiation over the small patch of the surface where the beam cuts through it; and this pressure must account for and just balance the rate of change of the quasi-momentum of the aether included inside the surface. Now where does this change of momentum come in? The propagation by waves is an alternating affair and so on the average the momentum in any part of the beam remains constant; the beam is however getting longer; a new region is being added in which there is momentum and so the change due to this added momentum per unit time must be equal to the pressure. The quasi-momentum per unit volume in the electromagnetic field in the general case is of amount

$$\frac{1}{c} \cdot [\mathbf{B} \cdot \mathbf{E}],$$

and is directed perpendicular to both vectors. In our case of plane propagation of wave motion in the aether this is

$$\frac{1}{c} \cdot \mathbf{B}_z \mathbf{E}_y,$$

and the general equations give

$$\mathbf{B}_z = \mathbf{E}_y,$$

so that the quasi-momentum per unit volume is

$$\frac{1}{c} \cdot \mathbf{B}_z^2,$$

or in the mean it is $$\frac{1}{2c} \mathbf{B}^2.$$

Thus on the average the momentum added in a time δt is

$$\frac{1}{2c} \mathbf{B}^2 c \delta t,$$

reckoned per unit area of cross-section in the beam. The rate of change of this is equal to $$\tfrac{1}{2}\mathbf{B}^2.$$

But the energy per unit volume in our wave is on the average equal to

$$\tfrac{1}{4}\mathbf{E}^2 + \tfrac{1}{4}\mathbf{B}^2 = \tfrac{1}{2}\mathbf{B}^2.$$

That is the average momentum per unit length in the beam is equal to the energy transmitted per unit time across any cross-section and this shows that it is balanced exactly by the pressure of radiation on the patch of the surface aforesaid.

The whole theory is thus consistent. It must however be noticed that the stresses and momentum involved in this discussion, about whose actual existence there may still be some doubts, are excessively small. The stress for example which gives rise to the phenomena of radiation pressure depends on the square of the vectors defining the field and is therefore nearly always smothered by the stress which propagates the wave and depends only on the first power of the vectors. A rough analogy is provided by the attraction of small objects by a vibrator like a tuning-fork in air. Very near such a vibrator in air the atmospheric pressure is less than that at a distance and so any object placed near the vibrator would have a greater pressure on its surface farthest from the vibrator and would therefore be impelled towards that body. But this resultant pressure depends on the square of the average pressure of the air whereas the sound propagation depends on the first order things. Thus in a body emitting light the reactions of the pressure of radiation would hardly ever be appreciable, being almost entirely swamped by the reaction to the setting up of the vibrations. This fact renders it almost impossible experimentally to test the existence of the 'momentum' force in free aether by testing for the reaction on a radiator, before the radiation from it has reached an absorber.

In these considerations the radiator has been considered as at rest, we must now calculate the effects due to motion.

273. In order to investigate whether the back pressure depends on the motion we proceed as before; and examine the reaction of an oblique beam emitted by a plane radiator travelling normally to itself with a velocity v. We consider any boundary drawn in the field as shown. The force acting on the path of the radiating surface which is inside this volume would then be balanced by the radiation pressure on the single patch of the geometrical surface were it not for the fact that the average 'momentum' of the field inside the surface is changing, owing to the fact that the beam is becoming shorter at a rate $v \cos \theta$, θ being the angle between the beam and normal to the surface. This rate of change of momentum together with the pressure along the ray from the radiator are balanced by the radiation pressure on the patch of the geometrical surface. Thus if E' is the energy density in the beam, the back push or radiation pressure is easily seen to be

$$p = \left(1 - \frac{v \cos \theta}{c}\right) E'.$$

But now we want E'. The question is whether the energy per unit volume in the radiation from a moving body is different from that of the same body at rest or does the nature of the radiation from a perfect radiator depend on its velocity?

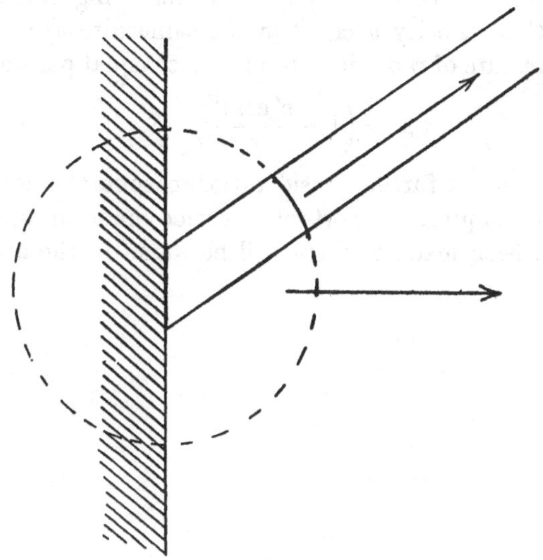

Fig. 40

A simple thermodynamic argument can be adopted to prove that the periods and amplitudes of the motions of the molecules or electrons in a body moving with uniform velocity do not depend on the velocity, so that the amplitude of the oscillation in the emitted wave train is the same but the wave length is necessarily altered by the motion according to the Doppler principle being shortened by the factor

$$\left(1 - \frac{v \cos \theta}{c}\right),$$

because if the radiator moves forward the waves are crowded up or shortened. The period of the wave is therefore shortened by a similar factor and thus the average energy per unit length is increased by the factor

$$\left(1-\frac{v\cos\theta}{c}\right)^{-2}.$$

Thus if we use E for the energy of the statical radiation we have

$$E'=\left(1-\frac{v\cos\theta}{c}\right)^{-2},$$

and thus

$$p=\left(1-\frac{v\cos\theta}{c}\right)^{-1}E.$$

Similarly for a perfect black body absorbing radiation and moving with a velocity $v'\cos\theta'$ in the same direction we should have the pressure of radiation on its surface equal per unit area to

$$p'=\left(1+\frac{v'\cos\theta'}{c}\right)^{-1}E.$$

We shall return to further considerations of some of these problems in the last chapter. Important thermodynamical applications which have been made of them will be found in the appropriate text-books.

CHAPTER VIII

THE ELECTRODYNAMICS OF MOVING MEDIA

274. The general equations of electrodynamic theory. We have up to the present confined our considerations mainly to the electromagnetic and electrodynamic phenomena of systems in which the ponderable matter is either actually at rest or is at least in a state of such slow motion that it may at any instant be regarded as at rest relative to the instantaneous electromagnetic field. We must now discuss certain aspects of the more general case of rapidly moving electromagnetic systems: such a discussion appears to be necessary not only because of its intrinsic theoretical interest but because all electrodynamic phenomena are concerned with more or less rapid motion of electrically charged bodies. Absolute rest is of course unknown by the human intelligence, and for instance all electrostatic fields created on this earth necessarily partake of the motion of the earth through space, so that they are of a type more general than that discussed in the earlier chapters of this book.

We shall begin by formulating the general equations of electrodynamic theory; these have already been set out in full on a previous occasion, but with a view to emphasising the point we may here briefly indicate their deduction on a purely dynamical basis.

It has been shown in the previous chapter that on the tentative assumption of appropriate forms for the potential and kinetic energies of an electrical system presumed to comprise merely a group of electrons or electrically charged particles in motion in the aether, the complete circumstances of the configuration and motion of the system can be described by means of the ordinary equations of dynamical theory: in such a mode of formulation of the theory the only effect of the interaction between the electromagnetic condition of the aether and the charge on the moving electron is completely specified as a force of ordinary mechanical type on the typical electron of vector amount

$$e\mathbf{F} = e\left(\mathbf{E} + \frac{1}{c}[\mathbf{vB}]\right),$$

where e is the charge on the electron and $\mathbf{v}$ its velocity; $\mathbf{E}$ is the aethereal electric force, defined in terms of the vector and scalar potentials by the relation

$$\mathbf{E} = -\frac{1}{c}\frac{d\mathbf{A}}{dt} - \nabla\phi,$$

$\mathbf{B}$ is the magnetic force vector and

$$\mathbf{B} = \text{curl}\,\mathbf{A}.$$

On such a theory the total effective current is

$$\mathbf{C} = \mathbf{C}_1 + \frac{d\mathbf{E}}{dt} + c\,\text{curl}\,\mathbf{I}_1 + \rho\mathbf{v} + \frac{d\mathbf{P}}{dt},$$

where $\mathbf{C}_1$ is the true current of conduction; $\frac{d\mathbf{E}}{dt}$ the fictitious current of aethereal displacement; $\frac{d\mathbf{P}}{dt}$ the true current of material polarisation; $c\,\text{curl}\,\mathbf{I}_1$ the current which in its magnetic aspects is the effective equivalent of the distribution of magnetic polarisation, including both the true magnetisation $\mathbf{I}$ and the quasi-magnetisation due to the convection of the polarised medium with velocity $\mathbf{v}$

$$\mathbf{I}_1 = \mathbf{I} + \frac{1}{c}[\mathbf{P}\mathbf{v}];$$

finally $\rho\mathbf{v}$ is the current due to the convection of the material medium charged to density ρ at any point.

275. All of these relations can be regarded either in the light of definitions or as relations of a purely dynamical nature. It follows from them that

$$\begin{aligned}\text{curl}\,\mathbf{F} &= \text{curl}\,\mathbf{E} + \frac{1}{c}\,\text{curl}\,[\mathbf{v}\mathbf{B}]\\ &= -\frac{1}{c}\frac{d}{dt}(\text{curl}\,\mathbf{A}) + \frac{1}{c}\,\text{curl}\,[\mathbf{v}\mathbf{B}]\\ &= -\frac{1}{c}\frac{d\mathbf{B}}{dt} + \frac{1}{c}\,\text{curl}\,[\mathbf{v}\mathbf{B}].\end{aligned}$$

Now the right-hand side of this equation, when multiplied by $-c$ and integrated as regards its normal component over any surface, expresses the time rate of change of the magnetic induction through the surface regarded as moving at each point with the charge system with velocity $\mathbf{v}$. Our equation is thus the complete analytical expression of Faraday's circuital relation which states that the line integral of the *electromotive force* $\mathbf{F}$ round any circuit which is *carried along with the matter* is equal to the time rate of diminution of the magnetic induction through it multiplied by $1/c$.

We have also, of course,

$$\text{curl}\,\mathbf{B} = +\frac{\mathbf{C}}{c},$$

or if it is preferred not to include the magnetism as molecular current whirls so that the total current is only

$$\mathbf{C} - c\mathbf{I}_1,$$

this relation becomes

$$\text{curl}\,\mathbf{H} = \text{curl}\,(\mathbf{B} - \mathbf{I}_1)$$

$$= \frac{1}{c}\,(\text{total current}),$$

where the magnetic induction vector **H** is defined by the relation

$$\mathbf{H} = \mathbf{B} - \mathbf{I}_1.$$

This is the expression of Ampère's circuital relation that the line integral of the magnetic force round any circuit, fixed or moving*, is at each instant equal to the flow of the Maxwellian total current through it multiplied by the factor $1/c$. It is important to notice that as the magnetic induction is here introduced into the theory it is a subsidiary quantity defined in terms of **B** and the magnetisation.

When the material medium, however heterogeneous, is at rest in the aether, these electrodynamic equations reduce precisely to Maxwell's original scheme

$$\text{curl}\,\mathbf{E} = -\frac{1}{c}\frac{d\mathbf{B}}{dt},$$

$$\text{curl}\,\mathbf{H} = \frac{1}{c}\,\mathbf{C},$$

with

$$\mathbf{C} = \mathbf{C}_1 + \frac{d\mathbf{D}}{dt},$$

$\mathbf{C}_1$ being the true conduction current.

When the material medium is in motion these equations are modified in the following respects; there is a term arising from convection of electric polarisation added to the magnetism, which changes $\mathbf{I}$ to $\mathbf{I}_1$, and there is the current arising from the convection

* Time differentials are not involved.

of electric charge which supplies the term $\rho\mathbf{v}$, a term which Maxwell in some connections temporarily overlooked but which has been fully restored by Fitzgerald and others.

276. The existence of a magnetic field due to the convection of electrically charged bodies and of polarised dielectrics has been experimentally verified by Roentgen* and Rowland†; doubts were subsequently thrown on the interpretation of their results by Cremieu‡ but the experiments have been repeated with much greater precision by Eichenwald§, with results which completely substantiate the verification of the theoretical predictions.

The arrangement finally adopted by Eichenwald consisted mainly of a parallel circular plate condenser with a uniform dielectric slab. The rapid motion was produced by rotating the whole condenser round an axis of symmetry perpendicular to the plates. If the charge on the plates of the condenser is of density σ at any point, there will be a convection current due to its being dragged on with the system which will be of amount

$$\sigma\mathbf{v},$$

$\mathbf{v}$ denoting the velocity of the system at that point.

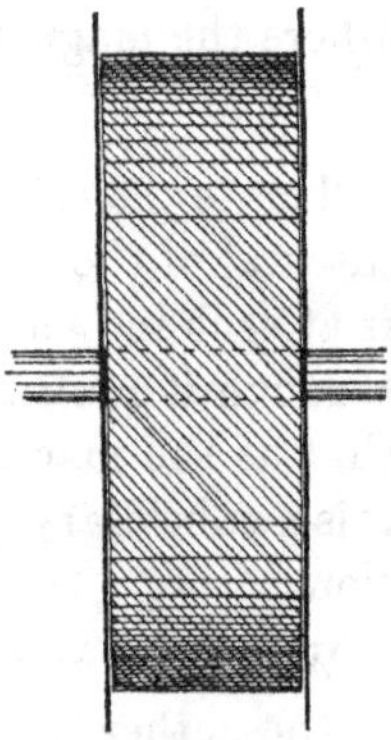

Fig. 41.

In addition the dielectric medium will be polarised to intensity $\mathbf{P}$ and the convection of this will also be equivalent to a current of intensity curl $[\mathbf{vP}]$. If we neglect the irregularity of the edges this field between the plates will be uniform right across and thus $\mathbf{P}$ will be constant throughout the interior of the slab and there will be no volume distribution of current of this latter type; but it exists as a surface distribution on the abrupt interfaces of the dielectric where the density is

$$[\mathbf{n}_1 \cdot [\mathbf{Pv}]],$$

if $\mathbf{n}_1$ is the unit normal vector whose direction is in the positive

* *Berlin. Ber.* (1885), p. 198.

† *American Jour. of S.* (3) 15 (1878), p. 30.

‡ *Paris C. R.* 130 (1900), p. 1544, 131 (1900), pp. 575, 797.

§ *Ann. d. Phys.* 11 (1904), p. 421.

direction of **P**, i.e. straight across between the plates in the present instance. This current thus appears as of magnitude $|\mathbf{P}.\mathbf{v}|$ and is directed parallel to the direction of **v** at each place, but in the opposite sense.

The total effective current in this arrangement is a surface current on the plates of the condenser and of density

$$(\sigma - \mathbf{P})\,\mathbf{v}.$$

But if **D** is the total electric displacement across the dielectric

$$\sigma = \mathbf{D} = \mathbf{E} + \mathbf{P},$$

and thus the surface current density is simply $|\mathbf{E}\mathbf{v}|$ in the direction of **v**, i.e. directed in circles round the axes of rotation.

The important point to notice is that this current and therefore also the magnetic field associated with it does not in any way depend on the dielectric material, but only on the potential difference between the plates: and this was exactly verified by Eichenwald.

The importance of this experiment is the confirmation which it provides for the fundamental hypothesis on which the present theory is based. The modern theory of electromagnetism is built on the idea of an aether permeated by a large number of electrons or electric point charges, either free or grouped together in material atoms, and it is with these charges and their general configuration and motion that we are alone concerned. The motion of a material medium is thus effectively accounted for in the motion of its constituent electrons. But what about the aether? Can this medium move also; and is it dragged along with the matter which is in motion through it? We have in our discussions tacitly neglected the possibility of any such motion of the fundamental medium and this course appears not only the simplest one but it is found to be more consistent with experimental facts. This is the original view of Fresnel, Lorentz and Larmor; but the opposite view has been strongly advocated by Stokes in optical theory and Hertz in electrical theory. According to their views the motion of a piece of matter through the aether necessarily produces by a sort of mechanical dragging action a convective motion of the aether itself in the neighbourhood of the piece of moving matter. That such a view is inconsistent with the result of Eichenwald's experiment is however easily seen, for according to it no distinction need be made between the separate parts of the total displacement current, and the whole effect summed up in the term **D** is presumed to be convected with the matter: the Roentgen current would—in such a theory—have, as is easily seen, a density

$$\text{curl}\,[\mathbf{Dv}],$$

and this adopted into the theory of Eichenwald's experiment would lead to the result that there should be no resultant current at all, the current due to the convection of the polarised dielectric and aether just balancing that due to the convection of the charges on the plates.

The experiment carried out by H. A. Wilson and described above* also in some respects affords another result in favour of the theory of a stationary aether. It was there shown that the effects of the rotation of a dielectric substance in a magnetic field can be fully and accurately explained on the assumption that it is merely the electrons in the dielectric atoms that are convected with that substance, the aether itself remaining absolutely at rest.

Thus it cannot but be admitted that the course adopted in the above exposition of the theory is at least perfectly consistent with our experience. We shall however refer to this point later and mention further and perhaps more exact evidence in its favour, and also some difficulties in the way of its acceptance. We may perhaps here mention a direct attempt made by Lodge to detect an aethereal drag accompanying the mass of a very large rapidly rotating flywheel, but with negative results.

277. The steady linear translation of an electrostatic system†. The general equations of the previous paragraph enable us to treat in detail the electrodynamic relations of an electrical system in steady uniform motion through the aether. In order that a steady electric state may be possible without permanent currents of conduction, it is necessary that the configuration of the matter shall be permanent and that its motion shall be the same at all times

* P. 250.

§ Cf. Larmor, *Phil. Trans.* A, 190 (1897), p. 226. The theory is due originally to J. J. Thomson, *Phil. Mag.* (5), 11 (1881), p. 229; *Phil. Mag.* (5), 28 (1889), p. 1; *Phil. Mag.* (5), 31 (1891), p. 149; *Recent Researches*, p. 16. Cf. also Heaviside, *Phil. Mag.* (5), 27 (1889), p. 324; G. F. C. Searle, *Phil. Trans.* 187, A (1896), p. 675; *Phil. Mag.* (5), 44 (1897), p. 329.

relative to this configuration and to the aether, and also to the extraneous magnetic field, if there is one: this confines it to uniform spiral motion on a definite axis fixed in the aether. We shall here confine our attention to the case when the motion is one of uniform translation and in which there is no extraneous field, electric or magnetic. Under these circumstances the magnetic induction through any circuit moving with the system being constant, the electromotive force **F** is derived from a potential Φ

$$\mathbf{F} = -\operatorname{grad}\Phi,$$

because its line integral round such a circuit vanishes. Inside a conductor the electromotive force **F** must vanish, otherwise electric separation would be going on; therefore Φ must be a constant over and inside any conductor in the system.

Φ is called after Searle the *convection potential* of the field of the moving system*.

If we refer the field to axes fixed in and moving with the material system and use **v** as the vector velocity of the system, then the total current density at any point in the system is

$$\mathbf{C} = \frac{\partial \mathbf{E}}{\partial t} + \rho\mathbf{v},$$

ρ being the density of the charge distribution at the point; it is presumed that the system consists entirely of conductors and free aether, no dielectrics being present. But on account of the steadiness of the motion

$$\frac{\partial \mathbf{E}}{\partial t} + (\mathbf{v}\nabla)\,\mathbf{E} = 0,$$

so that the current density may be written in the form

$$\mathbf{C} = -(\mathbf{v}\nabla)\,\mathbf{E} + \rho\mathbf{v},$$

or since

$$\rho = \operatorname{div}\mathbf{E},$$

$$\mathbf{C} = -\{(\mathbf{v}\nabla)\,\mathbf{E} - \mathbf{v}\operatorname{div}\mathbf{E}\}$$

$$= -\operatorname{curl}[\mathbf{E}\,.\,\mathbf{v}],$$

the velocity **v** being uniform throughout the system. We have therefore from Ampère's relation

$$\operatorname{curl}\mathbf{B} = \frac{1}{c}\mathbf{C} = -\frac{1}{c}\operatorname{curl}[\mathbf{E}\,.\,\mathbf{v}],$$

* Schwarzschild calls it the *electrokinetic potential.* Cf. *Gött Nachr.* (*math. phys. Kl.*) (1903), p. 125.

so that $$\text{curl}\left\{\mathbf{B} + \frac{1}{c}[\mathbf{E}\,.\,\mathbf{v}]\right\} = 0,$$

which implies that the vector

$$\mathbf{B} - \frac{1}{c}[\mathbf{v}\,.\,\mathbf{E}]$$

is the gradient of a potential function: we write

$$\mathbf{B} - \frac{1}{c}[\mathbf{vE}] = -\text{ grad }\psi,$$

ψ is an undetermined function which will be continuous as to itself and its gradient except at the surfaces of transition. The most general value of **B** consistent with the circuital relation is thus

$$\mathbf{B} = \frac{1}{c}[\mathbf{vE}] + \text{grad }\psi,$$

the part of it depending on ψ would include the extraneous magnetic field, if there were one, and also the field due to magnets, if any, that belong to the material system itself.

If there is no external applied magnetic field and the moving system itself contains no magnetic matter, the magnetic field of the moving charges will be sufficiently defined by the magnetic vector potential **A** so that

$$\mathbf{B} = \frac{1}{c}[\mathbf{vE}],$$

the function ψ being not now necessary, there being no external circumstances to be allowed for.

Combining the relation between **E** and **B** with the direct dynamical relation

$$\mathbf{F} = \mathbf{E} + \frac{1}{c}[\mathbf{vB}] = \mathbf{E} + \frac{1}{c}[\mathbf{v}\,.\,\mathbf{B}],$$

we get $$\mathbf{F} = \mathbf{E} - \frac{1}{c}[\mathbf{v}\,.\,\nabla]\,\psi + \frac{1}{c^2}[[\mathbf{E}\,.\,\mathbf{v}]\,.\,\mathbf{v}]$$

$$= \mathbf{E}\left(1 - \frac{\mathbf{v}^2}{c^2}\right) + \frac{\mathbf{v}}{c^2}(\mathbf{F}\,.\,\mathbf{v}) - \frac{1}{c}[\mathbf{v}\nabla]\,\psi,$$

wherein as above $$\mathbf{F} = -\text{ grad }\Phi,$$

and $$(\mathbf{vF}) = (\mathbf{vE}).$$

Again since the total current is always a stream we have

$$\text{div }\mathbf{E} = \rho,$$

so that $$\operatorname{div}\left(\mathbf{F}-\frac{\mathbf{v}}{c^2}(\mathbf{vF})\right)=\left(1-\frac{\mathbf{v}^2}{c^2}\right)\operatorname{div}\mathbf{E},$$

or $$\nabla^2\Phi=\frac{1}{c^2}(\mathbf{v}\nabla)^2\Phi-\rho\left(1-\frac{\mathbf{v}^2}{c^2}\right),$$

where now ψ has disappeared. This is the characteristic equation from which the single independent variable Φ of the problem is to be determined, subject to the condition that it is to be constant over each conductor.

For the interior of a conductor Φ is constant and the electromotive force $\mathbf{F}$ vanishes; but the aethereal displacement $\frac{1}{4\pi}\mathbf{E}$ does not vanish in the conductors, being now given by

$$\left(1-\frac{\mathbf{v}^2}{c^2}\right)\mathbf{E}=\frac{1}{c}[\mathbf{v}\nabla]\psi,$$

which makes it circuital so that there is no volume distribution of electrification.

278. In an investigation in detail of the field produced by the motion, it will conduce to brevity if we take $\mathbf{v}$ to be a velocity parallel to one of the axes of coordinates, say the x-axis. We shall also use the notation

$$\beta\equiv\frac{|\mathbf{v}|}{c},\quad \kappa^2=\left(1-\frac{|\mathbf{v}|^2}{c^2}\right)=1-\beta^2.$$

The characteristic equation for the convection potential Φ is then

$$\nabla^2\Phi=\beta^2\frac{\partial^2\Phi}{\partial x^2}-\kappa^2,$$

and this has to be solved subject to the condition that Φ is constant over each conductor of the system: as the change in the form of the equation arising from the motion depends on β^2, the differences thereby introduced will all be of the second order of small quantities.

We can restore the above characteristic equation for Φ, the potential of the electromotive force, to an isotropic form by a geometrical strain of the system and the surrounding space represented by

$$(x_0,y_0,z_0)=(\kappa^{-1}x,y,z),$$

where of course $\kappa^2=1-\beta^2$.*

Now let us compare our moving system, which we may generally describe as S, with the correlative system S_0 obtained by this

* This transformation was suggested by Thomson and Heaviside.

transformation and supposed at rest: we shall assume that the density ρ_0 of the charge distribution in S_0 is reduced from the corresponding value in S in the ratio $\kappa : 1$ so that corresponding elements of volume contain the same total charges: then if ϕ_0 is the electrostatic potential of these charges on S_0,

$$\nabla_0{}^2\phi_0 = -\rho_0$$

is the characteristic equation satisfied by ϕ_0 in this system. The general type of solution of this equation is, as we had it before,

$$\phi_0 = \frac{1}{4\pi}\int \frac{\rho_0\, dv_0}{r_0},$$

the integral being taken over the entire field and r_0 denoting the distance of the element dv_0 from the point in the field at which the function is calculated.

Now the potential Φ in the moving system satisfies the equation

$$\nabla_0{}^2\Phi = -\kappa^2\rho = -\kappa\rho_0,$$

so that

$$\Phi = \kappa\phi_0 = \frac{\kappa}{4\pi}\int \frac{\rho_0\, dv_0}{r_0},$$

and since

$$\rho\, dv = \rho_0\, dv_0$$

and

$$r_0{}^2 = \frac{1}{\kappa^2}(x - x_P)^2 + (y - y_P)^2 + (z - z_P)^2,$$

(x_P, y_P, z_P) denoting the coordinates of the point at which Φ is calculated, and (x, y, z) the coordinates of the position of dv, we may write

$$\Phi = \frac{\kappa}{4\pi}\int \frac{\rho\, dv}{r_0},$$

and this is the general type of solution for the convection potential in any moving system of the type under consideration.

Again, comparing the components of the electrostatic force

$$\mathbf{E}_0 = -\nabla\phi_0$$

in S_0 with the corresponding components of the electromotive force

$$\mathbf{F} = -\nabla\Phi$$

in S, we see that

$$\mathbf{F}_x = -\frac{\partial\Phi}{\partial x} = -\frac{\partial\phi_0}{\partial x_0} = \mathbf{E}_{0x},$$

$$(\mathbf{F}_y, \mathbf{F}_z) = -\left(\frac{\partial}{\partial y}, \frac{\partial}{\partial z}\right)\Phi = -\kappa\left(\frac{\partial}{\partial y_0}, \frac{\partial}{\partial z_0}\right)\phi_0$$

$$= \kappa\,[\mathbf{E}_{0y}, \mathbf{E}_{0z}].$$

Thus the forces on corresponding elements of charge in the two systems are equal as regards their components in the direction of motion, but the components in any direction at right angles to this direction are smaller in the moving system in the ratio $\kappa : 1$.

Thus if we have solved the electrostatic problem for any system S_0 at rest, i.e. if we have determined the equilibrium distribution of electricity on the conductors in the system under the influence of the rigid charge distribution, then we can immediately deduce the solution for the equilibrium distribution of electricity on the conductors in a uniformly moving system S obtained from S_0 by a uniform contraction in the direction of the motion in the ratio $\kappa : 1$.

Suppose, for example, that the system S_0 is represented by a uniform distribution of electricity of total amount q_0 throughout the thin homoeoidal ellipsoidal shell between two similar and similarly situated concentric ellipsoids and that there are no other bodies in the system. If a_{0_1}, a_{0_2}, a_{0_3} are the axes of the mean ellipsoid on which this shell lies, then we know that the appropriate form of the electrostatic potential ϕ_0 has the constant value

$$\frac{q_0}{8\pi}\int_0^\infty \frac{dt}{\sqrt{(a_{0_1}{}^2+t)\,(a_{0_2}{}^2+t)\,(a_{0_3}{}^2+t)}}$$

throughout the interior of the ellipsoid, whilst at external points the value is

$$\frac{q_0}{8\pi}\int_\lambda^t \frac{dt}{\sqrt{(a_{0_1}{}^2+t)\,(a_{0_2}{}^2+t)\,(a_{0_3}{}^2+t)}},$$

where q_0 is the total charge on the ellipsoid and in the last expression λ is the positive root of the cubic equation

$$\frac{x^2}{a_{0_1}{}^2+t}+\frac{y^2}{a_{0_2}{}^2+t}+\frac{z^2}{a_{0_2}{}^2+t}=1.$$

Moreover we have seen also that, since the potential ϕ_0 is constant throughout the interior of the ellipsoid, the distribution of charge thus specified is identical in the limit with the surface distribution of charge of the same total amount on the same ellipsoid when composed of conducting material.

Now by uniformly contracting this ellipsoid and its space in the ratio κ parallel to any definite line we obtain another ellipsoid with semi-axes (a_1, a_2, a_3). If this new ellipsoid is moved parallel to the chosen line with the velocity appropriate to the ratio κ, the original statical system and its field will exactly correspond in the manner just defined to the system it defines; on it the convection potential will therefore take the constant value

$$\Phi = \frac{\kappa q}{8\pi}\int_0^\infty \frac{dt}{\sqrt{(a_{0_1}{}^2 + t)(a_{0_2}{}^2 + t)(a_{0_3}{}^2 + t)}},$$

whilst at external points its value is

$$\frac{\kappa q}{8\pi}\int_\lambda^\infty \frac{dt}{\sqrt{(a_{0_1}{}^2 + t)(a_{0_2}{}^2 + t)(a_{0_3}{}^2 + t)}},$$

λ being the positive root of the equation

$$\frac{x_0{}^2}{a_{0_1}{}^2 + t} + \frac{y_0{}^2}{a_{0_2}{}^2 + t} + \frac{z_0{}^2}{a_{0_3}{}^2 + t} = 1.$$

The equilibrium distribution of electricity on a moving conductor is characterised by the fact that the electromotive force in its interior vanishes, i.e. the convection potential Φ is constant there. Thus the distribution on the moving ellipsoid obtained by contraction of the static distribution on the conducting ellipsoid in S_0 is identical with the distribution which would hold if the moving ellipsoid were conducting. But when it is remembered that the electric distribution in S_0 is the limit of a uniform distribution between two concentric, similar and similarly situated ellipsoids and that in the process of uniform contraction these ellipsoids remain concentric similar and similarly situated, it follows that the new distribution of charge on the moving ellipsoid (a_1, a_2, a_3) would be exactly the same as if it were in equilibrium. Thus the distribution of charge on a conducting ellipsoid is not disturbed by imparting a uniform translatory motion to it*.

Two particular cases of this general theorem have assumed special importance on account of the applications which have been made

* Mr H. S. Jones has suggested to me a modification of this proof. If it is assumed that the distribution on the conducting ellipsoid in motion which gives zero force inside it is the equilibrium one, we can argue exactly as in the statical case that the surface density varies as the central perpendicular on the tangent plane at the point, since we have shown that the electric force due to any moving point charge is radial and, *for any given direction*, varies inversely as the radius squared. Thus since $\sigma \propto p$ and the total charge is unaltered, the distribution must remain unaffected by the motion.

of them to illustrate the properties of a moving electron, which is nothing more nor less than a charged particle.

279. In the first case the conductor in motion is assumed to be spherical in form, say of radius a*. The conductor in the correlative static system will then be a prolate spheroid with axes $\left(\frac{a}{\kappa}, a, a\right)$, if the motion is along the direction of the x-axis. The appropriate form of the convection potential can then be written in the form

$$\Phi = \frac{\kappa q}{8\pi}\int_{\lambda}^{\infty}\frac{dt}{(a^2+t)\sqrt{\frac{a^2}{\kappa^2}+t}},$$

where λ is the positive root of the quadratic

$$\frac{x^2}{\kappa^2\left(\frac{a^2}{\kappa^2}+t\right)} + \frac{y^2+z^2}{a^2+t} = 1,$$

and reduces to the constant value

$$\Phi = \frac{\kappa q}{8\pi}\int_{0}^{\infty}\frac{dt}{(a^2+t)\sqrt{\frac{a^2}{\kappa^2}+t}}$$

on the surface of the sphere.

The integrals in these cases can be directly evaluated by the substitution

$$\tau^2 = \frac{a^2}{\kappa^2} + t,$$

so that it becomes

$$\Phi = \frac{\kappa q}{4\pi}\int_{\lambda}^{\infty}\frac{d\tau}{\tau^2 - a^2\frac{1-\kappa^2}{\kappa^2}}$$

$$= \frac{\kappa^2 q}{8\pi a\sqrt{1-\kappa^2}}\log\frac{\sqrt{a^2+\lambda\kappa^2}+a\sqrt{1-\kappa^2}}{\sqrt{a^2+\lambda\kappa^2}-a\sqrt{1-\kappa^2}},$$

which reduces to the value

$$\Phi = \frac{\kappa^2 q}{8\pi a\sqrt{1-\kappa^2}}\log\frac{1+\sqrt{1-\kappa^2}}{1-\sqrt{1-\kappa^2}}$$

$$= \frac{q}{8\pi a}\cdot\frac{1-\beta^2}{\beta}\log\frac{1+\beta}{1-\beta},$$

* M. Abraham, *Ann. d. Phys.* (IV.) X. p. 105 (1903)

on the surface of the sphere. The full details of the field can now be determined: the charge on the sphere is uniformly distributed over the surface.

280. In the second case* the moving surface is an oblate spheroid of axes

$$(a\kappa, a, a),$$

the first being in the direction of motion. The surface in the correlative static system which has axes

$$\left(\frac{a\kappa}{\kappa}, a, a\right)$$

is therefore a sphere of radius a. In this case the electrostatic potential ϕ_0 is

$$\phi_0 = \frac{q}{4\pi r} = \frac{q}{4\pi\sqrt{x^2 + y^2 + z^2}},$$

so that the convection potential Φ is

$$\Phi = \frac{\kappa^2 q}{4\pi\sqrt{x^2 + \kappa^2 (y^2 + z^2)}},$$

which educes to the constant value

$$\Phi = \frac{\kappa q}{4\pi a},$$

on the surface of the moving conductor of which the equation is

$$x^2 + \kappa^2 (y^2 + z^2) = a^2\kappa^2.$$

The field in the moving system is of a much simpler character than that of the previous example and we may therefore examine it in more detail with a view to illustrating some of the general features of these convection fields. The electromotive force $\mathbf{F}$ at any point in the field has components

$$\mathbf{F}_x, \mathbf{F}_y, \mathbf{F}_z = -\left(\frac{\partial}{\partial x}, \frac{\partial}{\partial y}, \frac{\partial}{\partial z}\right)\Phi$$

$$= \frac{\kappa^2 q/4\pi}{\{x^2 + \kappa^2 (y^2 + z^2)\}^{\frac{3}{2}}}\{x^2, \kappa^2 y, \kappa^2 z\},$$

the electric force at the same point, which in the most general case of translation along the axis Ox has components

$$(\mathbf{E}_x, \mathbf{E}_y, \mathbf{E}_z) = \left(\mathbf{F}_x, \frac{\mathbf{F}_y}{\kappa^2}, \frac{\mathbf{F}_z}{\kappa^2}\right),$$

* Cf. Lorentz, *The Theory of Electrons*, p. 210. This is also the solution for a point charge.

is therefore given by

$$\frac{\kappa^2 q/4\pi}{\{x^2 + \kappa^2 (y^2 + z^2)\}^{\frac{3}{2}}} (x, y, z).$$

The magnetic force in the field, given generally by the relation

$$\mathbf{B} = \frac{1}{c} [\mathbf{v} \,.\, \mathbf{E}],$$

has therefore the components

$$(\mathbf{B}_x, \mathbf{B}_y, \mathbf{B}_z) = \frac{\beta q \kappa^2/4\pi}{\{x^2 + \kappa^2 (y^2 + z^2)\}^{\frac{3}{2}}} (0, -z, y).$$

The formulae thus obtained for the electric and magnetic force vectors in the field are considerably simplified by a simple transformation in which account is taken of the fact that the conditions

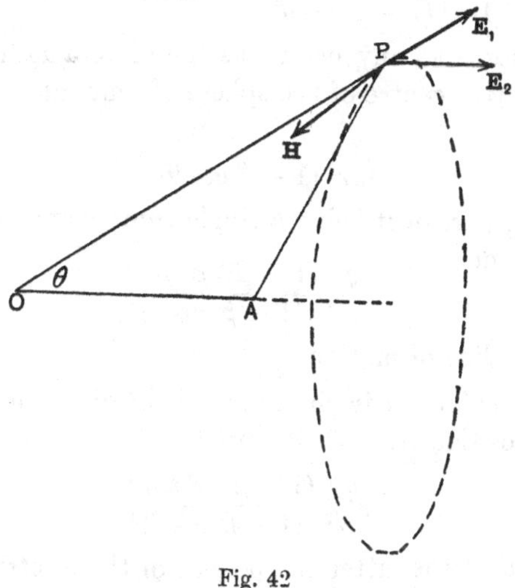

Fig. 42

in the field at any point have really originated from the motion of the sphere not at the instant at which they are examined but at a previous effective instant and that they have been propagated out with the velocity c from the position of the sphere at this instant. We now choose as the origin of a spherical polar coordinate system the centre O (see figure) of the sphere in its effective position for the special point P of the field at which the conditions are investigated: the polar axis is taken along the direction of the motion:

then if $OP = r$ and if A is the actual position of the centre of the sphere at the instant t (and this is the origin of the rectangular coordinate system), then

$$OA = \beta r,$$

and thus

$$x^2 = (r\cos\theta - \beta r)^2,$$

$$y^2 + z^2 = r^2 \sin^2\theta.$$

Thus

$$x^2 + \kappa^2(y^2 + z^2) = r^2 - 2\beta r \cos\theta + \beta^2 r^2 \cos^2\theta$$
$$= r^2(1 - \beta\cos\theta)^2,$$

and thus the electric force has the components

$$(\mathbf{E}_x, \mathbf{E}_y, \mathbf{E}_z) = \frac{\kappa^2 q/4\pi}{r^2(1-\beta\cos\theta)^3}\{\cos\theta - \beta, \sin\theta\cos\phi, \sin\theta\sin\phi\},$$

whilst the components of the magnetic force are

$$\frac{-\kappa^2\beta q \sin\theta}{4\pi r^2(1-\beta\cos\theta)^3}(0, \sin\phi, -\cos\phi).$$

The electric force at any point has therefore a radial component from the effective centre of the sphere of amount

$$\frac{q}{4\pi r^2}\frac{1-\beta^2}{(1-\beta\cos\theta)^3},$$

the remaining part of it being a single component in the meridian plane of amount

$$\frac{q}{4\pi r^2}\frac{(1-\beta^2)\beta\sin\theta}{(1-\beta\cos\theta)^3}$$

parallel to the line of motion.

The magnetic force is in circles round the direction of the motion and at any position (r, θ) its intensity is

$$\frac{q}{4\pi r^2}\frac{(1-\beta^2)\beta\sin\theta}{(1-\beta\cos\theta)^3},$$

which is equal to the latter component of the electric force*.

281. The dynamics of moving electrified systems. In the previous paragraph we have discussed the character of the field in the neighbourhood of an electrical system moving with uniform velocity along a straight line. It is of course assumed that the system thus discussed has been in motion in the manner specified for a sufficiently long time previous to the instant at which it is

* These formulae are due to Heaviside, *Phil. Mag.* 27 (1889), p. 332; *Electrician*, Dec. 7, 1888, p. 148.

examined: the conditions implied in this restriction will be fairly obvious when reference is made to the discussions of chapter VII where the mode of establishment of such steady fields is reviewed in detail. It was there seen that in the process of setting up the uniform motion, say from rest, a shell of disturbance is sent out into the surrounding electrostatic field: this shell travels out and away from the system with the velocity c of radiation leaving behind it the new steady field associated with the charges in motion and in which the field vectors are expressed by functions of position decreasing rapidly (like $1/r^2$) as the point is taken more and more distant from the charges themselves. Thus if the shell of disturbance has got to such a distance from the system that the field vectors in the uniform field are negligibly small we may regard the effective conditions for the uniform motion as practically established, and we have then no further concern with the expanding radiation field in which the energy has attained a constant value.

The practical aspect of the results thus obtained lies, of course, in their application to the explanation of the observed mechanical relations of such moving systems of charges and it is with these explanations that we shall now concern ourselves. The considerations of the present paragraph will be confined mainly to the translatory motion of such rigid electrical systems as have been examined in the previous paragraph.

282. In chapter VI we showed quite generally that the resultant force of electrodynamic origin on any material system contained within any closed surface f can be separated into two parts, the first of which can be represented simply as a stress across the surface f which per unit area is of intensity proportional to the square of the field vectors, whilst the second turns out to be expressible as the result of a distribution of bodily forces throughout the interior field of intensity (vectorial) per unit volume

$$-\frac{1}{c}\frac{d}{dt}[\mathbf{EB}].$$

This conclusion takes a remarkable form if the system under consideration is of finite dimensions and if the surface f is then extended indefinitely. For then the field vectors at points of the surface are so small that the first part of the total forcive just specified may be neglected. We conclude then that the total force on the whole system in this case can be represented as a wrench which when reduced to the origin of the rectangular coordinates as base point is specified by the linear and angular components **F** and **G** respectively, where

$$\mathbf{F} = -\frac{1}{c}\int \frac{d}{dt}[\mathbf{EB}]\,dv$$

$$= -\frac{d\mathbf{M}_e}{dt},$$

and

$$\mathbf{G} = -\frac{1}{c}\int\left[\mathbf{r}\,.\,\frac{d}{dt}[\mathbf{EB}]\right]dv$$

$$= -\frac{d\mathbf{N}_e}{dt} + [\mathbf{M}_e\,.\,\mathbf{v}],$$

wherein

$$\mathbf{M}_e = \frac{1}{c}\int [\mathbf{EB}]\,dv$$

and

$$\mathbf{N}_e = \frac{1}{c}\int [\mathbf{r}\,.\,[\mathbf{E}\,.\,\mathbf{B}]]\,dv,$$

all integrals being extended throughout the entire field: **r** denotes the vector coordinate of position of a point whose components are (x, y, z).

The forcive exerted by and through the aether on any electrical system without induced or intrinsic magnetisation is therefore equal and opposite to the change per unit time of the quantity which we have tentatively defined above as the electromagnetic momentum in the aethereal field of the system.

If the motion of the charges in the system is one of uniform translation in a fixed direction the field will, as we have just seen, be carried on with the system in a steady configuration. The total electromagnetic momentum in it will therefore be constant. The linear component of the reacting electromagnetic force in the system would then be zero but there will be a couple unless the field is symmetrical about the direction of motion.

In the more general case however whenever the motion of the system is accelerated there will arise a reaction on account of its charges specified as a wrench with linear and angular components

$$-\frac{d\mathbf{M}_e}{dt}, \quad -\frac{d\mathbf{N}_e}{dt} + [\mathbf{M}\,.\,\mathbf{v}].$$

There will of course in the general case be a reaction force to the acceleration of the motion of the system on account of its material momentum and this will be a wrench with linear and angular components

$$-\frac{d\mathbf{M}_0}{dt}, \; -\frac{d\mathbf{N}_0}{dt} + [\mathbf{N}_0\mathbf{v}].$$

Thus to maintain the accelerated motion external action of some kind is necessary; if the external force system reduces to the same base as a wrench with components

$$\mathbf{F}_0, \; \mathbf{G}_0,$$

then we shall have

$$\mathbf{F}_0 + \mathbf{F} = \frac{d\mathbf{M}_0}{dt}, \qquad \mathbf{F}_0 = \frac{d\,(\mathbf{M}_0 + \mathbf{M}_e)}{dt},$$

$$\mathbf{G}_0 + \mathbf{G} = \frac{d\mathbf{N}_0}{dt} + [\mathbf{v} \,.\, \mathbf{N}_0], \quad \mathbf{G}_0 = \frac{d}{dt}(\mathbf{N}_0 + \mathbf{N}_e) + [\mathbf{v} \,.\, \mathbf{N}_0 + \mathbf{N}_e].$$

From this point of view it would appear that the idea of an electromagnetic momentum is just as legitimate a conception as that of ordinary material momentum; in any case, of course, the conception, legitimate or not, provides a convenient mode of expressing a definite fact of theory.

283. In the practical application of these principles it is first necessary to determine the vectors $\mathbf{M}_e$ and $\mathbf{N}_e$ for the system under consideration, and this requires a knowledge of the complete electromagnetic field of the moving charges. Now we have already examined the details of the mode of generation and propagation of such electromagnetic fields: we saw that the conditions at any point of the field at a definite instant were made up by superposition of disturbances from all the separate electrons in the system, emitted at the appropriate effective previous time and place for each of them and transmitted thence with the velocity of radiation. The definition of the field at any instant in the most general case is therefore tremendously complicated and it is only in a very few restricted cases that it has been accomplished. But for the majority of the applications we shall have to make of these principles, such generality of procedure, however desirable it might be from the theoretical point of view, is not at all necessary. In fact it will be seen by reference back to the analysis of the previous chapter that the effective conditions in the field for the determination of the force on the system of electrodynamic origin are in reality merely the conditions which hold in the immediate neighbourhood of the charges in the system. Thus if the system of charges is of finite extent and the changes in its motion are not too rapid the conditions in the field throughout the extent of the field covering the charge distribution will at any instant be practically the same as those which exist in a system moving uniformly with the instantaneous configuration and velocity of the given system. Now the new conditions in the field are smoothed out with the velocity c of radiation so that the condition here implied is that the relative configuration and motion of the system must not change appreciably in the time taken by radiation to cross the system, so that the effective field is at each instant smoothed out to the type appropriate to the motions of the charges before these are appreciably altered. We shall of course in our calculations assume that the uniform field extends to an indefinite distance beyond the system so that we can avail ourselves of the separation of the forcive mentioned above. This procedure is legitimate because, since the force is in reality defined by the conditions of the field in the immediate neighbourhood of the system, the type of field to which this local field continues is irrelevant and may for the purposes of the detailed calculations be taken to be the simple uniform field of the instantaneous motion.

This is the essence of an equilibrium theory and it restricts our analysis to application only in the cases of so-called quasi-stationary motion. The importance in a dynamical theory of the restriction thus implied lies in the fact that the effective conditions in the field are then determined solely by the instantaneous configuration and motion of the charge system giving rise to it, so that instead of having to include in the analysis, in addition to the finite number of the coordinates of the charge system, an infinite number of coordinates to specify the conditions in the aethereal medium surrounding it, we have only to reckon with the former by themselves.

There is a mechanical analogy in the theory of the motion of a solid body through an ordinary elastic medium such as the air. If the motion of the solid is slow enough the conditions of the elastic medium adjust themselves at each instant to the state of motion then existing, and an equilibrium theory applies just as if the medium were incompressible. But in the other extreme case of rapid motion with large accelerations the whole of the circumstances in the surrounding medium are complicated by the continual generation of waves of compression starting out from the solid. In this case the resistance to the motion of the solid is practically all due to the elastic resistance of the surrounding medium to the setting up of waves in it.

284. Thus in the calculation of the quantities involved we may assume that at each instant the field is of the steady type appropriate to that instantaneous motion and position investigated in the previous paragraph; but in this case the motive forces of electromagnetic origin on the elements of charge in the system have a potential Φ, just as the ordinary static forces in electrostatics. We may therefore conclude as in the electrostatic theory that the mechanical forces of electrodynamic origin acting on the material bodies of the system carrying charges have a potential which in the general case is expressed by the integral

$$W = \tfrac{1}{2}\int \Phi\rho\, dv,$$

taken throughout the field, but which in simpler cases will reduce to the forms already discussed at length in the electrostatic theory. From another aspect this quantity may be regarded as potential energy stored up in the system; but it is to be noticed that it also includes a part, due to the magnetic field, which we have previously classed as kinetic energy and which therefore when reckoned as potential energy must have its sign changed: it is easy to verify this in the particular case of a charged conductor moving uniformly in a straight line such as discussed in the previous paragraph: for in that case the magnetic or kinetic energy of the system is

$$T = \tfrac{1}{2}\int \mathbf{B}^2 dv$$

and

$$\mathbf{B} = \frac{1}{c}[\mathbf{vE}],$$

so that if the motion is along the x-axis

$$2T = \beta^2 \int (\mathbf{E}_y{}^2 + \mathbf{E}_z{}^2)\, dv;$$

whilst the ordinary potential or electric energy of elastic strain in the aether is given by

$$2U = \int (\mathbf{E}_x^2 + \mathbf{E}_y^2 + \mathbf{E}_z^2)\, dv,$$

so that
$$W = U - T = \int \tfrac{1}{2} (\mathbf{E}_x^2 + \kappa^2\mathbf{E}_y^2 + \kappa^2\mathbf{E}_z^2)\, dv$$
$$= \int \frac{1}{2}\left(\mathbf{F}_x^2 + \frac{\mathbf{F}_y^2 + \mathbf{F}_z^2}{\kappa^2}\right) dv$$
$$= \frac{1}{2\kappa^2}\int (\kappa^2\mathbf{F}_x^2 + \mathbf{F}_y^2 + \mathbf{F}_z^2)\, dv$$
$$= \frac{1}{2\kappa^2}\int \left\{\kappa^2\left(\frac{\partial\Phi}{\partial x}\right)^2 + \left(\frac{\partial\Phi}{\partial y}\right)^2 + \left(\frac{\partial\Phi}{\partial z}\right)^2\right\} dv,$$

and this is easily transformed in the usual manner by Green's theorem and gives

$$W = U - T = -\frac{1}{2\kappa^2}\int \Phi\left(\kappa^2\frac{\partial^2\Phi}{\partial x^2} + \frac{\partial^2\Phi}{\partial y^2} + \frac{\partial^2\Phi}{\partial z^2}\right) dv,$$
$$W = +\tfrac{1}{2}\int \Phi\rho\, dv$$

as above*.

It must of course be remembered that this is not necessarily all the energy that is in the field, but it represents the variable part in the present case. The motion itself when under examination is of course of the quasi-stationary type so that the radiation field and the energies associated with it are negligible: but the motion must have been started by a non-stationary impulse in some remote past time and in this process a circular shell of disturbance of the wave motion type is sent out into the field and the energy in this part which however tends to a constant value is not necessarily negligible. But the more distant the time of generation, the farther is this wave shell away from the system and the nearer does the field inside it approximate to the actual field belonging to the quasi-stationary motion which the analysis maps out.

285. The above deduction of the form of W shows that the function
$$L = -W$$
also serves as a sort of Lagrangian function. The fact that the force

* Cf. Schwarzschild, "Zwei Formen des Prinzips der kleinsten Aktion in der Elektronentheorie," *Gött. Nachr.* (*math. phys. Kl.*), 1903, p. 125.

function and Lagrangian function with the sign changed agree in the cases of quasi-stationary motion is not necessarily confined to the present problem: in fact the general equations of motion of any system are of the type

$$\frac{d}{dt}\left(\frac{\partial L}{\partial \dot{\theta}}\right) - \frac{\partial L}{\partial \theta} = \frac{\partial W}{\partial \theta},$$

and if the changes in the motion only take place very slowly, if the accelerations, that is, are small, this equation practically reduces to

$$-\frac{\partial L}{\partial \theta} = \frac{\partial W}{\partial \theta},$$

so that

$$W = -L,$$

except perhaps for a constant.

This interpretation of the force function as a Lagrangian function has another significance: in fact we see from the form given for L, viz.

$$L = -\tfrac{1}{2}\int \{\mathbf{E}_x^2 + \kappa^2(\mathbf{E}_y^2 + \mathbf{E}_z^2)\}\, dv,$$

that

$$\frac{\partial L}{\partial \beta} = \beta \int (\mathbf{E}_y^2 + \mathbf{E}_z^2)\, dv - \int \left\{\mathbf{E}_x \frac{\partial \mathbf{E}_x}{\partial \beta} + \kappa^2\left(\mathbf{E}_y \frac{\partial \mathbf{E}_y}{\partial \beta} + \mathbf{E}_z \frac{\partial \mathbf{E}_z}{\partial \beta}\right)\right\} dv.$$

Now the former of the integrals on the right is

$$= \int (\mathbf{E}_y\mathbf{H}_z - \mathbf{E}_z\mathbf{H}_y)\, dv$$

$$= c\mathbf{M}_e:$$

the second is on the other hand equal to zero for it is

$$-\int\left(\frac{\partial \Phi}{\partial x}\cdot\frac{\partial \mathbf{E}_x}{\partial \beta} + \frac{\partial \Phi}{\partial y}\cdot\frac{\partial \mathbf{E}_y}{\partial \beta} + \frac{\partial \Phi}{\partial z}\cdot\frac{\partial \mathbf{E}_z}{\partial \beta}\right) dv,$$

and this transforms in the usual way by Green's theorem and becomes

$$\int \Phi \frac{\partial}{\partial \beta}(\operatorname{div} \mathbf{E})\, dv$$

$$= \int \Phi \frac{\partial \rho}{\partial \beta}\, dv = 0^*.$$

It follows therefore that

$$\frac{\partial L}{\partial \beta} = c\mathbf{M}_e,$$

* In the case of deformable systems this integral does not in general vanish and the simple relation between L and $\mathbf{M}_e$ does not subsist. See however below, p. 355.

or since
$$\beta = \frac{|\mathbf{v}|}{c},$$

$$\frac{\partial L}{\partial |\mathbf{v}|} = \frac{1}{c}\frac{\partial L}{\partial \beta} = \mathbf{M}_e,$$

and this relation again exhibits the analogy between the so-called electromagnetic momentum of the system and an ordinary momentum in dynamical theory: it also provides us with the simplest means of calculating $\mathbf{M}_e$.

It is important to notice either as a deduction from this last equation or as a consequence of the particular type of field determined that the vector of linear momentum of a rigid system of charges moving in a straight line is directed entirely along that line: there is no tendency to motion across the direction of translation.

286. Next let us turn to another fundamental aspect of these matters. In ordinary dynamical theories the existence of momentum implies the presence of a material mass moving with a velocity. Now we have seen that for instance an electrostatic system in uniform motion in a straight line would possess something akin to momentum, viz. what we have called electromagnetic momentum, in the direction of its motion, even if its material mass were negligibly small. Thus if we are prepared to adopt the analogy between electromagnetic and material momentum, it appears as at least convenient to extend the analogy still farther and to say that the existence of electromagnetic momentum implies also an electromagnetic mass moving with a velocity. All that it is intended to imply in such a statement is that an electrical system possesses a certain amount of inertia on account of the field and charges in it, just as if it had an additional mass of ordinary type: the inertia offered by this additional mass of the system to accelerated motion exactly accounts for the reaction which according to our theories arises from the electromagnetic field surrounding the system.

We can put this point in another way. The equations of linear translation of an electrical system were obtained in the form

$$\mathbf{F} = \frac{d}{dt}(\mathbf{M}_0 + \mathbf{M}_e),$$

$\mathbf{F}$ being the applied force vector: now suppose that the motion is of the quasi-stationary type so that both $\mathbf{M}_0$ and $\mathbf{M}_e$ are functions of geometrical configuration and velocities only, and are both parallel to the direction of the velocity $\mathbf{v}$ of the system. If then we use $\mathbf{R}$ for the radius of curvature of the path of the system, and $\mathbf{R}_1$ a unit vector along $\mathbf{R}$, then

$$\mathbf{F} = \frac{d}{d|\mathbf{v}|}(\mathbf{M}_0 + \mathbf{M}_e)\frac{d|\mathbf{v}|}{dt} + \left|\frac{\mathbf{M}_0 + \mathbf{M}_e}{\mathbf{v}}\right| \cdot \frac{\mathbf{R}_1|\mathbf{v}|^2}{\mathbf{R}},$$

and now we recognise the ordinary definition of inertia mass in dynamics on the basis of Newton's second law of motion. Thus even if $\mathbf{M}_0$, the material mass of the body, is zero there will still be an apparent mass of electrodynamic origin which for accelerations in the direction of the line of motion is of amount

$$\frac{d|\mathbf{M}_e|}{d|\mathbf{v}|} = \frac{1}{c}\frac{d|\mathbf{M}_e|}{d\beta} = \frac{1}{c^2}\frac{d^2L}{d\beta^2},$$

while it is

$$\left|\frac{\mathbf{M}_e}{\mathbf{v}}\right| = \frac{1}{c}\frac{|\mathbf{M}_e|}{\beta} = \frac{1}{c^2}\frac{1}{\beta}\frac{dL}{d\beta}$$

for accelerations perpendicular to the direction of motion.

Thus the quasi-stationary motion of an electromagnetic system is effectively modified on account of the charges on it just as if it possessed additional mass of an amount however depending on the relative direction of its main velocity and its acceleration.

These results may be further illustrated by application to the two special cases examined in the previous paragraph. When the moving system consists solely of a uniformly charged sphere, the convection potential assumes the constant value

$$\Phi = \frac{q}{8\pi a}\frac{1-\beta^2}{\beta}\log\frac{1+\beta}{1-\beta}$$

on the surface of the sphere: the force function or, if we prefer it, the Lagrangian function of the motion with the sign changed is therefore

$$\tfrac{1}{2}\int \Phi\rho\, dv = -L = \frac{q^2}{16\pi a}\frac{1-\beta^2}{\beta}\log\frac{1+\beta}{1-\beta}.$$

Thus we have

$$\mathbf{M}_e = \frac{1}{c}\frac{dL}{d\beta}$$

$$= \frac{q^2}{8\pi ac\beta}\left\{\left(\frac{1+\beta^2}{2\beta}\right)\log\frac{1+\beta}{1-\beta} - 1\right\},$$

and then the longitudinal and transverse masses turn out to be*

$$m_e = \frac{q^2}{8\pi a c^2} \cdot \frac{1}{\beta^2} \left\{ \frac{2}{1-\beta^2} - \frac{1}{\beta} \log \frac{1+\beta}{1-\beta} \right\},$$

$$m_t = \frac{q^2}{8\pi a c^2} \cdot \frac{1}{\beta^2} \left\{ -1 + \frac{1+\beta^2}{2\beta} \log \frac{1+\beta}{1-\beta} \right\}.$$

These masses, functions of the velocity, increase indefinitely as $\beta = \frac{|\mathbf{v}|}{c}$ is increased and become infinite for $\beta = 1$, i.e. when the velocity of the electron attains the velocity of light. This means of course that such a velocity would never be attained. They have the common limit

$$m_0 = \frac{q^2}{6\pi a c^2}$$

when β is small.

In the second case the moving system is of variable dimensions, having the form of an oblate spheroid with axes $(\kappa a, a, a)$ when the motion is with velocity $\mathbf{v}$, where

$$\kappa^2 = 1 - \frac{|\mathbf{v}|^2}{c^2} = 1 - \beta^2.$$

In this case the convection potential assumes the constant value

$$\Phi = \frac{\kappa q}{4\pi a} = (1-\beta^2)^{\frac{1}{2}} \frac{q}{4\pi a},$$

on the surface of the body where the charge is confined: and thus now

$$L = -\frac{\kappa q^2}{8\pi a}.$$

It follows therefore, just as above, that

$$\mathbf{M}_e = \frac{1}{c} \frac{dL}{d\beta} = \frac{q^2}{8\pi a c} \cdot \frac{\beta}{\sqrt{1-\beta^2}},$$

so that†

$$m_e = \frac{q^2}{8\pi a c^2} \cdot \frac{1}{(1-\beta^2)^{\frac{3}{2}}},$$

$$m_t = \frac{q^2}{8\pi a c^2} \cdot \frac{1}{(1-\beta^2)^{\frac{1}{2}}},$$

which again become infinite as the velocity approaches that of light, starting from the same limit

$$m_0' = \frac{q^2}{8\pi a c^2} = \frac{4}{3} m_0.$$

* Abraham, *Ann. d. Phys.* 10 (1903), p. 105; *Die Theorie der Elektrizität*, II. p. 181.

† Lorentz, *Theory of Electrons*, p. 210.

287. There is however a difficulty in the second case. In fact if instead of calculating $\mathbf{M}_e$ from the Lagrangian or force function we deduce it directly from the definition

$$\mathbf{M}_e = \frac{1}{c}\int [\mathbf{EB}]\, dv,$$

which is equal to

$$\frac{1}{c^2}\mathbf{v}_x \int (\mathbf{E}_y{}^2 + \mathbf{E}_z{}^2)\, dv,$$

in our case we find that

$$\mathbf{M}_e = -\frac{4\mathbf{v}}{3\kappa^2 c^2} L = \frac{q^2}{6\pi a c^2}\frac{\mathbf{v}}{(1-\beta^2)^{\frac{1}{2}}}.$$

This value is larger than the previous one in the ratio 4 : 3 and it leads to formulae for the masses which are also larger in the same ratio and which therefore in the limit have the same value as in the previous case.

The fact is that in a system of the present type where the dimensions of the conductors are varying the internal or intrinsic elastic energy of the material of the conductor which is consequent on its charged condition is varying and the complete electromagnetic energy reckoned with in the calculation of L is not all dynamically available. Part of it is in fact supplied from the store of elastic energy to which the elastic forces maintaining the form of the conductors are due and this part can not be used for external mechanical purposes.

This point can be illustrated by a detailed calculation for the present case. The rate at which work is done by the field on the conductor is

$$\int_f \left(\mathbf{v}\,.\,\mathbf{E} + \frac{1}{c}[\mathbf{vB}]\right)\sigma df$$

integrated over its surface. But the conductor contracts whilst in accelerated motion and therefore its fore end moves at each instant slower than its centre and its back end faster. If the velocity of the centre is denoted by v and is presumed to be directed along the axis OX the velocity of the points distant x from the centre is

$$v + x\dot{\kappa}, \quad \kappa = \left(1 - \frac{v^2}{c^2}\right)^{\frac{1}{2}}.$$

The integral for the rate of work thus splits into two parts. The first or

$$\int_f v\left(\mathbf{E}_x + \frac{1}{c}[\mathbf{vB}]_x\right)\sigma df$$

represents the work done by the field on the conductor as a whole. The second or

$$\int_f x\dot{\kappa}\left(\mathbf{E}_x + \frac{1}{c}[\mathbf{vB}]_x\right)\sigma df$$

is the work done by the field forces in the deformation; with sign reversed this part may be regarded as work supplied by the internal elastic forces of other than electric nature which must necessarily act to balance the tendency of the field to deform the conductor. If we insert the usual stationary values it is easy to verify that this last integral is

$$\frac{q^2}{24\pi a}\frac{\beta}{\kappa}\frac{d\beta}{dt},$$

which is the time rate of change of the function

$$A - \frac{q^2\kappa}{24\pi a},$$

A being a constant, usually taken as $q^2/6\pi a$. This then is the part of the total energy of the electron which is concerned with the deformation forces: the remainder or

$$\frac{q^2\kappa}{8\pi a} - \left(\frac{q^2}{6\pi a} - \frac{q^2\kappa}{24\pi a}\right) = \frac{q^2}{6\pi a}(1-\kappa)$$

is the part which alone corresponds to the mechanically effective forces exerted by the field on the conductor and it therefore represents the work function which must be used in place of the L-function given above. No inconsistency is then introduced by the independent method of calculating the momentum.

The value of A in the above formula is chosen so that in the limit for small motions the mechanically effective energy reduces to

$$\frac{1}{2}\frac{q^2}{6\pi ac^2}v^2 = \frac{1}{2}m_0 v^2.$$

288. The great interest in these results lies in their application to the explanation of the properties of the electron. It is impossible to charge an ordinary piece of matter with sufficient electricity to produce an electromagnetic inertia at all comparable with its ordinary inertia: but an electron carries a charge which is so enormous, compared with its mass, that its electromagnetic inertia is far greater than its material inertia if it has any. Again pieces of matter even so small as the individual molecules are much too cumbersome to get up a speed comparable with that of light, whereas the electrons are often found under natural circumstances travelling with such speeds, so that they should exhibit the additional characteristic of the electromagnetic mass in increasing with the speed. Some time after Thomson's determination of the ratio of the charge to the mass, Kaufmann* proceeded by the same method to determine the functional form of this ratio in terms of the velocity, with a view to testing the effectiveness of the theoretical formulae. He worked with the negative electrons which are thrown off as the β-rays from radium with velocities ranging up to ·95 c. Now it was found that while the velocity increases from about ·5 c. to the higher value the corresponding value of e/m considerably diminishes, and exactly in the way that we might expect if the charge remains constant and the mass increases according to the theoretical formulae. The first experiments were unable to distinguish between the two types of formulae given, the first by Abraham and the second by Lorentz, but later and more exact methods tend to the view that the simpler functional forms given by Lorentz's method are the more exact of the two.

The conclusion to be drawn from these facts is that at all events the electromagnetic mass of an electron has an appreciable influence; but the experiments went further, the type of function obtained pointed to the conclusion that the electromagnetic mass greatly preponderates over any material mass that the electrons may have. Indeed Kaufmann's numbers show no trace of an influence of the material mass at all, his ratio of the effective masses for two different velocities agreeing within the limits of experimental error with the theoretical ratio of the electromagnetic masses themselves.

If the material mass of an electron is inappreciable under all circumstances it may be treated as absolutely zero. It is no use talking of something that cannot be traced. On such a view the electron consists merely of an element of negative electricity.

* *Gött. Nachr.* 1 (1901), 5 (1902), 3 (1903); *Phys. Zeitschr.* 1902, p. 55; *Ann. d. Phys.* 19 (1906), p. 487.

Of course by the negation of a material mass, the electron loses much of its substantiality because our theory has been interpreted entirely in terms of charged matter and the ordinary mechanical relations of such matter. In speaking therefore of the electron as having no material mass, we must nevertheless leave it sufficient substantiality to enable us to speak of forces acting on it. In terms of a pure aether theory we might say that the electron, which is necessarily a centre of an electric field of strain in the aether, is nothing more nor less than the centre or nucleus of the strain thus depicted, a knot, so to speak, which can however move freely from one point of the medium to another. This point of view has been advocated by Larmor and its simplicity is a great point in its favour.

289. It is natural to enquire whether there is similarly no material mass for the elements of positive electricity. Unfortunately no decisive answer is yet possible to this question, but the view is now generally held that not only is there no material mass in the ordinary sense of the word in the positive particles, but that the material mass of any ordinary piece of matter is nothing but the inertia mass of the electric charges involved in it. This is the 'electromagnetic theory of matter' and although it is not possible for us to enter here into any further details, it must be admitted that the evidence of both the experimental and theoretical researches of recent years is gradually tending to confirm the substantiality of the view.

The purely electromagnetic theory is of course not entirely free from difficulties even as regards the electrons themselves, especially if we are to retain the Lorentzian idea of these particles. We have seen that his idea is that of a charged particle which changes its shape as its velocity alters and we have also noticed that such change of shape must necessarily be controlled by intra-electronic forces of other than purely electric type. We are thus at the outset forced to complicate the purely electrical constitution of the particle by the introduction of other types of force—and energy—holding them together and controlling their deformation; having gone so far there is no special reason why we should not depart even wider from the purely electric view.

Latterly since the development of the theory of relativity to its full extent this question of whether there is ultimately more than one type of mass has fallen into the background. It is found that in fact even ordinary inertial mass must in reality also be a function of the velocity of precisely the same type as the electrical mass, so that attempts to discriminate between the two will probably fail in any case.

In this theory even another point of view is emerging. It appears that all forms of energy must have associated with them a proportional amount of inertia and therefore energy is coming to be regarded as more fundamental even than matter, which then becomes in its smallest parts—electrons or positive particles—simply intense collocations of energy of different types. Such a view is of course not free from difficulties, but it is enticing, and may perhaps in the near future be more successfully elaborated than it has been at present, when its possibilities as a consistent theory may perhaps emerge.

290. Optical phenomena in moving media. We now leave these interesting questions concerning moving charges to turn to another aspect of electrodynamic theory of moving systems in general. We have so far merely treated of steady electromagnetic fields convected with a system of bodies having a motion of translation, but it is equally important to consider the other extreme case of the propagation of very rapid electromagnetic disturbances across the field between such moving systems. The practical aspect of these considerations lies in their application to the explanation of the optical phenomena in systems having a motion of translation, as for instance all terrestrial bodies have by the annual motion of the earth.

We consider in the first place the propagation of electric waves across a dielectric medium, which is moving with uniform velocity $\mathbf{v}$ parallel to a definite direction which we shall take to be the x-axis of a system of rectangular coordinates. According to the general scheme of equations formulated above we have the electromotive force expressed in terms of the vector and scalar potentials and the magnetic induction by the relation

$$\mathbf{F} = -\frac{1}{c}\frac{d\mathbf{A}}{dt} - \operatorname{grad}\phi + \frac{1}{c}[\mathbf{vB}],$$

where $$\mathbf{B} = \operatorname{curl} \mathbf{A},$$

and if $\mathbf{C}$ is the total current of Maxwell's theory

$$\frac{1}{c}\mathbf{C} = \operatorname{curl} \mathbf{H},$$

and $$\mathbf{B} = \mathbf{H} + \frac{1}{c}[\mathbf{Pv}],$$

if there are no magnetic substances about; $\mathbf{P}$ is the dielectric polarisation: thus*

$$\nabla^2 \mathbf{A} = -\frac{1}{c}\mathbf{C} - \operatorname{curl}[\mathbf{Pv}].$$

These equations are satisfied by the propagation of a train of transverse waves along the x-axis, in which $\mathbf{F}_x$, $\mathbf{H}_x$, $\mathbf{C}_x$, $\mathbf{A}_x$ are all null, while ϕ is a function of y, z.

Thus for such a wave train

$$\begin{aligned}[\mathbf{vB}] &= [\mathbf{v}[\nabla\mathbf{A}]]\\ &= \nabla(\mathbf{Av}) - (\mathbf{v}\nabla)\mathbf{A}\\ &= -v\frac{\partial \mathbf{A}}{\partial x},\end{aligned}$$

and thus $$\mathbf{F}_y = -\frac{1}{c}\left(\frac{d}{dt} + v\frac{\partial}{\partial x}\right)\mathbf{A}_y - \frac{\partial\phi}{\partial y},$$

$$\mathbf{F}_z = -\frac{1}{c}\left(\frac{d}{dt} + v\frac{\partial}{\partial x}\right)\mathbf{A}_z - \frac{\partial\phi}{\partial z}.$$

Now since $\operatorname{div} \mathbf{A} = 0$ any theory that makes

$$\operatorname{div} \mathbf{F} = 0$$

will also make $\nabla^2\phi = 0$, that is, will make ϕ merely the static potential of the electric charges in the field. We shall then have

$$\nabla^2 \mathbf{F} = \frac{1}{c^2}\left(\frac{d}{dt} + v\frac{\partial}{\partial x}\right)\{\mathbf{C} + c \operatorname{curl}[\mathbf{Pv}]\},$$

and the remainder of the analysis depends on the relation between $\mathbf{C}$ and $\mathbf{F}$.

291. Now we have seen that the total current $\mathbf{C}$ is composed of two parts: the first, an aethereal part, of density

$$\frac{d\mathbf{E}}{dt},$$

* We are using Maxwell's instantaneous vector potential.

which is not convected, presumably because the aether does not participate in the motion of the matter; the second part depends on the material polarisation and is measured by its time rate of change

$$\frac{\partial \mathbf{P}}{\partial t}.$$

Again $$\operatorname{curl}[\mathbf{Pv}] = (\mathbf{v}\nabla)\,\mathbf{P} - \mathbf{v}\,(\nabla\mathbf{P})$$

and as far as concerns the two components under consideration this is the same as

$$v\,\frac{\partial \mathbf{P}}{\partial x}$$

whilst $$\mathbf{P} = \epsilon'\mathbf{F},$$

where $\epsilon = 1 + \epsilon'$ is the dielectric constant of the medium*. Thus

$$\mathbf{C} + c\operatorname{curl}[\mathbf{Pv}] = \frac{d\mathbf{E}}{dt} + (\epsilon - 1)\left(\frac{\partial}{\partial t} + v\frac{\partial}{\partial x}\right)\mathbf{F}.$$

Now putting ϕ null for purely transverse waves, which will be found to cause no discrepancy, we have

$$\mathbf{E} = -\frac{1}{c}\frac{d\mathbf{A}}{dt},$$

and thus keeping $\mathbf{A}$ as the more convenient independent variable

$$\nabla^2\mathbf{A} = \frac{1}{c^2}\left[\frac{\partial^2}{\partial t^2} + (\epsilon - 1)\left(\frac{\partial}{\partial t} + v\frac{\partial}{\partial x}\right)^2\right]\mathbf{A}.$$

If the period of the waves propagated is $\frac{2\pi}{p}$ all the functions may be taken to depend on the time and position coordinates by the factor

$$e^{ip(c't-x)},$$

where c' is the velocity of propagation; substituting this form we find that

$$c^2 = c'^2 + (\epsilon - 1)(c' - v)^2,$$

whence since v/c' is very small we have approximately

$$c' = \frac{c}{\sqrt{\epsilon}}\left\{1 + \left(1 - \frac{1}{\epsilon}\right)\frac{v}{c'}\right\}$$

$$= \frac{c}{\sqrt{\epsilon}} + \left(1 - \frac{1}{\epsilon}\right)v.$$

* It is to be understood that the constant here involved is that appropriate to rapidly alternating fields and is a function of the frequency of alternation.

Thus the velocity of propagation of radiation in the moving medium is increased from its value in the same medium at rest by the amount

$$\left(1 - \frac{1}{\epsilon}\right) v^*.$$

This result was formulated by Fresnel† long before the electromagnetic theory had been formulated, and it has been experimentally verified first by Fizeau but much more accurately by Michelson and Morley‡ in connection with the propagation of light in flowing water. In the experiments water was made to flow in opposite directions through two parallel tubes placed side by side and closed at both ends by glass plates: the two interfering beams of light were passed through the tubes in such a manner that throughout their course one went with the water and the other against it.

292. These experiments provide the most decisive, as well as the most accurate, test of Fresnel's hypothesis of a stagnant aether, on which the above theory has been based, and they definitely exclude the more complicated hypothesis advanced by Stokes. This is easily seen because if the motion of a piece of matter through the aether drags that medium with it, all that is contained in the flowing column of water in the experiments will share the translatory motion of flow: the propagation of light in the water will then go on in its interior in exactly the same manner as if it were at rest; so that its actual velocity in space will be $\left(\frac{c}{\sqrt{\epsilon}} + v\right)$§.

A further important point is also emphasised by the result obtained above. It essentially involves not only the distinction between aethereal and material polarisation currents but also the distinction between the forces producing these currents, which is

* For dispersing media this formula requires correction by the insertion in the bracket of a term $\frac{p}{2\epsilon}\frac{d\epsilon}{dp}$. Cf. Lorentz, *Versuch einer Theorie u.s.w.* p. 101.

† *Paris C. R.* 33 (1851), p. 349; *Ann. d. Phys.* (1853), p. 377.

‡ *Amer. Jour. of Sc.* 31 (1885).

§ The whole question has however quite recently been re-opened by certain experiments by Miller in America. He has obtained positive results in a repetition of the Michelson-Morley experiment which point to there being a flow of aether caused by the earth's motion.

brought out in the general dynamical theory of the previous chapter. It was there shown that the force of electrodynamic origin which acts on the electrons and produces the true current of material displacement is the electromotive force **F**, whereas the force which strains the aether and produces the aethereal displacement current is the electric force, differing from the electromotive force **F** by the dynamical part $\frac{1}{c}[\mathbf{vB}]$.

The fundamental hypotheses on which the theory is based are therefore the only ones which are consistent with the results of experiments made especially with a view to testing them: we must now follow the theory through into some of its more important consequences in order to see whether it is entirely sufficient or whether it still requires modification and amplification. Before proceeding at once to the optical theory it will however be convenient to elaborate at the present stage a few further details of a question already dealt with at some length on a previous occasion.

293. The ponderomotive force which an electromagnetic field exerts on any interface separating a perfectly conducting or absorbing medium from the surrounding free space is determined mainly in the form of a traction per unit area

$$\mathbf{T} = \tfrac{1}{2}\{2\mathbf{E}\mathbf{E}_n + 2\mathbf{B}\mathbf{B}_n - \mathbf{n}_1(\mathbf{E}^2 + \mathbf{B}^2)\},$$

$\mathbf{n}_1$ being the unit vector along the normal to the surface.

The resultant force exerted by the field on any such body is obtained by integration of this surface force over the outer surface of the body.

Now let us find how this surface force is altered when the body moves in a vacuum: the above expression is then increased for momentum is taken up as a result of the motion. If **v** is the velocity of the typical element df of the surface, the momentum picked up by the body on account of the motion of df is then

$$\frac{1}{c}\mathbf{v}_n[\mathbf{EB}]\,df.$$

It follows that the pressure per unit area on the moving surface is of the form

$$\mathbf{T}' = \left\{\mathbf{E}\mathbf{E}_n + \mathbf{B}\mathbf{B}_n - \frac{1}{2}\mathbf{n}(\mathbf{E}^2 + \mathbf{B}^2) + \frac{\mathbf{v}_n}{c}[\mathbf{EB}]\right\}.$$

We can transform this into a more convenient form: to do this we start from the identity

$$\mathbf{v}_n[\mathbf{EB}] + \mathbf{E}_n[\mathbf{B}\,.\,\mathbf{v}] + \mathbf{B}_n[\mathbf{vE}] = \mathbf{n}_1(\mathbf{v}\,.\,[\mathbf{EB}]),$$

so that we may write

$$2\mathbf{T}' = 2\mathbf{E}'\mathbf{E}_n + 2\mathbf{B}'\mathbf{B}_n - \mathbf{n}\left\{\mathbf{E}^2 + \mathbf{B}^2 - \frac{2}{c}(\mathbf{v}\,.\,[\mathbf{EB}])\right\},$$

where we have written $\mathbf{E}'$ and $\mathbf{B}'$ for the vectors

$$\mathbf{E}' = \mathbf{E} + \frac{1}{c}[\mathbf{vB}], \quad \mathbf{B}' = \mathbf{B} + \frac{1}{c}[\mathbf{E}\,.\,\mathbf{v}].$$

We may notice also that

$$(\mathbf{EE}') = \mathbf{E}^2 - \frac{1}{c}(\mathbf{v}\,.\,[\mathbf{EB}]), \quad (\mathbf{BB}') = \mathbf{B}^2 - \frac{1}{c}(\mathbf{v}\,.\,[\mathbf{EB}]),$$

so that $$\mathbf{T}' = \left\{\mathbf{E}'\mathbf{E}_n + \mathbf{B}'\mathbf{B}_n - \frac{\mathbf{n}}{2}((\mathbf{EE}') + (\mathbf{BB}'))\right\}$$

is the general expression for the electromagnetic surface on a moving surface.

If the problem concerns the pressure of radiation on a perfectly absorbing surface in motion, then **E**, **B** are simply the vectors of the field of the incident radiation, there being no reflexion. But with a moving mirror things are different; here the reflexion takes place so that at all points of the surface the tangential components of the electromotive force ($\mathbf{E}'$) and the normal components of magnetic induction (**B** and therefore also of the vector $\mathbf{B}'$) must vanish*. Thus at the surface

$$\mathbf{B}_n = 0, \quad [\mathbf{nE}'] = 0,$$

so that $$0 = [\mathbf{E}\,.\,[\mathbf{E}'\mathbf{n}]] = \mathbf{E}'\mathbf{E}_n - \mathbf{n}(\mathbf{EE}'),$$

so that we have now

$$\mathbf{T}' = \tfrac{1}{2}\mathbf{n}_1(\mathbf{EE}' - \mathbf{BB}')$$

$$= \frac{\mathbf{n}_1}{2}(\mathbf{E}^2 - \mathbf{B}^2).$$

Each of these two last formulae determine the surface forces exerted by the electromagnetic field on the moving mirror. Since $\mathbf{n}_1$ is a unit vector parallel to the normal to the surface, $\mathbf{T}'$ is always

* Provided of course the substance of the mirror is perfectly reflecting, so that there is no penetration.

perpendicular to this surface. The radiation pressure gives no tangential forces on the perfectly reflecting surface.

294. Now let us return to our purely optical considerations in moving media.

The fundamental conception of optical science is the relation between the direction of the ray and that in which the radiant waves are travelling. When the material medium is at rest in the aether, the ray, or the path of the radiant energy, is the same relative to the matter as to the aether. In the circumstances of moving matter there are however two kinds of rays to be distinguished, one of them being the paths of the radiant energy with respect to the particles of the moving matter, the other the absolute path of the radiant energy in the stagnant aether, or as we may say in space. As radiation is revealed to us wholly by its action on matter it is the former type of rays that is of objective importance.

Let us suppose that a ray from a fixed source passes through a small opening in a screen at O* and after traversing the distance OP in the free aether arrives at P. An observer situated at P and the screen at O may be supposed to have a common velocity $\mathbf{v}$. The opening O has then arrived at the new position O' by the time that the radiation which passed through it at O has reached the observer at P and this observer, supposed ignorant of the motion, would regard $O'P$ as the direction of the ray: this is therefore the direction of the *relative ray* and it is parallel to the direction of the relative velocity of the actual light and the observer

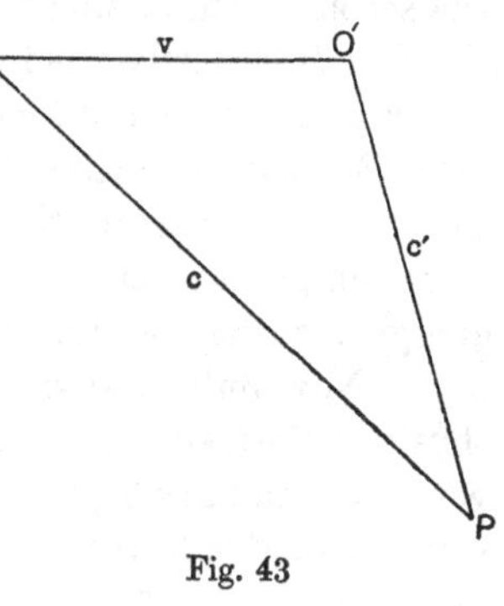

Fig. 43

$$\mathbf{c}' = \mathbf{c} - \mathbf{v}.$$

Bradley explained by this simple construction the aberration of the starlight which arises from the yearly motion of the earth: the periodic motion of the earth round the sun gives rise to a periodic change in the direction of the relative ray of light from a star and

* For example the object glass of a telescope.

therefore also to yearly periodic change in the apparent position of the star.

295. This simple definition of the relative ray can also be obtained by purely electromagnetic reasoning without introducing the conceptions of geometrical optics.

The absolute ray of light is determined by the Poynting vector

$$\mathbf{S} = c\,[\mathbf{EB}],$$

and it gives the flux of energy per unit area through a small surface placed perpendicular to its direction. If this surface be regarded as perfectly black, all the energy which is thus transferred is absorbed by the surface.

The relative flux of energy per unit area through a small surface perpendicular to the direction denoted by $\mathbf{n}$ and moving with the velocity $\mathbf{v}$ is

$$\mathbf{S}_n - \tfrac{1}{2}\mathbf{v}_n\,(\mathbf{E}^2 + \mathbf{B}^2),$$

the second term arising from the fact that the energy provided in the flux $\mathbf{S}$ is partly used up in establishing the energy in the ray which gets longer and longer on account of the motion of the surface. Suppose now that the moving surface is perfectly absorbing; then we may regard the component of the relative ray in the direction perpendicular to this surface as defined by the amount of energy absorbed per unit time by the surface and transformed into heat. Now only part of the energy which reaches the surface is thus transformed, the remainder being used up in doing mechanical work as a radiation pressure on the surface: moreover the rate of working of this latter force per unit area is

$$-\frac{1}{2}\left\{2\mathbf{E}_n\,(\mathbf{vE}) + 2\mathbf{B}_n\,(\mathbf{vB}) - \mathbf{v}_n\,(\mathbf{E}^2 + \mathbf{B}^2) + \frac{2\mathbf{v}_n}{c}\,(\mathbf{v}\,.\,[\mathbf{EB}])\right\},$$

so that the component of the relative ray perpendicular to the direction of the surface is determined by

$$\mathbf{S}_n + \left\{\mathbf{E}_n\,(\mathbf{vE}) + \mathbf{B}_n\,(\mathbf{vB}) - \mathbf{v}_n\,(\mathbf{E}^2 + \mathbf{B}^2) + \frac{\mathbf{v}_n}{c}\,(\mathbf{v}\,.\,[\mathbf{EB}])\right\},$$

which may be regarded as the component in the same direction of the vector

$$\mathbf{S}' = \mathbf{S} + \left\{\mathbf{E}\,(\mathbf{vE}) + \mathbf{B}\,(\mathbf{vB}) - \mathbf{v}\,(\mathbf{E}^2 + \mathbf{B}^2) + \frac{\mathbf{v}}{c}\,(\mathbf{v}\,.\,[\mathbf{EB}])\right\},$$

which determines the relative ray vector. If we write

$$\mathbf{E}' = \mathbf{E} + \frac{1}{c}[\mathbf{vB}], \quad \mathbf{B}' = \mathbf{B} - \frac{1}{c}[\mathbf{vE}],$$

then we see that

$$[\mathbf{E}'\mathbf{B}'] = [\mathbf{EB}] - \frac{1}{c}[\mathbf{E}\,.\,[\mathbf{vE}]] - \frac{1}{c}[\mathbf{B}\,.\,[\mathbf{vB}]] + \frac{1}{c^2}[[\mathbf{vE}]\,.\,[\mathbf{vB}]],$$

and

$$[\mathbf{E}\,.\,[\mathbf{vE}]] = \mathbf{v}\mathbf{E}^2 - \mathbf{E}\,(\mathbf{vE}),$$

$$[\mathbf{B}\,[\mathbf{vB}]] = \mathbf{v}\mathbf{B}^2 - \mathbf{B}\,(\mathbf{vB}),$$

$$[[\mathbf{vE}]\,.\,[\mathbf{vB}]] = \mathbf{v}\,([\mathbf{vE}],\ \mathbf{B}) - \mathbf{B}\,([\mathbf{vE}]\,.\,\mathbf{v})$$

$$= \mathbf{v}\,(\mathbf{v}\,.\,[\mathbf{EB}]),$$

so that finally

$$\mathbf{S}' = c\,[\mathbf{E}'\mathbf{B}'],$$

which is the simplest form of the general expression of the relative ray vector: its component in any direction determines the amount of heat developed per unit time per unit area in a perfectly absorbing surface moving with the appropriate velocity and placed perpendicular to this direction.

It remains to show that for plane waves this definition of the relative ray direction agrees with the elementary rule given in the earlier part of the paragraph.

For a plane polarised wave train the absolute vector velocity **c** of propagation, and the vectors of electric and magnetic force intensities are mutually perpendicular to one another: and since the intensities of these forces are equal to one another we have

$$\mathbf{B} = \frac{1}{c}[\mathbf{cE}], \quad \mathbf{E} = \frac{1}{c}[\mathbf{Bc}].$$

Thus

$$\mathbf{E}' = \frac{1}{c}[\mathbf{Bc}'], \quad \mathbf{B}' = \frac{1}{c}[\mathbf{c}'\mathbf{E}],$$

so that

$$\mathbf{S}' = c\,.\,\frac{1}{c}[[\mathbf{Bc}']\,.\,[\mathbf{c}'\mathbf{E}]]$$

$$= c\,.\,\frac{\mathbf{c}'}{c^2}(\mathbf{c}'\,.\,[\mathbf{EB}])$$

$$= \frac{\mathbf{c}'}{c^2}(\mathbf{c}'\,.\,\mathbf{S}),$$

so that **S′** is parallel to the relative velocity of the light to the receiving surface: the elementary rule for the construction of the path is therefore completely in accordance with the electromagnetic theory of these things.

296. These considerations are confined to cases where the source of light is at rest in the aether and they therefore only apply to such problems as the aberration of light from fixed stars. We have next to enquire whether the motion of a system has any influence on optical phenomena taking place inside it. Such a question is of practical importance in considerations respecting the effect of the motion of the earth on optical phenomena taking place on its surface. Can we determine the motion of the earth by optical measurements in a laboratory? Let us imagine that at the time $t = 0$ a light signal is emitted from the point O. At the instant t it will have arrived at P, the absolute direction of its path being OP, which we can denote by the vector $\mathbf{r}$. In the interval thus occupied by the actual radiation in travelling from O to P the source of light will have moved to O' with the velocity $\mathbf{v}$ of the system to which it and the observer at P belong. The radius $O'P$ in the vector sense is given by

$$\mathbf{r}' = \mathbf{r} - \mathbf{v}t,$$

and of course

$$\mathbf{r} = \mathbf{c}t,$$

so that

$$\mathbf{r}' = (\mathbf{c} - \mathbf{v})\frac{r}{c} = \frac{r}{c}\mathbf{c}', \quad r = |\mathbf{r}|.$$

Thus the observed direction of the ray is identical with the actual direction of the line joining the instantaneous positions of source and observer. Thus in a uniformly moving system the source of light is seen just where it happens to be at the instant. The common motion of source and observer cannot therefore be detected by an observation of the direction of the relative ray. We might however still hope to detect this motion by a measurement of the length of the path of the ray, for the surface through P of constant absolute light dis-

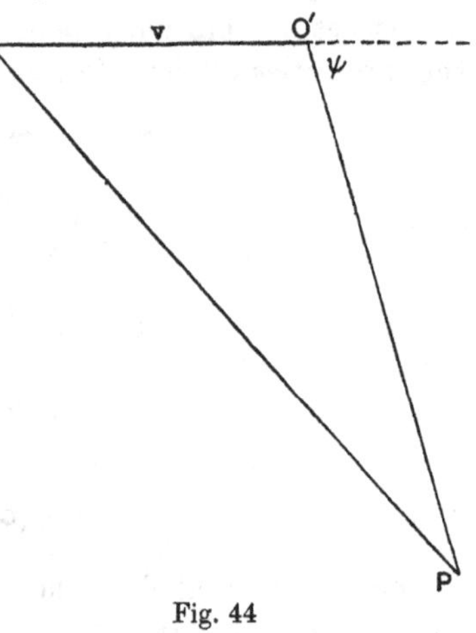

Fig. 44

tances is a sphere of centre O and the point O' from which the relative ray starts lies eccentric to this sphere, so that given lengths r of the absolute light path would correspond to different lengths r' of the relative light path $O'P$ depending on its direction. Now we see from the figure as given

$$r^2 = \beta^2 r^2 + r'^2 + 2\beta r r' \cos\psi,$$

where β is used for $\frac{|\mathbf{v}|}{c}$ just as before so that again using $\kappa^2 = 1 - \beta^2$

$$\kappa r = \frac{\beta r' \cos\psi}{\kappa} + \sqrt{r'^2 + \frac{\beta^2 r'^2 \cos^2\psi}{\kappa^2}}.$$

If now we choose moving axes with origin at P and the x-axis in the direction of motion of the observer and source we can write

$$\kappa r = \frac{\beta x'}{\kappa} + \sqrt{\frac{x'^2}{\kappa^2} + y'^2 + z'^2},$$

where (x', y', z') are the coordinates of O': this relation determines the absolute length of the path in terms of the relative coordinates of source and observer.

Now let us compare with this moving system a stationary system obtained from it by a uniform expansion in the ratio $1 : \kappa$ parallel to the direction of motion, i.e. the x-axis. The analytical transformation is expressed by the relations

$$x_0' = \frac{x'}{\kappa}, \quad y_0' = y', \quad z_0' = z',$$

giving the coordinates of the point P_0' corresponding to P'. If we use

$$r_0'^2 = x_0'^2 + y_0'^2 + z_0'^2,$$

then

$$\kappa r = r_0' + \beta x_0'.$$

The coordinates of the absolute effective position of the source in the moving system relative to P are therefore connected with the coordinates of the relative position in the stationary system by the relations

$$\kappa x_0' = x - \beta r, \quad y_0' = y, \quad z_0' = z,$$

and we have also

$$\kappa r_0' = r - \beta x,$$

$$\kappa x = x_0' + \beta r_0', \quad y = y_0', \quad z = z_0'.$$

297. Let us now examine the question as to whether a measurement of the length of the relative light path by an observer moving with the system will enable a determination of the motion of the system to be made. The method is to send two parts of the same beam along different paths and then unite them and determine by an interference method whether the one has had to traverse a longer path. Suppose that light is sent along the relative path $O'P$ and then reflected back by a mirror to O'. The length of the absolute path of the incident beam is

$$r_1 = \kappa^{-1} r_0' + \beta\kappa^{-1} x_0',$$

and that of the reflected beam is

$$r_2 = \kappa^{-1} r_0' - \beta\kappa^{-1} x_0',$$

the sum of the two being

$$r_1 + r_2 = 2\kappa^{-1} r_0'.$$

Now draw a sphere round O' as centre and of radius r'. For any point P on this sphere the path $O'PO'$ would be the same if the system were at rest, but when it is in motion this is by no means the case. In this case the absolute path is determined not by the relative radius in S but by the corresponding length r_0' in the expanded system S_0,

$$r_0'^2 = \frac{x'^2}{\kappa^2} + y'^2 + z'^2,$$

whilst

$$r'^2 = x'^2 + y'^2 + z'^2.$$

Thus a given total relative path corresponds to different lengths of the absolute path. If the relative path is parallel to the direction of motion

$$r_0'^2 = \frac{r'^2}{\kappa^2},$$

so that

$$r_1 + r_2 = \frac{2r'}{\kappa^2},$$

whilst if the relative path is perpendicular to the direction of motion

$$r_0' = r',$$

and

$$r_1 + r_2 = \frac{2r'}{\kappa}.$$

Thus with equal relative paths the ratio of the absolute light paths in the two cases is $1 : \kappa$ greater in the first than in the second. The difference of the light paths is then Δl when

$$\frac{\Delta l}{l} = (1 - \beta^2)^{-\frac{1}{2}} - 1 = \tfrac{1}{2}\beta^2,$$

if we neglect higher powers of β.

This difference in length of the absolute light paths for a beam propagated over a given length in a moving system parallel and perpendicular to the motion was first noticed by Maxwell. It formed the object of investigation in a now famous experiment conducted by Michelson and Morley* with a view to determining the motion of the earth. Two pencils of light coming from the same source are brought to inference after traversing perpendicular paths of equal length, the one perpendicular to the direction of motion of the earth and the other parallel to it. Although the precision attained in the measurements was such that a difference equal to $\frac{1}{20}$ of that expected would have been observed and measured, no difference whatever was observed at all which was not within the very close limits of experimental error. The conclusion must therefore be drawn that no effect of the kind mentioned can be experimentally detected.

298. Thus if we retain the view tacitly assumed throughout that the dimensions of a body are unaltered by imparting a uniform motion to it, the theory as we have expounded it is inconsistent with experience. In order however to surmount this difficulty it is suggested by Fitzgerald† and Lorentz‡ that we make the further not unnatural assumption that the dimensions of a body in motion are altered in such a way that its diameter parallel to the direction of motion are contracted in the corresponding ratio

$$\kappa : 1 \text{ or } \left(1 - \frac{|\mathbf{v}|^2}{c^2}\right)^{\frac{1}{2}} : 1.$$

This of course disposes of the difficulty of the Michelson-Morley experiment for it throws the points in the stationary system S_0 which correspond to loci of constant relative paths round O' on to a sphere, and constant relative paths then correspond to constant absolute paths.

However extraordinary this hypothesis may appear at first sight, it must be admitted that it is by no means gratuitous, if we assume that the intermolecular forces act through the mediation of the aether in a manner similar to that which we know to be the case in regard to electric and magnetic forces. If that is so, the trans-

* *Phil. Mag.* Dec. 1887. Cf. also Morley and Miller, *Phil. Mag.* 9 (1905).

† Cf. Lodge, *Phil. Trans.* 184 A (1893), p. 727.

‡ "Versuch einer Theorie der elektrischen u. optischen Erscheinungen in bewegten Körpen," Leiden (1895).

lation of the matter will most likely alter the action between two molecules or atoms in a manner similar to that in which it alters the attraction or repulsion between electrically charged particles. As then the form and dimensions of a solid body are determined in the last resort solely by the intensity of the molecular forces, an alteration of the dimensions cannot well be left out of consideration.

In its theoretical aspect there is thus nothing to be urged against the hypothesis. As regards its experimental aspect we at once notice that the elongation or contraction which it implies is extraordinarily minute. It would involve a shortening in the diameter of the earth of about $6\frac{1}{2}$ centimetres. The only experimental arrangements in which it could come into evidence would be just of the type of this one of Michelson's which first suggested it*.

299. The Lorentz transformation of the electromagnetic equations. Before finally closing this work reference must be made to a fundamental property of the electromagnetic equations which is of great importance in the discussions relating to the conditions in moving electrical systems and which coordinates in one principle many of the properties of these systems treated separately above. We shall simplify our discussion by assuming that the whole electrodynamic properties of matter can be explained on the basis of a stationary aether and electrons, and we shall so far substantiate these latter elements as to assume that they are extremely small but finite entities consisting essentially and solely of electricity distributed with a finite volume density throughout their small extent, where the fundamental equations are also presumed to hold. In this case the only constituents of the complete current of the theory are its purely aethereal part

$$\frac{d\mathbf{E}}{dt}$$

and a part due to the convection of the electrons which, at any place, is of density

$$\rho\mathbf{u},$$

ρ being the density of the charge in the point of the electron passing over it.

* Miller has published quite recently brief details of certain results obtained in repetitions of these experiments at low and at high altitudes. While he confirms Michelson's result at the low altitudes he obtains definite positive results at high altitudes; but the significance of these is not yet very clear.

The equations of the theory referred to a frame of reference fixed in the aether thus assume the form

$$\operatorname{curl} \mathbf{B} = \frac{1}{c}\frac{d\mathbf{E}}{dt} + \rho\mathbf{u},$$

$$\operatorname{curl} \mathbf{E} = -\frac{1}{c}\frac{d\mathbf{B}}{dt}$$

with $$\operatorname{div} \mathbf{E} = \rho, \ \operatorname{div} \mathbf{B} = 0.$$

It is now at least of philosophical interest to determine how the electrical phenomena in such a system would behave to an observer moving uniformly relative to the fixed frame of reference with the velocity v, which we may, for simplicity, take to be parallel to the x-axis of the fixed coordinate system. The first step naturally will be to take a set of axes moving with the observer, but parallel to the old ones. When this is done the equations lose their original simplicity; but it is found that the original form can be recovered by a slight modification of the variables*.

300. We write $\kappa^2 = 1 - \frac{v^2}{c^2}$ and then put

$$x' = \frac{1}{\kappa}(x - vt), \ y' = y, \ z' = z,$$

$$t' = \frac{1}{\kappa}\left(t - \frac{vx}{c^2}\right),$$

and then if ϕ is any function of (x, y, z, t)

$$\frac{\partial\phi}{\partial x} = \kappa^{-1}\left(\frac{\partial\phi}{\partial x'} - \frac{v}{c^2}\frac{\partial\phi}{\partial t'}\right), \quad \frac{\partial\phi}{\partial y} = \frac{\partial\phi}{\partial y'}, \quad \frac{\partial\phi}{\partial z} = \frac{\partial\phi}{\partial z'}$$

and $$\frac{\partial\phi}{\partial t} = \kappa^{-1}\left(\frac{\partial\phi}{\partial t'} - v\frac{\partial\phi}{\partial x'}\right).$$

Thus in terms of the new variables the fundamental equations of Ampère assume the form

$$\frac{\kappa^{-1}}{c}\left(\frac{\partial\mathbf{E}_x}{\partial t'} - v\frac{\partial\mathbf{E}_x}{\partial x'}\right) + \frac{\rho\mathbf{u}_x}{c} = \frac{\partial\mathbf{H}_z}{\partial y'} - \frac{\partial\mathbf{H}_y}{\partial z'}$$

$$\frac{\kappa^{-1}}{c}\left(\frac{\partial\mathbf{E}_y}{\partial t'} - v\frac{\partial\mathbf{E}_y}{\partial x'}\right) + \frac{\rho\mathbf{u}_y}{c} = \frac{\partial\mathbf{H}_x}{\partial z'} - \kappa\left(\frac{\partial\mathbf{H}_z}{\partial x'} - \frac{v}{c^2}\frac{\partial\mathbf{H}_z}{\partial t'}\right)$$

$$\frac{\kappa^{-1}}{c}\left(\frac{\partial\mathbf{E}_z}{\partial t'} - v\frac{\partial\mathbf{E}_z}{\partial x'}\right) + \frac{\rho\mathbf{u}_z}{c} = \kappa\left(\frac{\partial\mathbf{H}_y}{\partial x'} - \frac{v}{c^2}\frac{\partial\mathbf{H}_y}{\partial t'}\right) - \frac{\partial\mathbf{H}_x}{\partial y'}.$$

* Cf. Lorentz, "Versuch einer Theorie der elektrischen u. optischen Erscheinungen," etc.; Larmor, *Aether and Matter*, chaps. x and xi; Lorentz, *Amsterdam Proc.* (1904), p. 809; Einstein, *Annalen d. Phys.* 17 (1905).

The second and third of these can be written in the form

$$\frac{1}{c}\frac{\partial \mathbf{E}_y'}{\partial t'} + \frac{\rho \mathbf{u}_y}{c} = \frac{\partial \mathbf{H}_x'}{\partial z'} - \frac{\partial \mathbf{H}_z'}{\partial x'},$$

$$\frac{1}{c}\frac{\partial \mathbf{E}_z'}{\partial t'} + \frac{\rho \mathbf{u}_z}{c} = \frac{\partial \mathbf{H}_y'}{\partial x'} - \frac{\partial \mathbf{H}_x'}{\partial y'},$$

where $\mathbf{E}_y' = \kappa^{-1}\left(\mathbf{E}_y - \frac{v}{c}\mathbf{H}_z\right), \quad \mathbf{E}_z' = \kappa^{-1}\left(\mathbf{E}_z + \frac{v}{c}\mathbf{H}_y\right),$

$$\mathbf{H}_y' = \kappa^{-1}\left(\mathbf{H}_y + \frac{v}{c}\mathbf{E}_z\right), \quad \mathbf{H}_z' = \kappa^{-1}\left(\mathbf{H}_z - \frac{v}{c}\mathbf{E}_y\right).$$

These last equations are equivalent to

$$\mathbf{E}_y = \kappa^{-1}\left(\mathbf{E}_y' + \frac{v}{c}\mathbf{H}_z'\right), \quad \mathbf{E}_z = \kappa^{-1}\left(\mathbf{E}_z' - \frac{v}{c}\mathbf{H}_y'\right),$$

$$\mathbf{H}_y = \kappa^{-1}\left(\mathbf{H}_y' - \frac{v}{c}\mathbf{E}_z'\right), \quad \mathbf{H}_z = \kappa^{-1}\left(\mathbf{H}_z' + \frac{v}{c}\mathbf{E}_y'\right).$$

If now we substitute these values in the first equation and use also

$$\mathbf{E}_x' = \mathbf{E}_x, \quad \mathbf{H}_x' = \mathbf{H}_x,$$

it becomes

$$\frac{\kappa^{-1}}{c}\left\{\frac{\partial \mathbf{E}_x'}{\partial t'} - v\left(\frac{\partial \mathbf{E}_x'}{\partial x'} + \frac{\partial \mathbf{E}_y'}{\partial y'} + \frac{\partial \mathbf{E}_z'}{\partial z'}\right)\right\} + \frac{\rho \mathbf{u}_x}{c} = \kappa^{-1}\left(\frac{\partial \mathbf{H}_z'}{\partial y'} - \frac{\partial \mathbf{H}_y'}{\partial z'}\right),$$

while the equation $\qquad \operatorname{div} \mathbf{E} = \rho$

becomes under similar circumstances

$$\kappa^{-1}\left(\frac{\partial \mathbf{E}_x'}{\partial x'} - \frac{v}{c^2}\frac{\partial \mathbf{E}_x'}{\partial t'}\right) + \kappa^{-1}\left(\frac{\partial \mathbf{E}_y'}{\partial y'} + \frac{v}{c}\frac{\partial \mathbf{H}_z'}{\partial y'}\right) + \kappa^{-1}\left(\frac{\partial \mathbf{E}_z'}{\partial z'} - \frac{v}{c}\frac{\partial \mathbf{H}_y'}{\partial z'}\right)$$
$$= \rho.$$

Multiplying this by $\frac{v}{c}$ and adding it to the first equation and remembering that

$$\kappa^2 = 1 - \frac{v^2}{c^2},$$

we get $\qquad \frac{1}{c}\left\{\frac{\partial \mathbf{E}_x'}{\partial t'} + \frac{\rho}{\kappa}(u_x - v)\right\} = \frac{\partial \mathbf{H}_z'}{\partial y'} - \frac{\partial \mathbf{H}_y'}{\partial z'}$

and also $\qquad \frac{\partial \mathbf{E}_x'}{\partial x'} + \frac{\partial \mathbf{E}_y'}{\partial y'} + \frac{\partial \mathbf{E}_z'}{\partial z'} = \frac{\rho}{\kappa}\left(1 - \frac{v u_x}{c^2}\right).$

301. On treating Faraday's equations in the same way we find that they are equivalent to

$$-\frac{1}{c}\frac{\partial \mathbf{H}_x'}{\partial t'} = \frac{\partial \mathbf{E}_z'}{\partial y'} - \frac{\partial \mathbf{E}_y'}{\partial z'},$$

$$-\frac{1}{c}\frac{\partial \mathbf{H}_y'}{\partial t'} = \frac{\partial \mathbf{E}_x'}{\partial z'} - \frac{\partial \mathbf{E}_z'}{\partial x'},$$

$$-\frac{1}{c}\frac{\partial \mathbf{H}_z'}{\partial t'} = \frac{\partial \mathbf{E}_y'}{\partial x'} - \frac{\partial \mathbf{E}_x'}{\partial y'},$$

together with the conditional equation

$$\frac{\partial \mathbf{H}_x'}{\partial x'} + \frac{\partial \mathbf{H}_y'}{\partial y'} + \frac{\partial \mathbf{H}_z'}{\partial z'} = 0.$$

Thus finally if we write

$$\mathbf{u}_x' = \frac{\mathbf{u}_x - v}{1 - \dfrac{v\mathbf{u}_x}{c}} = \frac{dx'}{dt'},$$

$$\mathbf{u}_y' = \frac{\kappa \mathbf{u}_y}{1 - \dfrac{v\mathbf{u}_x}{c}} = \frac{dy'}{dt'},$$

$$\mathbf{u}_z' = \frac{\kappa \mathbf{u}_z}{1 - \dfrac{v\mathbf{u}_x}{c}} = \frac{dz'}{dt'},$$

and
$$\rho' = \rho\kappa^{-1}\left(1 - \frac{v\mathbf{u}_x}{c}\right) = \rho\frac{dt'}{dt},$$

the original form of the equations is completely restored.

Moreover since the determinant of the transformation is unity and the transformation itself is an orthogonal one, we must have

$$dx'dy'dz'dt' = dx\,dy\,dz\,dt$$

so that also
$$\rho' dx'dy'dz' = \rho\,dx\,dy\,dz$$

or the charges in corresponding elements of volume in the two systems are the same.

Thus if we transform any electrical system S in the manner specified above into another system S' in such a way that the charges in corresponding volume elements in the two systems are the same, the electromagnetic equations defining the sequence of events in the system remain unaltered, so long as the vectors **E**$'$, **H**$'$ in S' correspond respectively to the vectors **E**, **H** in S.

302. According to the transformation as a mere geometrical correspondence, the length of a line in the direction of the axis of x, measured in the coordinates (x', y', z'), is greater than its length measured in the coordinates (x, y, z) in the ratio $1 : \kappa$. Thus the Fitzgerald-Lorentz hypothesis of the reduction in the dimensions of a body when it moves relatively to an observer is reduced by this geometrical transformation to the assumption that in the variables associated with it the shape of the body is unaltered.

A similar result holds as regards the time. Suppose we have a clock moving with the origin of the (x', y', z') system of coordinates: then $x' = 0$ and so $x = vt$. Thus if t_1', t_2' are the times of two consecutive events as recorded on the moving clock and t_1, t_2 the times for the same events as registered in the fixed system, then

$$\kappa^{-1}(t_2' - t_1') = t_2 - t_1.$$

We may for instance take t_2', t_1' to represent the two consecutive hour strokes of the clock. It is then clear from the equation above that the moving clock as observed from the fixed system will appear to have its periodic time increased in the ratio $1 : \kappa$. The frequency will be decreased in the inverse ratio.

303. Let us* now imagine an observer, whom we shall call A and to whom we shall assign a fixed position in the aether, to be engaged in the study of the phenomena in the stationary electromagnetic system S. We shall suppose him to be provided with a measuring rod and a clock. By these means he will be able to determine the coordinates (x, y, z) of any point and the time t for any instant, and by studying the electromagnetic field as it manifests itself to him at different places he will be led to the introduction of the vectors **E** and **H** and the fundamental equations connecting them.

Let A' be a second observer, whose task it is to examine the phenomena in the system S', and who himself also moves through the aether with the velocity v, without being aware of his motion or that of the system S'. Let this observer also be provided with a similar measuring rod, which will however be contracted on account of its possessing the motion v, and a clock working properly to his time. If this observer study the electromagnetic phenomena

* Cf. Lorentz, *Theory of Electrons*, ch. v.

in the system S' he will introduce the vectors $\mathbf{E}'$, $\mathbf{H}'$ and find that they are connected by exactly the same equations as were the vectors used by the observer A in the stationary system.

Thus if both A and A' were to keep a record of their observations and the conclusions they draw from them, these records would, on comparison, be found to be identical. In other words neither observer can detect, by an examination of circumstances in his own system, whether he is in motion or at rest.

304. Thus far it has been the task of each observer to examine the conditions in his own system. Let us now suppose that each observer is able to see the system to which the other belongs and to study the phenomena in it. The observer A, in studying the electromagnetic field in S', will be led to introduce the new variables x', y', z', t'; $\mathbf{E}'$, $\mathbf{H}'$, etc., and so will establish our relative equations which are exactly those deduced directly by A'. On the other hand A', in studying the field of S, would introduce new variables (x, y, z, t) defined by the reversed relations (the v to him is now negative) and eventually he would deduce the original equations found by A.

Thus the impressions received by the two observers in examining their own or any other electromagnetic system would be exactly identical, so that in reality neither could be sure of his own or the other's state of motion: all that they know is that their relativity velocity is v. We conclude that if the transformation mathematically defined above has any truth in fact then no measurements of electromagnetic conditions will ever enable anyone to determine whether he is in motion or not, except relatively to other bodies. The idea of absolute motion, frequently employed in our discussions, thus proves to be immaterial to our physical theories.

The conclusion just stated can hardly be avoided on purely philosophical grounds, and if it is granted as intuitive then a sufficient scientific reason is provided for the general validity of the transformation on which it was based. In fact Einstein*, starting from the philosophical principle that absolute motion is physically indeterminate, derives the transformation by purely mathematical considerations. As a consequence of the principle he assumes that

* *Ann. d. Phys.* 17 (1905).

the velocity of propagation of light in any direction referred to any axes must be the same and then proceeds to determine what amount of arbitrariness in the space and time variables is consistent with this assumption. If the changes between the variables from the fixed to a moving system is linear and such that a point at rest in the fixed system corresponds to a point moving with a uniform velocity v parallel to the direction of the x-axis in the other, then we are limited to transformations of the type

$$x' = k\,(x - vt), \quad y' = ly, \quad z' = lz,$$
$$t' = \alpha x + \beta y + \gamma z + \delta t.$$

The analytical criterion is that the equation

$$dx^2 + dy^2 + dz^2 - c^2 dt^2 = 0$$

must lead to the equation

$$dx'^2 + dy'^2 + dz'^2 - c^2 dt'^2 = 0$$

and it is easy to verify that the only forms of the above transformation are those in which

$$k = l\kappa^{-1} = l\left(1 - \frac{v^2}{c^2}\right)^{-\frac{1}{2}},$$
$$\alpha = -\frac{lv}{c^2\kappa}, \quad \beta = \gamma = 0, \quad \delta = l\kappa^{-1}.$$

The transformation thus only differs from that adopted above by the presence of the arbitrary constant factor l; this can however easily be reduced to unity by a proper choice of units.

305. The four-dimensional symmetry of the electromagnetic equations. In the discussions of the various points raised in the previous paragraph an even more remarkable property of the electromagnetic equations has appeared. They appear to possess almost perfect symmetry with respect to the four variables of space and time, a conclusion which is not without its physical significance.

To obtain the symmetry in its most obvious form it is necessary to introduce an entirely new notation in the equations, adopting therein the conventions of four-dimensional algebraic geometry. We use (x_1, x_2, x_3) for the space coordinates (x, y, z) and x_4 in place of the time variable, but with the imaginary factor ic $(i = \sqrt{-1})$; thus we write

$$x = x_1, \quad y = x_2, \quad z = x_3, \quad ict = x_4.$$

Vectors in four-dimensional space may be of two kinds; if they are of the nature of and can be represented by a line they will be sufficiently defined by their four components, the projections of this line on the coordinate axes. There is however a second type of vector—called a *six-vector*—which is of the nature of a surface element; the typical example is the parallelogram formed with the two line elements (dx_1, dx_2, dx_3, dx_4) and $(\delta x_1, \delta x_2, \delta x_3, \delta x_4)$ drawn from any point as adjacent sides which has the six-components —the projections of the area on the six coordinate planes— which are

$$(dx_2\delta x_3 - \delta x_2 dx_3,\ dx_3\delta x_1 - \delta x_3 dx_1,\ dx_1\delta x_2 - \delta x_1 dx_2,$$
$$dx_1\delta x_4 - \delta x_1 dx_4,\ dx_2\delta x_4 - \delta x_2 dx_4,\ dx_3\delta x_4 - \delta x_3 dx_4).$$

We now define the fundamental force-vector $\mathbf{F}$ of the theory by its six-components $(F_{23}, F_{31}, F_{12}, F_{41}, F_{42}, F_{43})$ which are related to the electric and magnetic force intensities by the following equations

$$(F_{23}, F_{31}, F_{12}) = (\mathbf{B}_x, \mathbf{B}_y, \mathbf{B}_z),$$
$$(F_{41}, F_{42}, F_{43}) = i\,(\mathbf{E}_x, \mathbf{E}_y, \mathbf{E}_z).$$

The associated components of the vector are such that

$$F_{rs} = -F_{sr},\quad F_{ss} = 0;$$

the former is a general property characteristic of six-vectors.

With these conventions the equations of Faraday which in their ordinary form are

$$\operatorname{curl}\mathbf{E} = -\frac{1}{c}\dot{\mathbf{B}}$$

with the characteristic equation for $\mathbf{B}$

$$\operatorname{div}\mathbf{B} = 0$$

are then

$$\frac{\partial F_{23}}{\partial x_4} + \frac{\partial F_{34}}{\partial x_2} + \frac{\partial F_{42}}{\partial x_3} = 0,$$
$$\frac{\partial F_{34}}{\partial x_1} + \frac{\partial F_{41}}{\partial x_3} + \frac{\partial F_{13}}{\partial x_4} = 0,$$
$$\frac{\partial F_{12}}{\partial x_4} + \frac{\partial F_{24}}{\partial x_1} + \frac{\partial F_{41}}{\partial x_2} = 0,$$
$$\frac{\partial F_{12}}{\partial x_3} + \frac{\partial F_{23}}{\partial x_1} + \frac{\partial F_{31}}{\partial x_2} = 0.$$

Further if we introduce a four-vector $\mathbf{s}$ defined by its components

$$s_1 = \rho\mathbf{v}_x,\quad s_2 = \rho\mathbf{v}_y,\quad s_3 = \rho\mathbf{v}_z,\quad s_4 = i\rho,$$

then the Ampèrean equations, in the simple form to which we have restricted ourselves in this chapter, become simply

$$\frac{\partial F_{12}}{\partial x_2}+\frac{\partial F_{13}}{\partial x_3}+\frac{\partial F_{14}}{\partial x_4}=s_1,$$

$$\frac{\partial F_{23}}{\partial x_3}+\frac{\partial F_{24}}{\partial x_4}+\frac{\partial F_{21}}{\partial x_1}=s_2,$$

$$\frac{\partial F_{34}}{\partial x_4}+\frac{\partial F_{31}}{\partial x_1}+\frac{\partial F_{32}}{\partial x_2}=s_3,$$

$$\frac{\partial F_{41}}{\partial x_1}+\frac{\partial F_{42}}{\partial x_2}+\frac{\partial F_{43}}{\partial x_3}=s_4,$$

the last equation being the appropriate form of the associated equation

$$\operatorname{div}\mathbf{E}=\rho.$$

306. The equations here given in their four-dimensional form can also be written in an integral form corresponding to Gauss's surface integral theorem in electrostatics. There is in fact a four-dimensional analogue of Green's lemma which, if $\mathbf{P}$ is any six vector, takes the form

$$\int_f \sum_{i,k} P_{ik}\,df_{ik}=\int_v \sum_i \left(\sum_s \frac{\partial P_{is}}{\partial x_s}\right) dv_{jkl},$$

wherein the former integral is taken over the bounding surface f —of which the typical element has components df_{ik}—of the three-dimensional volume v which has the four components dv_{jkl}. In each case the Σ denotes a sum over all possible values (1, 2, 3, 4) of the different indices underneath it.

In this way we see that the two sets of electromagnetic equations are the differential equivalents of the integral equations

$$\int_f \sum_{i,k} F_{ik}\,df_{ik}=\int_v \sum_i s_i\,dv_{jkl}$$

and

$$\int_f \sum_{i,k,l,m} F_{ik}\,df_{lm}=0 \quad (l, m \neq i, k).$$

These integral properties have been applied by Whittaker to extend the notion of tubes of force—he calls the four-dimensional ones *calamoids*. The geometrical idea of such tubes in four dimensions is not only possible but it appears that the tubes have fundamental physical properties which are the analogues of the simpler displacement properties of tubes in static fields.

The lateral surfaces of the tubes are surfaces satisfying the equations

$$\sum_{i,k} F_{ik} df_{ik} = 0, \quad \sum_{i,k,l,m} F_{lm} df_{ik} = 0, \quad (l, m \neq i, k)$$

but the ends may be chosen in two different ways. We may either take $df_{ik} = \lambda F_{ik}$, wherein λ is defined by the relation

$$df^2 = \lambda^2 \sum_{i,k} F_{ik}{}^2 = \lambda^2 (\mathbf{E}^2 - \mathbf{B}^2),$$

or we may take $df_{ik} = \lambda F_{lm}$, wherein λ has the same value if df is the same in both cases. In the first case the integral theorems are equivalent to the statements that

(i) $\dfrac{(\mathbf{EB})}{(\mathbf{E}^2 - \mathbf{B}^2)^{\frac{1}{2}}} df$ is constant along a tube,

and (ii) $(\mathbf{B}^2 - \mathbf{E}^2)^{\frac{1}{2}} df$ is equal to the total amount of $\mathbf{s}$ inside. In the second case the conclusions are similar but the fundamental quantities involved are interchanged.

307. This is however by no means the end of these symmetrical relations, for it can be shown that every fundamental equation of the theory has its symmetrical form. Let us introduce the four-vector which is related to the scalar and vector potentials by the equations

$$(\phi_1, \phi_2, \phi_3) = (\mathbf{A}_x, \mathbf{A}_y, \mathbf{A}_z), \quad \phi_4 = i\phi.$$

Then the six equations

$$\mathbf{B} = \text{curl}\, \mathbf{A}, \quad \mathbf{E} = -\frac{1}{c}\frac{d\mathbf{A}}{dt} - \nabla\phi$$

connecting the electric and magnetic force with these potentials appear as members of a set of six symmetrical equations of type

$$F_{ik} = \frac{\partial \phi_k}{\partial x_i} - \frac{\partial \phi_i}{\partial x_k}.$$

The potential in this theory is an auxiliary mathematical function without direct physical significance. It is in fact, like the scalar and vector potentials from which it is derived, not completely defined in the mathematical sense for if ϕ_i is a solution of the above equations then so also is

$$\phi_i + \frac{\partial \psi}{\partial x_i} \qquad i = 1, 2, 3, 4,$$

where ψ is any scalar function of the four variables x_i. To obtain the specific functions called the retarded potentials we restrict ϕ and **A** so that they satisfy the equation

$$\operatorname{div} \mathbf{A} + \frac{1}{c}\frac{\partial \phi}{\partial t} = 0,$$

which in our notation is simply

$$\sum_i \frac{\partial \phi_i}{\partial x_i} = 0.$$

The characteristic equations satisfied by the component potentials are then simply

$$\sum_i \frac{\partial^2 \phi_k}{\partial x_i^2} = -s_k \qquad k = 1, 2, 3, 4,$$

of which the solutions are also symmetrical in the four variables.

308. The dynamical equations also possess their appropriate symmetry. The force per unit volume on a charge distribution of density ρ moving with velocity **v** is given by the formula

$$\mathbf{F} = \rho\left(\mathbf{E} + \frac{1}{c}[\mathbf{vB}]\right).$$

The three components of this force are obviously the first three components of the composite four-vector

$$F_i = \sum_k F_{ik} s_k \qquad i = 1, 2, 3\ 4.$$

The fourth component of this same vector also has a physical significance for it is

$$F_4 = \frac{i\rho}{c}(\mathbf{vE}),$$

so that it measures the rate at which the electrodynamic forces are doing work.

The equations of motion (cf. p. 354) of a typical charge nucleus (electron) with charge q and momentum

$$\mathbf{M}_e = m\mathbf{v} = \frac{m_0 \mathbf{v}}{(1-\beta^2)^{\frac{1}{2}}},$$

are

$$\frac{d}{dt}(m\mathbf{v}) = q\left(\mathbf{E} + \frac{1}{c}[\mathbf{vB}]\right).$$

Let us now introduce ds as the real four-dimensional element of the path of the electron which is defined so that

$$\begin{aligned} ds^2 &= c^2 dt^2 - dx^2 - dy^2 - dz^2 \\ &= -(dx_1^2 + dx_2^2 + dx_3^2 + dx_4^2), \end{aligned}$$

then
$$\left(\frac{ds}{dt}\right)^2 = c^2 - \mathbf{v}^2 = c^2(1-\beta^2).$$

The components of the momentum $m\mathbf{v}$ are the three quantities

$$\frac{m_0}{\sqrt{1-\beta^2}}\left(\frac{dx_1}{dt}, \frac{dx_2}{dt}, \frac{dx_3}{dt}\right) = \frac{m_0}{c}\left(\frac{dx_1}{ds}, \frac{dx_2}{ds}, \frac{dx_3}{ds}\right),$$

so that the equations of motion are the three equations

$$m_0 \frac{d^2x_i}{ds^2} = K_i \qquad i = 1, 2, 3,$$

wherein K_1, K_2, K_3 are the components of the vector

$$\frac{q}{(1-\beta^2)^{\frac{1}{2}}}\left(\mathbf{E} + \frac{1}{c}[\mathbf{vB}]\right).$$

The fourth symmetrical equation of this set is then simply

$$m_0 \frac{d^2x_4}{ds^2} = \frac{q}{(1-\beta^2)^{\frac{1}{2}}}(\mathbf{Ev}),$$

which in the older notation is

$$\frac{d}{dt}\left\{\frac{m_0c^2}{(1-\beta^2)^{\frac{1}{2}}}\right\} = q(\mathbf{Ev}),$$

the equation of energy, that is if we define the quantity in the bracket as the kinetic energy of the moving charge, or at least to an additive constant.

309. The new definition for the kinetic energy of a moving charge whose mass is

$$m = \frac{m_0}{(1-\beta^2)^{\frac{1}{2}}}$$

shows that it is equal to

$$T = \frac{m_0c^2}{(1-\beta^2)^{\frac{1}{2}}} + \alpha.$$

For small values of $\mathbf{v}$ (or β) this is approximately equal to

$$T = \tfrac{1}{2}m_0\mathbf{v}^2 + (\alpha + m_0c^2),$$

so that if $\alpha = -m_0c^2$, that is if

$$T = m_0c^2\left(\frac{1}{(1-\beta^2)^{\frac{1}{2}}} - 1\right),$$

then the definition is not inconsistent with the ordinary definition for the small velocities which in practice are alone realised.

Further in terms of the mass we have

$$T = mc^2 - m_0c^2 = (m - m_0)\, c^2,$$

so that the kinetic energy of the moving electron is simply proportional to the excess of its mass over the purely static mass of the charge.

The proportionality between mass and energy obtained in this result has been accepted as a fundamental physical fact and all types of energy are now presumed to be associated with a proportional amount of mass. It is of course implied, as regards the electromagnetic energies, in the similarity in the vectors determining the momentum of the field on the one hand and the energy flux throughout it on the other, as the following analysis of a special case will show.

We have seen that when radiant energy, the only type which is not properly associated with a nuclear charge, falls on and is absorbed by a surface it exerts on that surface a force in the direction of propagation and equal to

$$F = \frac{1}{c}\frac{dW}{dt},$$

c being the velocity of radiation and dW/dt the rate of absorption of energy by the surface. This formula can be written in the form

$$Fdt = \left(\frac{dW}{c^2}\right) c.$$

Thus the quantity dW of radiant energy when absorbed by the body imparts to it an amount of momentum equal to Fdt which in this case is equal to

$$\left(\frac{dW}{c^2}\right) c.$$

Thus the associated momentum is the same as if the energy (which travels with velocity c) had a mass

$$m = \frac{dW}{c^2},$$

which is the same result as above.

310. The dynamical equations of the previous paragraph are special cases of the general momentum and energy equations. We first assume these equations in the form in which they involve the Maxwellian stress and the Poynting vector for the flux of energy.

The general stress-tensor of Maxwell may in our notation be expressed by its six components

$$T_{ik} = (\mathbf{E}_i\mathbf{E}_k - \tfrac{1}{2}\mathbf{E}^2\delta_{ik}) + (\mathbf{B}_i\mathbf{B}_k - \tfrac{1}{2}\mathbf{B}^2\delta_{ik}) \qquad (i_k = 1, 2, 3),$$

where δ_{ik} is such that $\delta_{ik} = 1$ if $i = k$

$= 0$ if $i \neq k$.

The vector of momentum per unit volume in the field is given by

$$\mathbf{G} = \frac{1}{c^2}\mathbf{S} \qquad \mathbf{S} = c[\mathbf{EB}].$$

The force per unit volume comprised in the field is then

$$\mathbf{F} = \sum_{i=1}^{3} \frac{\partial T_{ik}}{\partial x_i} - \frac{d\mathbf{G}}{dt} \qquad (i_k = 1, 2, 3),$$

and the energy equation is

$$\frac{dW}{dt} + \operatorname{div}\mathbf{S} = -(\mathbf{Fv}),$$

where $W = \tfrac{1}{2}(\mathbf{E}^2 + \mathbf{B}^2)$.

Now consider the six-vector

$$S_{ik} = \sum_{r=1}^{4} F_{ir}F_{kr} - \tfrac{1}{4}\sum_{r,s=1}^{4} F_{rs}{}^2\delta_{ik}.$$

Obviously $S_{ik} = -T_{ik} \quad (i, k = 1, 2, 3)$.

Further S_{14}, S_{24}, S_{34}

are the components of the vector

$$\frac{i}{c}\mathbf{S} = ic\mathbf{G},$$

whilst $S_{44} = -W$.

That is the six space components of the tensor S are the Maxwell stress components; the three space time components are proportional to the components of the Poynting vector or its equivalent, the momentum density, and the time component is the energy density. The equations of energy and momentum, four in number, are then simply the members of the one symmetrical set

$$F_i = -\sum_k \frac{\partial S_{ik}}{\partial x_k} \qquad (i = 1, 2, 3, 4),$$

wherein F_i is the vector $\sum_k F_{ik}s_k$

introduced above.

The vector S is called the *stress-energy tensor.*

311. The alternative forms of the theory in which different forms for W, the energy density, and S, the flux vector, are employed are also susceptible of similar treatment. In fact it is obvious from the relations by which these quantities are defined that the most general possible expressions for these quantities are of the form

$$W = - S_{44} - 2\Psi \frac{\partial^2 \Psi}{\partial {x_4}^2} + \chi \delta_{44} + C,$$

$$T_{ik} = - S_{ik} - 2\Psi \frac{\partial^2 \Psi}{\partial x_i \partial x_k} + \chi \cdot \delta_{ik},$$

$$(\mathbf{S}_x, \mathbf{S}_y, \mathbf{S}_z) = c^2 (\mathbf{G}_x, \mathbf{G}_y, \mathbf{G}_z)$$
$$= (S_{14}, S_{24}, S_{34}) - 2\Psi \frac{\partial}{\partial x_4} \left(\frac{\partial}{\partial x_1}, \frac{\partial}{\partial x_2}, \frac{\partial}{\partial x_3} \right) \Psi,$$

where S_{ik} has the specific value implied above, Ψ is any scalar function of the four variables x_1, x_2, x_3 and x_4, and χ is defined so that

$$\chi = \left(\frac{\partial \Psi}{\partial x_1}\right)^2 + \left(\frac{\partial \Psi}{\partial x_2}\right)^2 + \left(\frac{\partial \Psi}{\partial x_3}\right)^2 + \left(\frac{\partial \Psi}{\partial x_4}\right)^2.$$

The arbitrary constant C is inserted in W to ensure that this quantity never becomes negative.

It is also possible to express in this manner the more general results which include an account of the dielectric and magnetic properties of the general continuous field, but it is hardly necessary to dwell any further on such extensions.

312. Finally the connection of the electromagnetic equations with the Principal of Least Action also has its four-dimensional equivalent form.

Apart from the intrinsic term depending on the structure of the nuclei the action integral after the introduction of the undetermined multipliers is, in our notation,

$$\int \left\{ \sum_{i,k} \left({F_{ik}}^2 - 2\phi_i \frac{\partial F_{ik}}{\partial x_k} \right) - 2 \sum_i \phi_i s_i \right\} dV,$$

where $dV = dx_1 dx_2 dx_3 dx_4$ is the element of four-dimensional volume and

$$s_1, s_2, s_3 = \Sigma q (\mathbf{v}_x, \mathbf{v}_y, \mathbf{v}_z), \quad s_4 = \Sigma iq,$$

the summations in these last equations being taken per unit volume over the nuclei.

We can introduce the intrinsic terms in the integral either in the previous way by inserting a generalised Lagrangian function depending on the nuclei and leaving its determination to the subsequent analysis, or we can accept the conclusions of the previous detailed analysis for the single electron and take for each one a term

$$L = m_0 (1 - \beta^2)^{\frac{1}{2}} \qquad \beta = \frac{(\mathbf{v})}{c}.$$

Thus this part of the function is now

$$\Sigma \int L\,dt = \Sigma \int m_0 (1 - \beta^2)^{\frac{1}{2}}\,dt = \Sigma \frac{m_0}{c} \int ds,$$

and is also symmetrical in the four variables.

The form of the last result for the part of the Lagrangian function depending on the nuclei seems to suggest a purely geometrical interpretation for the whole problem. The general action integral has to be stationary for variations which leave the four-dimensional paths of the separate particles unaltered. And when we further realise that the conditions for a stationary value of an integral of this type are simply the conditions that the value of the integral itself should be invariant for a differential change of variables, we may be tempted further towards the geometrical and away from the dynamical significance of the whole theory. However, we must not impart too great importance to the geometrical as distinct from the physical form of the theory. It is of course not surprising that the purely geometrical interpretation of the relations of the theory is possible; it would rather have been surprising if it were otherwise, for all the equations have been interpreted in terms of vectors, which are defined so that they can be represented in magnitude and direction (but not in physical significance) by geometrical elements.

Whatever interpretation we place on the relations of our theory however we see that the whole of our equations are expressible in a form which involves the four independent variables of space and time in a perfectly symmetrical manner. This conclusion taken in conjunction with the results of the previous section is usually interpreted as implying that there is in reality no fundamental distinction between the two types of variables; but in this connection it is as well to remember that mere analytical or geometrical equivalence of the measures of time and distance is not necessarily all that we are concerned with; the numerical measures must always have dimensions associated with them.

Finally we must notice a further question raised by these discussions as to the reality of the aether, or indeed of any of the more substantial bases of the foregoing theory. Since the form of our equations is precisely the same whether they are referred to fixed or moving axes, we can draw no conclusion from such form—or in fact from any phenomena described by them—as to whether we ourselves are moving relatively to the aether or not. But if motion of the aether relative to matter has no influence on physical phenomena connected with that matter, is it likely that there can be other characteristics of such an aether which insist on being taken more seriously?

Further, it is impossible to determine the absolute motion of any body—that is, in this theory, its motion relative to the aether—and therefore we cannot know its correct normal mass or even its normal dimensions; and when two bodies are in relative motion we have the curious paradox that each is contracted when viewed from a frame of reference on the other. We must not however pretend to be capable of getting outside of our own scheme of these things in order to view them from an independent angle. All observations of physical theory, and more particularly our descriptions of them, are made in terms of ordinary matter with which we are all now familiar, and our theories—that is our correlations of facts—necessarily have this relative significance, and they can have no other. As necessity demands we extend and shift our frame of reference and alter our measuring instruments, but they remain in essence material. It is in fact quite impossible for us to attempt to examine *space*, for example, apart from its material content, for we have no sense which is capable of perceiving it. We may fit on to any distribution of matter any theoretically possible kind of space, be it curved or flat, that we may like, but we must not pretend that it is anything more than a mere fiction of the imagination. And so it is with our aether. All that we really recognise in this and in other theories are the different related performances of pieces of matter, and we can only describe them in relation to a material framework (real or imaginary) plotted out with material rulers. What happens in the space between (space is that something which can be occupied by matter) we do not and cannot know. But in order to simplify our description of the observed relations it is convenient and helpful to imagine the aether existing throughout our framework and behaving in the requisite manner, and our description is to a certain extent saved by the fact, which we have frequently emphasised, that no special constitution is essential for the development of the ideas.

If a stage is reached in our mental development when we can *all* think clearly in terms of four-dimensional tensors and the complex differential equations associated with them, we may be in a position to dispense with our present material aids to description; but we shall merely replace them by others which we shall then consider more suitable and appropriate.

APPENDIX I

ON THE MECHANISM OF MAGNETIC INDUCTION

313. Paramagnetism and diamagnetism. The theory of induced magnetism developed above is based largely on a linear relation between the induced polarisation intensity **I** and the inducing magnetic force **B** which in isotropic media is of the form

$$\mathbf{I} = \mu'\mathbf{B} = (1 - \mu)\,\mathbf{B}.$$

If the susceptibility μ' is positive then the permeability $\mu < 1$ and the magnetisation has the same sign as the magnetising force, that is, it is in the same direction. If however μ' is negative the permeability is greater than unity and the two are in opposite directions. In practice both cases occur. Those substances for which μ' is positive are called *paramagnetic* substances, whilst those for which μ' is negative are called *diamagnetic* substances.

In all diamagnetic substances μ' is very small: it is largest in Bismuth where however it only amounts to $2{\cdot}5 \times 10^{-6}$. In paramagnetic substances it is always equally small except in the substances of the iron group, which form a special magnetic group of their own known as the *ferromagnetic* substances, the typical ones being Iron, Nickel and Cobalt.

In general then we may say that the above theory with the linear relation is applicable under all circumstances except when ferromagnetic substances are present in the field.

The distinction between paramagnetic and diamagnetic magnetism is fundamental in this theory, because the two phenomena wherever they occur vary in quite a different manner with changing physical conditions of the body. The diamagnetic coefficient is apparently nearly independent of the temperature and of changes in the chemical state of the material. On the other hand in all feebly paramagnetic bodies the coefficient μ' of magnetisation varies inversely as the absolute temperature (ϑ) with a degree of accuracy which tends to perfection at high temperatures. This simple law, first formulated by Curie, may be written in the symbolic form

$$\mu' \vartheta = \text{const.},$$

and its accuracy has been confirmed by recent investigation for certain substances down to the lowest attainable temperatures. There are however certain substances whose behaviour does not conform to the rule; some of these appear to obey a law of the form

$$\mu' (\vartheta + \kappa) = \text{const.},$$

where κ is constant.

Curie himself drew the obvious inference from this physical difference. Diamagnetism is an affair of the internal constitution of the molecules, having only slight relation to their bodily motions on which the temperature depends. On the other hand paramagnetism is an affair of orientation of the molecules in space without change of internal conformation, so that alteration of the mean state of translational motion is involved in it and we should expect a temperature effect.

314. We can illustrate the point of view in more detail by the analysis for the simple case of a gaseous medium. If we assume that the molecules of the gas are small charged particles possessing rotatory degrees of freedom, we can imagine that the revolving of the charge elements provides the current whirls which are—on Ampère's hypothesis—equivalent to the magnetism observed, and then we can get the temperature effect by imagining that the motions in the rotatory degrees of freedom are in thermodynamic equilibrium with the ordinary to and fro motions of the molecules. We assume that the rotation of a molecule about an axis with angular momentum M is equivalent to a magnetic moment along that axis equal to μM.

If then we assume an axis of symmetry and use the Eulerian coordinates (θ, ϕ, ψ) for the axes of principal inertia with corresponding moments A, A, C, the kinetic energy of the rotatory motions, which are dynamically independent of the translatory motions, is of the form

$$T = \tfrac{1}{2} [A (\dot{\theta}^2 + \sin^2 \theta \dot{\psi}^2) + C (\dot{\psi} + \dot{\phi} \cos \theta)^2].$$

In addition there is the magnetic potential energy which, if the applied magnetic force intensity B is along the polar axes of the coordinates, is

$$-\mu B\,[A\dot{\phi}\sin^2\theta + C\cos\theta\,(\dot{\psi} + \dot{\phi}\cos\theta)],$$

the term in the square brackets representing the component of the resultant momentum in the direction of B.

The difficulty of the present method is in the application of the statistical method. According to the general theory the distribution function defining the number of molecules with their coordinates θ, ϕ, ψ and velocities (or momenta) $\dot{\theta}$, $\dot{\phi}$, $\dot{\psi}$ in the usual differential range of these variables is of the form $e^{-E/R\vartheta}$; but the E in this expression, which represents the total energy of the system, is not necessarily the sum of the kinetic and potential energies. For this sum is not conserved on the collision or interaction of the molecules. To find a more suitable function which is properly conserved we have to proceed on more general lines*. In dynamical theory it is shown that for a dynamical system of more general type defined by the coordinates $q_1, q_2, \ldots q_n$ the quantity

$$E = -L + \sum_{i=1}^{n} \frac{\partial L}{\partial \dot{q}_i}\dot{q}_i,$$

wherein $L = T - V$, is such a function. When L involves the velocities $\dot{q}_i$ only as a homogeneous quadratic function this expression for E reduces to $T + V$, but in other cases, as the present, where L contains terms linear in the velocities, no such reduction to the ordinary energy expression is possible. In such cases it is usual to call E, defined as above, the energy of the system, this being the only expression to which definite physical conception, involving conservation, is possible.

In the present case

$$L = \tfrac{1}{2}\,[A\,(\dot{\theta}^2 + \sin^2\theta\dot{\phi}^2) + C\,(\dot{\psi} + \dot{\phi}\cos\theta)^2] \\ + \mu B\,[A\dot{\phi}\sin^2\theta + C\cos\theta\,(\dot{\psi} + \dot{\phi}\cos\theta)],$$

from which it is easy to see that

$$E = \tfrac{1}{2}A\,(\dot{\theta}^2 + \sin^2\theta\dot{\phi}^2) + \tfrac{1}{2}C\,(\dot{\psi} + \dot{\phi}\cos\theta)^2.$$

The statistical distribution in its invariant form is, after Liouville, expressed not in terms of the velocities but of the corresponding

* I am indebted to Prof. Larmor for the correct form of this argument.

momenta Θ, Φ, Ψ. These are defined generally from the Lagrangian function L by the equations

$$\Theta = \frac{\partial L}{\partial \dot{\theta}} = A\dot{\theta},$$

$$\Phi = \frac{\partial L}{\partial \dot{\phi}} = A \sin^2 \theta \dot{\phi} + C \cos\theta\,(\dot{\psi} + \dot{\phi}\cos\theta) + \mu B[A \sin^2\theta + C\cos^2\theta],$$

$$\Psi = \frac{\partial L}{\partial \dot{\psi}} = C\,(\dot{\psi} + \dot{\phi}\cos\theta) + \mu BC\cos\theta,$$

and in terms of them E can be written in the form

$$E = \frac{\Theta^2}{2A} + \frac{(\Phi - \Psi\cos\theta - \mu BA\sin^2\theta)^2}{2A\sin^2\theta} + \frac{(\Psi - \mu BC\cos\theta)^2}{2C}.$$

The statistical theory now tells us that the number of elements per unit volume for which the variables Θ, Φ, Ψ, θ, ϕ, ψ lie within the small differential range $d\Omega$ is

$$dn = a e^{-\frac{E}{R\vartheta}}\, d\Omega,$$

where $$d\Omega = \sin\theta\, d\theta\, d\phi\, d\psi\, d\Theta\, d\Phi\, d\Psi.$$

The constant a has to be determined from the fact that the total number n of the molecules per unit volume is specified so that

$$n = a\int_{-\infty}^{+\infty} e^{-\frac{E}{R\vartheta}}\, d\Omega.$$

Now each of the molecules in dn contributes the same amount to the magnetisation along the direction of B, viz.

$$\mu\,[A\dot{\phi}\sin^2\theta + C\cos\theta\,(\dot{\psi} + \dot{\phi}\cos\theta)],$$

so that on the whole we have

$$I = a\mu\int_{-\infty}^{+\infty} [A\dot{\phi}\sin^2\theta + C\cos\theta\,(\dot{\psi} + \dot{\phi}\cos\theta)]\, e^{-\frac{E}{R\vartheta}}\, d\Omega$$

$$= a\mu\int_{-\infty}^{+\infty} [\Phi - \mu B\,(A\sin^2\theta + C\cos^2\theta)]\, e^{-\frac{E}{R\vartheta}}\, d\Omega.$$

If the limits for Φ and Ψ are both from $-\infty$ to $+\infty$ the result of the integration for I is obviously zero. If the range of variation of the motion in the rotatory coordinates is unrestricted the resultant effect of the magnetic field will be zero, no resultant polarity either of diamagnetic or paramagnetic character will be induced. To get a finite result we must therefore restrict the variation in one of the rotatory coordinates ϕ, ψ. Let us examine for example the special case when Ψ is restricted to a single constant value, the same for all molecules, and let us for simplicity also take $A = 0$. We get then in effect

$$I = \alpha\mu \int_0^\pi x(\Psi - x)\, e^{-\frac{x^2}{2RC\vartheta}}\, dx,$$

where we have written

$$x = \Psi - \mu BC \cos\theta.$$

The polarisation is therefore in part diamagnetic and in part paramagnetic, the latter arising from the permanent angular momentum Ψ in the molecules and the former from the induced angular momentum $\mu BC\cos\theta$. That depending on Ψ is mainly a matter of thermal agitation of the molecules, but the other part is a purely internal molecular phenomenon.

315. In the special case examined by Langevin, who was the first to develop this mode of attack, $C = 0$ and $\mu = \infty$, but in such a way that

$$m = \mu\Psi$$

is a finite constant. It is then found that the magnetisation is entirely paramagnetic and its intensity is

$$I = nm\left[\coth\left(\frac{mB}{R\vartheta}\right) - \frac{R\vartheta}{mB}\right].$$

Thus I, the intensity of magnetisation induced, is a function of (B/ϑ). Moreover for small values of the inducing force the functional relation becomes a mere proportionality or

$$I = \frac{m^2nB}{2R\vartheta},$$

so that the susceptibility is

$$\mu' = \frac{nm^2}{2R\vartheta},$$

and it varies inversely as the absolute temperature in accordance with Curie's law.

In the more general case the two types of magnetism will be present and superposed so that circumstances will always be more complex than those examined in Langevin's case. In certain circumstances the two types may even just balance one another when the medium is apparently non-magnetic.

316. The relation between paramagnetism and temperature is so simple that it must be the expression of a theoretical principle. The following considerations in fact derive it from Carnot's principle on the assumption that the internal energy associated with the magnetisation is purely thermal in character. Consider then a mass of paramagnetic material, moved up at temperature $\vartheta + \delta\vartheta$ from a place where the intensity of the field is B to a place where it is $B + \delta B$ and then moved back again at a temperature ϑ. The aggregate per unit volume of the internal energy of the medium is increased by $-(B\delta I)$ in the move up and decreased by the same amount in the move back. Thus if $h + dh$ is the thermal energy per unit volume which it must receive from without at the higher temperature, and h that which it must return at the lower, in order to perform the amount of external work which is $\frac{\partial I}{\partial \vartheta}\delta B\delta\theta$, then by Carnot's principle

$$\frac{\partial I}{\partial \vartheta}\delta B = \frac{h}{\vartheta},$$

but by assumption $h = -B\delta I$ so that

$$\frac{\partial I}{\partial \vartheta}\delta B = -\frac{B\delta I}{\vartheta},$$

or

$$\frac{\partial I}{\partial \vartheta} = -\frac{B}{\vartheta}\frac{\partial I}{\partial B},$$

whence

$$I = f\left(\frac{B}{\vartheta}\right),$$

f being some arbitrary function. When the magnetising force B is very small or the temperature ϑ very large this relation is approximately equivalent to

$$I = \mu' B, \quad \mu' = \frac{a}{\vartheta},$$

which is again Curie's law.

Thus if the magnetisation arises from the effort of the magnetic field to orientate the molecules which are spinning about as the result of gaseous encounters, and if therefore the whole of the internal energy of the medium associated with its magnetisation is to be regarded as added thermal energy of these rotations and not as internal energy of any regular elastic type, then Curie's law is a natural result. If only the proportion $\frac{\vartheta}{\vartheta+\kappa}$ of the internal energy is converted into heat then we should derive the modified law

$$\mu'(\vartheta+\kappa) = \text{const.},$$

and in any case we may regard any such deviation from Curie's law as indicating an incomplete conversion of the internal energy into heat.

317. Ferromagnetism. The ferromagnetic substances, which with their compounds are the only substances to show large magnetic effects, are not only exceptional in respect to the magnitude of these effects, but also as to the manner in which they vary. The linear relation between the polarisation induced and the polarising force is far from being valid except perhaps when the inducing force is small, in which case it is necessarily sufficient, and even for a given field strength the variation of the susceptibility which is still defined by the relation

$$\mu' = I/B$$

with temperature does not even approximately obey Curie's law.

In the absence of adequate analytical means of expressing it the general relation between I and B is usually represented graphically

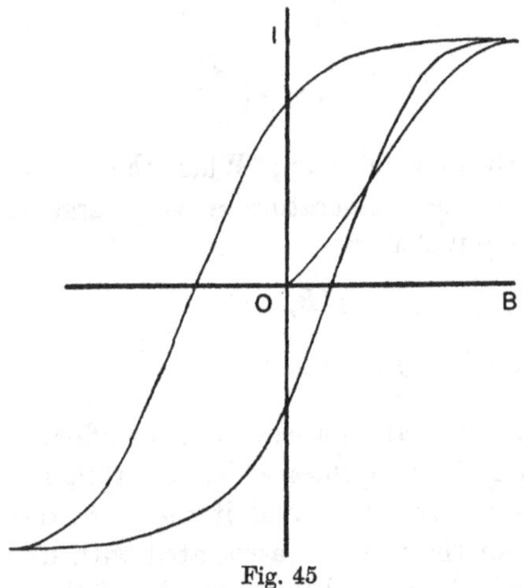

Fig. 45

and the type of curve obtained is always similar to that exhibited in Fig. 45, where the ordinates represent the values of I, the polarisation, and the abscissa the values of B, the force. For small values of B the curve is straight, indicating that the susceptibility is independent of the force. Beyond the point where $B = \cdot 02$ unit the curve begins to rise more rapidly and the value of μ' quickly approaches its constant value unity which it maintains for a considerable range. Eventually however it begins to decrease to the value zero which it has for large fields. This means that the intensity does not further increase as the force increases—the substance is *saturated*. For a specimen of soft iron Ewing found that saturation was practically obtained when $B = 16000$ units; for steel even larger forces are required.

If the applied field is gradually removed from a saturated piece of iron it will retain a considerable portion of its magnetisation—the property of magnetic *retentiveness* exhibited in particular by permanent magnets. If the process is further continued and a field of increasing strength in the opposite direction is applied, the magnetisation is gradually destroyed and then created in the opposite direction and ultimately to its full saturation value again. If at this stage the field is again gradually reduced and replaced slowly by the original one, a similar set of changes in the magnetisation occur but in the reverse order. The result is the familiar magnetic cycle figure already referred to.

318. Various more or less successful attempts have been made to account for this special complexity in the magnetic properties of the ferromagnetic substances. The matter is of the utmost importance from the technical point of view and has consequently received a good deal of attention in theoretical discussions, but it is only during the last few years that any substantial progress has been made. The mere fact that the ferromagnetic substances are exceptions to Curie's law suggests that their magnetism is a constitutional phenomena involved with the elastic and therefore structural properties of the media. Structurally these substances consist of an immense number of minute and irregular crystals—they all belong to the cubic system—whose axes are uniformly distributed in all directions. In each of these crystals we may imagine that the molecular magnets are arranged in a regular space lattice, and if an external field act on the substance all the elementary magnets in each crystal will tend to turn with their axes in the direction of the field: they are either wholly or partially prevented from so doing by the action of the mutual magnetic or other forces tending to hold them in their original stable configuration. With the increase of the field the molecules will turn more and more in the direction of the field and consequently the intensity of magnetisation becomes greater, tending however to its maximum or saturation value when all the magnets are in line with the field.

Detailed numerical calculations from this point of view have been made by Ewing and more extensively recently by K. Honda and his pupils. The latter in particular has had good success in fitting the actual results into a detailed theory. In any case, however, the ideas seem of too general a nature to be offered as a complete solution of the constitution problem.

319. More relevant information towards a solution is provided by the behaviour of one or two of the crystals containing iron. Pyrhotite for example, which is a sulphide of iron, has specially simple magnetic properties. This substance possesses three mutually perpendicular planes of magnetic symmetry and is much more easily magnetised in a direction perpendicular to one of these planes than in any other direction. Moreover if an attempt is made to magnetise the crystal in this one direction it is found that it remains unmagnetised until the magnetic force reaches the definite critical value B_s when the intensity of magnetisation assumes its saturation value I_s, which it will retain perfectly constant until the magnetic force reaches the value $-B_s$ when the magnetisation itself suddenly reverses to the value $-I_s$ and so on. The hysteresis curve showing the relation between I and B is thus a parallelogram (Fig. 46). Obviously in a substance of this type the molecular magnets have only two positions of equilibrium which are stable under the action of the internal constitutive forces, viz. the two positions in which they are perpendicular to the one principal plane. Moreover it appears that either of these positions ceases to be a position of stable equilibrium as soon as the opposing magnetic force reaches its critical value. This explanation is still further supported by the fact that the magnetic behaviour of the substance in any direction other than along its one principal axis is perfectly continuous and in complete accord with the theoretical consequences of such a view.

Broadly speaking the other ferromagnetic crystalline minerals exhibit the same general features: there are important differences in detail, but they all possess different magnetic properties along the different axes of symmetry and usually one axis of conspicuously easy magnetisation. Moreover the similarity in the general shape of the induction curves in all cases with that obtained for say iron shows that probably the only difference between iron and its crystals is that iron itself is probably constituted of an irregular conglomeration of small crystals of the simpler type. This view is supported by the fact that carefully prepared specimens of iron can be obtained which give an almost exact parallelogram curve of induction.

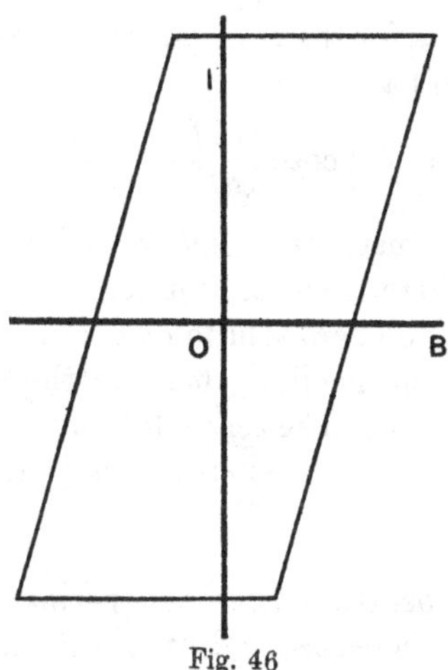

Fig. 46

320. A bold attempt has been made by Weiss at a detailed analytical theory of the ferromagnetic phenomena. Regarding the non-applicability of Curie's law as indicating the at least partial control by other than thermal agitations of the orientation processes induced by the field, he starts by assuming in a general manner that when the medium is distorted, that is when the molecules are so orientated that the intensity of magnetisation is I, then the elastic forces acting on the magnetic molecules are the equivalent of a magnetic force aI in the direction of I and acting on the magnetic element in the molecule*. We can, if we like, regard this force as a local force of magnetic origin, but this is not necessary for the argument and probably only represents the partial truth in any case.

On this idea we may regard the molecular magnets as in the gaseous case but orientated under the action of a complete force $B + aI$. If we take the simple case given by Langevin this gives, as the equation for I,

$$I = mn\left[\coth\left\{\frac{m}{R\vartheta}(B + aI)\right\} - \frac{R\vartheta}{m(B + aI)}\right],$$

or in terms of $I_s = mn$ the saturation value of I,

$$I = I_s\left[\coth\left\{\frac{I_s(B + aI)}{nR\vartheta}\right\} - \frac{nR\vartheta}{I_s(B + aI)}\right].$$

This formula accounts for the phenomena of permanent magnetisation, for if $B = 0$ we have

$$I = I_s\left[\coth\left(\frac{aII_s}{nR\vartheta}\right) - \frac{nR\vartheta}{aII_s}\right],$$

and under ideal circumstances n and a are both independent of ϑ and I, and there is therefore a solution for I which is different from zero. Moreover this non-zero solution when it exists represents the stable condition of the medium, because the magnetic potential energy in it is a minimum. The zero solution corresponds to a maximum value of the potential energy and is therefore in general unstable.

321. As the temperature rises the permanent magnetisation intensity gradually decreases, and at a certain temperature θ_s it vanishes altogether. Beyond this temperature the only possible real solution of the equation for I is the zero one. The temperature θ_s obtained from the equation may therefore be interpreted as the temperature at which the ferromagnetic quality disappears, that is the so-called critical temperature. In the neighbourhood of this temperature the magnetisation is always small, so that using

$$x = \frac{aII_s}{nR\vartheta},$$

* a is a constitutive constant of the medium.

we have approximately

$$\frac{1}{I_s}\frac{dI}{dx} = \frac{d}{dx}\left[\coth x - \frac{1}{x}\right] = 1 - \operatorname{cosech}^2 x + \frac{1}{x^2} = \tfrac{1}{3}.$$

But

$$\frac{dI}{dx} = \frac{nR\vartheta}{aI_s},$$

so that at the actual critical temperature

$$\frac{I_s}{3} = \frac{nR\vartheta_s}{aI_s},$$

or

$$\theta_s = \frac{aI_s^2}{3nR}.$$

If now we introduce this temperature into the relation determining the intensity of permanent magnetisation it assumes the form

$$I = I_s\left[\coth\left(\frac{3I\vartheta_s}{\vartheta I_s}\right) - \frac{\vartheta I_s}{3I\vartheta_s}\right],$$

so that the ratio I/I_s is a function of ϑ/ϑ_s and the function is the same for all substances.

Thus if we express the intensity of permanent magnetisation in terms of the maximum possible intensity I_s and the absolute temperature θ in terms of the absolute critical temperature, we obtain a characteristic equation for the intensity of permanent magnetisation at a given temperature, which is identical for all ferromagnetic substances. This relation has been tested by Weiss* in the case of magnetite and he finds that it is satisfied with great accuracy except at very low temperatures (− 79° C.) and in the neighbourhood of the critical temperature, where however the deviations are not large. When it is observed that there are no disposable constants in the formula this agreement between the observed facts and what at first sight appears to be merely a provisory theory can only be regarded as remarkable.

322. The theory proves equally successful in explaining the observed facts of the phenomenon of induced magnetisation. Of course, as we have already noticed, any simple theory of the present type is hardly likely to be directly applicable to the ferromagnetic metals, the magnetic behaviour of which is complicated by various secondary causes. It is however found that the simple phenomena

* *Jour. de Phys.* VI. p. 665 (1907).

accompanying the magnetisation of the various crystalline ferromagnetic minerals when placed in a field parallel to their principal magnetic axis fit in admirably with the theoretical conclusions to be drawn from the theory.

The general relation between I and B is expressed by the equation

$$I = I_s\left[\coth\left\{\frac{I_s(B+aI)}{nR\vartheta}\right\} - \frac{nR\vartheta}{I_s(B+aI)}\right].$$

This requires that in general I should be a continuous function of B when θ is maintained constant, which is just the opposite to what was actually found to be the case, the magnetisation being practically independent of the magnetic field and equal to its saturation value at the given temperature. This could be the case only when the internal local field of intensity aI far exceeds in actual magnitude any field that we can experimentally produce. Now this actually appears to be the case, as we shall soon see; so that under all circumstances except when I is very small, that is in the neighbourhood of the critical temperature, the intensity of magnetisation is determined by

$$I = I_s\left[\coth\left(\frac{aII_s}{nR\vartheta}\right) - \frac{nR\vartheta}{aII_s}\right],$$

and is equal to the saturation intensity in all fields.

In the neighbourhood of the critical temperature, when I is small, the previous argument is invalid. But then the general relation may be simplified by approximation, and for all practical purposes it is equivalent to

$$\frac{I}{I_s} = \frac{I_s(H+aI)}{3nR\vartheta},$$

or introducing the critical temperature

$$(\vartheta - \vartheta_s)\,I = \frac{H\vartheta_s}{a},$$

in which form it has been satisfactorily verified in numerous cases, even with iron. This latter agreement is not surprising because in the neighbourhood of the critical temperature the constitutional irregularities are becoming very unstable.

In all cases where the last relation is experimentally verified, a determination of the constant a may be made, and hence an estimate of the strength of the local magnetic fields. From measurements of this kind Weiss finds that the intensity of the local field aI in iron is 6·56 . 10^6 absolute units and it is of a similar order of magnitude in other substances. This is very much greater than the strongest magnetic field (5 . 10^4 units) which can be produced in the laboratory, so that the peculiar result that the intensity of induced magnetisation I_s does not, under the simplest circumstances, appreciably alter with the external field is satisfactorily accounted for.

APPENDIX II

ON THE MECHANISM OF METALLIC CONDUCTION

323. Many attempts have been made to develop into further detail the idea of the electric current as a process of diffusion, but before the introduction of the electron theory these attempts could hardly be described as very successful. The difficulty arises in the absence of any sign of transport of matter or of any chemical change accompanying the transport of electricity. Every current circuit must consist of two or more portions composed of different materials which may contain no element in common and we must suppose either that the particles which carry the current can pass from one material to the other or that they cannot. Either supposition lands us in enormous difficulties. If the particles cannot cross the boundary between the different metallic conductors they must remain piled up at these boundaries, even if they are identical with the atoms of the metal in which they move, and this piling up must alter the distribution of the mass in the conductor, while, if they are composed of some substance different from that of the rest of the material, some sort of chemical separation should occur. But the most careful experiments on pure metals and alloys have failed to show the slightest change in any properties of a metallic conductor induced by the passage of a current through it. On the other hand if we suppose that the particles can pass freely from one material to another new difficulties arise. We know that the atoms and groups of atoms of different elements differ markedly in their properties: and we could certainly detect the presence in one substance of atoms derived from a foreign element. Since the properties of the materials forming a non-uniform circuit are unchanged by the passage of the current, if the electricity is conveyed at all by diffusing particles, these particles must be of a nature common to all elements, or at least to all elements forming metallic conductors. Previous to the discovery of the electron however no such electrical elements were know to exist.

The discovery of the electron and the consequent formulation of the so-called 'electron theory' removed in one stroke all the difficulties thus inherent in the earlier theories. According to this theory* there are in every metal a very large number of (negative) electrons freely movable in the spaces between the atoms, and it is the diffusion of these electrons through the metal under the action of the electric force in the external field that is the essence of a conduction current. If there is no external field the velocities of these free electrons will be distributed equally in all directions: there will be no tendency for them to move in one direction rather than in any other; but if the metal is placed in an electric field the electrons are subject to a force in a definite direction (that of the force in the field) in virtue of their charge, and those moving in this one direction will have their velocities increased whilst those moving in the opposite direction will have theirs decreased, there will thus be a slight drift of the electrons in a definite direction and this constitutes a current of electricity: the velocity of drift is however kept in check by the continued encounters of the electrons with the metal molecules and with one another when additional forces come into play tending to deflect the electrons from their forward motion: the essential conditions for a diffusion flux are thus satisfied.

324. Problems relating to the motion of the innumerable number of electrons in a piece of metal are best treated by the statistical method which Maxwell introduced into the kinetic theory of gases, and which may be represented in a simple geometrical form so long as we are concerned only with the motion of translation of the electrons. Indeed it is clear that, if we construct a diagram in which the velocity of each electron is represented in direction and magnitude by a vector OP drawn from a fixed point O, the distribution of the ends of these vectors, the velocity points as we shall say, will give us an image of the state of the motion of the electrons.

If the positions of the velocity points are referred to axes of coordinates parallel to those that have been chosen in the metal itself, the coordinates of a velocity point are equal to the components ξ, η, ζ of the velocity of the corresponding electron.

* Cf. the original papers by Riecke, *Wied. Ann.* 66 (1898), pp. 353, 545, 1199; Drude, *Ann. der Phys.* (1900), 1, p. 566; 3, p. 369; Thomson, *Rapports de Congrès de Physique* (Paris, 1900), 3, p. 318. The treatment here given follows that given by Lorentz, *The Theory of Electrons*, p. 266, or *Proc. Amsterdam Academy*, 7 (1905), pp. 438, 585.

Let dv be an element of volume in the diagram, situated at the point (ξ, η, ζ) so small that we may neglect the changes of (ξ, η, ζ) from one of its points to another, but yet so large that it contains a great number of velocity points. Then this number may be reckoned to be proportional to dv: representing it by

$$f(\xi, \eta, \zeta)\, dv$$

per unit volume of the metal, we may say that, from a statistical point of view, f determines completely the motion of the swarm of electrons.

It is clear that the integral

$$\int f(\xi, \eta, \zeta)\, dv,$$

extended over the whole space of the diagram, gives the total number of electrons per unit of volume. In like manner

$$\int \xi f(\xi, \eta, \zeta)\, dv$$

represents the stream of electrons through a plane perpendicular to Ox: i.e. the excess of the number passing through the plane towards the positive side over the number of those which go in the opposite direction, both numbers being referred to unit of area and unit of time. This is seen by considering first a group of electrons having their velocity points in an element dv; these may be regarded as moving with equal velocities, and those of them which pass through an element df of area in the said direction between the moments t and $t + dt$ have been situated at the beginning of this interval in a certain cylinder having df for its base and the height ξdt. The number of these particles is found if one multiplies the volume of the cylinder by the number per unit volume.

Hence if $\int_+$ means an integration over the part of the diagram on the positive side of the η-ζ plane, and $\int_-$ an integration over the part on the opposite side, the number of the electrons which go to one side is

$$df\,dt \int_+ \xi f(\xi, \eta, \zeta)\, dv,$$

and that of the particles going the opposite way

$$df\,dt \int_- - \xi f(\xi, \eta, \zeta)\, dv.$$

The expression given above is the difference between these values divided by $dfdt$.

If all the electrons have equal charges e, the excess of the charge that is carried towards the positive side over that which is transported in the opposite direction is given by

$$C = e\int \xi f dv,$$

and it is easily seen that if we use

$$u^2 \equiv \xi^2 + \eta^2 + \zeta^2$$

for the square of the absolute velocity of an electron, then

$$H = \tfrac{1}{2}m\int \xi u^2 f dv$$

is the expression for the difference between the amounts of energy that are carried through the plane in the opposite direction.

The function f is to be determined by an equation that is to be regarded as the fundamental equation of the theory, and which we now proceed to establish on the assumption that the electrons are subject to a force giving them an acceleration $\mathbf{F}$ equal for all the electrons in one of the groups considered.

325. Let us fix our attention on the electrons lying at the time t in an element of volume dv of the metal and having their velocity points in the element dv of the diagram. If there were no encounters of the electrons, neither with other electrons nor with the metallic atoms, these electrons would be found, at the time $t + dt$, in an element dv' equal to dv and lying at the point $(x + \xi dt, y + \eta dt, z + \zeta dt)$. At the same time their velocity points would have been displaced to an element dv' equal to dv and situated at the point $(\xi + \mathbf{F}_x dt, \eta + \mathbf{F}_y dt, \zeta + \mathbf{F}_z dt)$. We should have therefore

$$f(\xi + \mathbf{F}_x dt, \eta + \mathbf{F}_y dt, \zeta + \mathbf{F}_z dt, x + \xi dt, y + \eta dt, z + \zeta dt, t + dt)\, dv' dv' = f(\xi, \eta, \zeta, x, y, z, t)\, dv dv.$$

The encounters or impacts which take place during the interval of time considered require us to modify this equation. The number of electrons constituting at the time $t + dt$, the group specified by $dv'dv'$, is no longer equal to the number of those which at the time

t belonged to the group $dv\,dv$, the latter number having to be diminished by the number of impacts which the group of electrons under consideration undergoes during the time dt and increased by the number of impacts by which an electron, originally not belonging to the group, is made to enter it. Writing $a\,dv\,dv\,dt$ and $b\,dv\,dv\,dt$ for these two numbers we have, after division by $dv\,dv = dv'dv'$,

$$f(\xi + \mathbf{F}_x dt, \eta + \mathbf{F}_y dt, \zeta + \mathbf{F}_z dt, x + \xi dt, y + \eta dt, z + \zeta dt, t + dt) = f(\xi, \eta, \zeta, x, y, z, t) + (b - a),$$

or what is practically the same thing

$$\mathbf{F}_x \frac{\partial f}{\partial \xi} + \mathbf{F}_y \frac{\partial f}{\partial \eta} + \mathbf{F}_z \frac{\partial f}{\partial \zeta} + \xi \frac{\partial f}{\partial x} + \eta \frac{\partial f}{\partial y} + \zeta \frac{\partial f}{\partial z} + \frac{\partial f}{\partial t} = b - a.$$

This is the general equation of which we have spoken. It remains to calculate a and b or at least the difference $(a - b)$ and here the difficulties begin: we can however simplify the problem if we neglect the mutual encounters of the electrons, considering only their impacts against the metallic atoms, whose masses are so great that they may be regarded as immovable.

Now in the absence of any extraneous forces a steady state of perfectly chaotic motion will soon be established among the electrons and one in which

$$f(\xi, \eta, \zeta, x, y, z, t) \equiv f_0(\xi, \eta, \zeta, x, y, z, t),$$

and on the assumptions just made it seems reasonable to suppose that this distribution will be similar to that which exists under similar circumstances in gas theory and is specified by Maxwell's law so that we may take

$$f = f_0 = Ae^{-qu^2},$$

where $$A = N\sqrt{\frac{q^3}{\pi^3}}, \quad q = \frac{3}{2u_m{}^2},$$

where N denotes the number of free electrons per unit volume and $u_m{}^2$ the mean square of their velocities. This is the perfectly chaotic distribution of motions and any departure from it arises as a result of the external forces or condition gradients tending to organise the perfect irregularity which this law specifies.

Now since the velocity of any electron after a collision is independent of that before collision, it follows that the distribution of velocities among any set of electrons when taken each just after its next collision succeeding the instant t will be wholly independent of the distribution at the time t and will in general be different from it unless indeed this latter distribution is that specified by Maxwell's law which is specially defined so as to remain unaltered by collision. It follows therefore also that the distribution of velocities among any set of electrons, each taken immediately after its next collision after the instant t, will in fact be precisely that specified by Maxwell's law and is therefore the same independently of the state of motion that may exist at the instant t: in other words the collisions completely obliterate any regularity which existed in the electronic motions before collision. Thus the number ($b\,dv\,dt$) of electrons which enter the specified group during the small time dt is precisely the same as the number which would enter the same group if Maxwell's law specified the distribution both before and after the collision and it might be calculated on this basis. The number $a\,dv\,dt$ of electrons leaving the group in the same time would then be exactly the same as $b\,dt$ if there were no external forces or condition gradients to modify the distribution established by the collisions. In the more general case it is however at once obvious that the number

$$(a - b)\, dv\, dt$$

can be calculated as the number of electrons removed by collision during the time dt from among the partial group of electrons contained in the specified group at the instant t, which is the excess of the number in this group over and above the number required by Maxwell's law: that is the number

$$(f - f_0)\, dv\, dt.$$

We thus want to find the number of collisions which the electrons of this group undergo in the small time dt. As will be seen later however we are only dealing with small variations from the distribution specified by Maxwell's law and therefore the number of electrons removed from the group under consideration will be proportional to the small excess of the number in this group over the number according to Maxwell's law; or in other words

$$b - a = -\frac{f - f_0}{\tau_m}.$$

327. The factor τ_m is a constant for the group under consideration but will in all probability vary from group to group. It will be a function of the resultant velocity of the electrons in the group and of the law of their interaction with the molecules on collision. Our ignorance of the precise nature of the action between a molecule and an electron precludes our probing the nature of this constant any further, but in a tentative theory we might assume that

$$\tau_m = \mu u^p.$$

This is in fact the type of law that has been derived by various writers on certain simple assumptions as to the nature of the collision. When it is as between hard elastic spheres Lorentz finds that $p = -1$, and if it is due to an inverse nth power-law repulsion $p = \frac{4}{n-1} - 1$. These values are however all difficult to reconcile with the experimental results, which seem to require that $p = -2$ or some higher negative power. The value $p = -2$ makes the mean time in a free path inversely proportional to the energy of the electron, a result which is probably not without its physical significance; it seems to indicate however that the atomic nuclei have a much greater scattering power than they would possess if they were simple force centres, especially as regards the more rapidly moving electrons.

328. It follows therefore that the equation for the function f, in the most general possible case of the present type, is

$$\mathbf{F}_x \frac{\partial f}{\partial \xi} + \mathbf{F}_y \frac{\partial f}{\partial \eta} + \mathbf{F}_z \frac{\partial f}{\partial \zeta} + \xi \frac{\partial f}{\partial x} + \eta \frac{\partial f}{\partial y} + \zeta \frac{\partial f}{\partial z} + \frac{\partial f}{\partial t} = \frac{f_0}{\tau_m} - \frac{f}{\tau_m}.$$

Now the difference between the functions f and f_0 may be shown to be extremely small in any real case, at least compared with f_0 itself, and we may therefore use $f = f_0$ on the left-hand side of this equation so that we get

$$f = f_0 - \tau_m \left(\mathbf{F}_x \frac{\partial f_0}{\partial \xi} + \mathbf{F}_y \frac{\partial f_0}{\partial \eta} + \mathbf{F}_z \frac{\partial f_0}{\partial \zeta} + \xi \frac{\partial f_0}{\partial x} + \eta \frac{\partial f_0}{\partial y} + \zeta \frac{\partial f_0}{\partial z} + \frac{\partial f_0}{\partial t} \right),$$

where of course f_0 has the value quoted above: on inserting this we find that

$$f = Ae^{-qu^2} \left[1 + 2q\tau_m (\xi \mathbf{F}_x + \eta \mathbf{F}_y + \zeta \mathbf{F}_z) - \frac{\tau_m}{A} \left(\xi \frac{\partial A}{\partial x} + \eta \frac{\partial A}{\partial y} + \zeta \frac{\partial A}{\partial z} \right) \right.$$
$$\left. + u^2 \tau_m \left(\xi \frac{\partial q}{\partial x} + \eta \frac{\partial q}{\partial y} + \zeta \frac{\partial q}{\partial z} \right) \right],$$

or in vector notation, using $\mathbf{u}$ as the vector velocity of the electron

$$f = Ae^{-qu^2}\left[1 + 2q\tau_m(\mathbf{uF}) - \frac{\tau_m}{A}(\mathbf{u}\nabla)A + \tau_m u^2(\mathbf{u}\nabla)q\right].$$

The density of the electric flux is then determined by its components

$$(\mathbf{C}_x, \mathbf{C}_y, \mathbf{C}_z) = e\iiint_{-\infty}^{+\infty}(\xi, \eta, \zeta) f\, d\xi\, d\eta\, d\zeta,$$

whilst the flux of kinetic energy is determined by the vector with components

$$(\mathbf{H}_x, \mathbf{H}_y, \mathbf{H}_z) = \iiint_{-\infty}^{+\infty}\frac{m}{2}u^2(\xi, \eta, \zeta) f\, d\xi\, d\eta\, d\zeta.$$

The integrals in each of these cases can be directly evaluated by the spherical polar transformation

$$\xi = u\cos\theta, \quad \eta = u\sin\theta\cos\phi, \quad \zeta = u\sin\theta\sin\phi,$$

and it is in this way found that

$$\mathbf{C} = \frac{4\pi Ae}{3}\left[i_4\left(2q\mathbf{F} - \frac{1}{A}\operatorname{grad} A\right) + i_6 \operatorname{grad} q\right],$$

whilst

$$\mathbf{H} = \frac{2\pi Am}{3}\left[i_6\left(2q\mathbf{F} - \frac{1}{A}\operatorname{grad} A\right) + i_8 \operatorname{grad} q\right],$$

wherein we have used

$$i_s = \int_0^\infty \tau_m u^s e^{-qu^2}du.$$

When $\tau_m = \mu u^p$ we have

$$i_s = \mu\int_0 u^{s+p}e^{-qu^2}du$$

$$= \frac{\mu}{2q^{\frac{s+p+1}{2}}}\Gamma\left(\frac{s+p+1}{2}\right),$$

and with these values we have

$$\mathbf{C} = \frac{2\pi\mu eA\Gamma\left(\frac{p+5}{2}\right)}{3q^{\frac{p+5}{2}}}\left[\left(2q\mathbf{F} - \frac{1}{A}\nabla A\right) + \frac{p+5}{2q}\nabla q\right],$$

and

$$\mathbf{H} = \frac{\pi\mu mA\Gamma\left(\frac{p+7}{2}\right)}{3q^{\frac{p+7}{2}}}\left[\left(2q\mathbf{F} - \frac{1}{A}\nabla A\right) + \frac{p+7}{2q}\nabla q\right].$$

329. We now consider the conduction of electricity in a homogeneous bar of the metal which is kept at the same temperature in all its parts: let this metal be acted on by an electric force E in the direction of its length which we may take to be directed along the x-axis. The force acting on each electron will then be eE so that

$$\mathbf{F}_x = \frac{eE}{m}, \quad \mathbf{F}_y = \mathbf{F}_z = 0,$$

and since the physical conditions of the metal are the same throughout its mass, the quantities A and q which depend on these will be constants so that

$$\text{grad } A = \text{grad } q = 0.$$

We have thus in this case a current of electricity defined by its flow per unit area across a section of the bar

$$C = \frac{4\pi\mu e A \Gamma\left(\frac{p+5}{2}\right)}{3mq^{\frac{p+3}{2}}} E$$

from which we may conclude that the conductivity of the metal under the specified conditions is

$$\kappa = \frac{4\pi\mu e^2 A \Gamma\left(\frac{p+5}{2}\right)}{3mq^{\frac{p+3}{2}}}.$$

This formula was first given by Lorentz for the case when $p = -1$.

In the simplest cases it appears that the conductivity at ordinary or room temperatures varies inversely as the absolute temperature. But on the Maxwell-Boltzmann theory q also varies inversely as the temperature and so to secure agreement in these cases we must assume that κ varies as q. Now

$$A = N\sqrt{\frac{q^3}{\pi^3}},$$

and therefore as mentioned above we must take $p = -2$. In this case the conductivity is simply

$$\kappa = \frac{\mu e^2 N}{3m} q = \frac{\mu e^2 N}{6R} \cdot \frac{1}{\vartheta}.$$

This simple result is obtained on the assumption that the applied electric field is the only effective driving field for the current. If there were local fields in the metal, either intrinsic or induced, this value of κ would have to be modified; it is therefore not difficult to see why the result is not generally verified in practice.

330. In order however to exhibit all the beauties of this theory it is necessary to consider the question of the conduction of heat in the metal. A bar of metal whose ends are maintained at different temperatures may be likened to a column of gas, placed for example in a vertical position and having a higher temperature at its top than at its base. The process by which the gas conducts heat consists in a kind of diffusion between the upper part of the column in which we find larger, and the lower one in which we find smaller, molecular velocities; the amount of this diffusion and the intensity of the flow of heat that results from it depend on the mean distance over which a molecule travels between two successive encounters. Now in the present theory of metals it seems at least plausible to assume that the conduction of heat goes on in a way that is exactly similar to that just described, only the carriers by which the heat is transformed from the hotter towards the colder parts of the body are now the free electrons, and the length of their free path is limited, not, as in the case of a gas by mutual encounters, but by the impacts against the metallic atoms, which we have supposed to remain at rest on account of their comparatively large mass. Of course this idea seems to imply that some sort of thermodynamical equilibrium exists between the metal molecules and the electrons, so that the latter are partaking of a real heat motion. It is usual to assume that this equilibrium is of the simple type that exists between different kinds of molecules in a compound gas, so that the mean kinetic energy of the electron expressed in the above notation by

$$\tfrac{1}{2}mu_m^2 = \frac{3m}{4q}$$

is determined by the simple law of the equality of mean energies. This means that if R is the universal gas constant and θ the absolute temperature of the metal we may write

$$\tfrac{1}{2}mu_m^2 = \frac{3R\theta}{2},$$

and then the relation

$$q = \frac{m}{2R\theta}$$

determines the dependence of q on the temperature. Such a simple relation however seems to be satisfactory only at ordinary or very high temperatures; in order to bring the theory of specific heats of solids into line with the experimental data obtained at low temperatures it has in fact become necessary to use a more complex relation. The theoretical basis of the modification need not here detain us however as our results are independent of any special theory. All we need to notice is that

$$Q = \tfrac{1}{2}Nmu_m^2$$
$$= \frac{3Nm}{4q}$$

is the total heat contained by the electrons in the body at the specified temperature.

331. In the general case it is not possible to determine a definite value for the quantity of energy of the irregular or chaotic part of the electronic motions which is transferred during their average flux, because the energy of each separate electron being in part kinetic energy and in part potential energy relative to the metal molecules and the external field is known only to an additive constant. In one special case however when the aggregate flux of the electrons vanishes, so that there is no flow of electricity, will this indefiniteness disappear and the flux of energy through the metal is completely determined by the vector **H** given above. As in this case the electric current is zero we must have the condition

$$2q\mathbf{F} - \frac{1}{A}\nabla A + \frac{p+5}{2q}\nabla q = 0,$$

and thus
$$\mathbf{H} = \frac{2\pi\mu m A\Gamma\left(\frac{p+7}{2}\right)}{3q^{\frac{p+9}{2}}}\nabla q.$$

We have however from above

$$\frac{1}{q^2}\nabla q = -\frac{4}{3Nm}\nabla Q = -\frac{4}{3Nm}\frac{dQ}{d\theta}\nabla\theta,$$

so that
$$\mathbf{H} = -\frac{8\pi\mu A\Gamma\left(\frac{p+7}{2}\right)}{q^{\frac{p+5}{2}}}\frac{\nabla Q}{d\theta}\nabla\theta,$$

and the conductivity for heat, as usually defined, is therefore in this case given by

$$\gamma = \frac{8\pi\mu A\Gamma\left(\frac{p+7}{2}\right)}{9Nq^{\frac{p+5}{2}}}\frac{dQ}{d\theta} = \frac{8\Gamma\left(\frac{p+7}{2}\right)\mu}{9q^{\frac{p+2}{2}}\pi^{\frac{1}{2}}}\frac{dQ}{d\theta},$$

and it depends only on the nature and physical conditions at the point in the metal under consideration.

Since Q is the total heat content of the electrons in the metal, $\frac{dQ}{d\theta}$ is what we may call, after Thomson, the specific heat of the electricity in the metal. In the particular case when $p = -2$ the coefficient γ is proportional to this specific heat, a result which without a great stretch of imagination may be regarded as substantiated by the fact that in a large number of cases the thermal conductivity varies with the temperature just as does the specific heat of the metal as a whole, even at every low temperatures where the specific heat becomes somewhat irregular.

Of course, at ordinary temperatures Q is proportional to the absolute temperature ϑ and then γ is a constant (if $p = -2$) independent of the temperature.

332. The ratio of the conductivities of heat and electricity is

$$\frac{\gamma}{\kappa} = \frac{2(p+5)}{9N^2e^2} Q \frac{\partial Q}{\partial \theta},$$

and at ordinary temperatures where

$$Q = \frac{3Nm}{4q} = \tfrac{3}{2} NR\vartheta$$

this reduces to

$$\frac{p+5}{2}\frac{R^2}{e^2}\vartheta,$$

and is therefore the same in all metals at the same temperature. This is the well-known Wiedemann-Franz law which has been successfully verified in numerous cases*: a few typical ones are exhibited below:

Aluminium	$\cdot 706 \,.\, 10^{-10}$
Copper	$\cdot 738 \,.\, 10^{-10}$
Silver	$\cdot 760 \,.\, 10^{-10}$
Iron	$\cdot 890 \,.\, 10^{-10}$

and seeing that the specific electrical resistances of these substances range from $\cdot 3 \,.\, 10^{-4}$ to $64 \,.\, 10^{-4}$ the agreement is remarkable.

* Full details of the experimental results to 1911 are given by K. Baedeker, *Die elektrischen Erschienungen in metallischer Leitern* (Braunschweig, 1911). Cf. also L. L. Campbell, *Galvanomagnetic and thermomagnetic effects* (Longmans, 1923).

The ratio of the two conductivities varies on any form of the theory as the absolute temperature, a law empirically formulated by Lorenz: the temperature coefficients of the ratio in the four cases quoted are respectively 4·37, 3·95, 3·77, 4·32.

But not only are these general qualitative results satisfactorily verified by our theory: the agreement is quantitative as well: in fact it is known from gas theory that

$$\left(\frac{R}{e}\right)^2 = 8{\cdot}26 \times 10^{-14},$$

where e is the electrostatic charge on the electron, and thus

$$\frac{\gamma}{\kappa} = \left(\frac{p+5}{2}\right) 8{\cdot}26 \times 10^{-14},$$

or taking an absolute temperature of 300° (i.e. 27° C.) this gives

$$\frac{\gamma}{\kappa} = \left(\frac{p+5}{2}\right) 24{\cdot}8 \times 10^{-12},$$

which is of the same order of magnitude as those quoted above, especially in the case when $p = 1$ or some number of this order of magnitude.

This agreement between the theory and experiment, first noticed by Drude, is one of the most beautiful results of this theory and points distinctly to the conclusion that the assumptions that both electricity and heat are carried through the metal by the electrons, and that these electrons are in a simple mechanical heat equilibrium with the metal molecules, are completely justified.

At low temperatures where Q is no longer proportional to the temperature the two simple laws just mentioned no longer hold as our formula would indicate. That there are discrepancies at ordinary temperatures is due to the uncertainty in the electrical coefficient already referred to.

333. In the preceding paragraphs we have considered two special cases of the transfer of heat and electricty in a homogeneous piece of metal. We shall finally consider briefly the more general case which leads to an explanation of the full circumstances in the thermoelectric phenomena discussed above. For the sake of generality we shall introduce the notion of molecular forces of one kind or another exerted by the atoms of the metal on the electrons and producing for each electron a resulting force along the direction in which the metal is not homogeneous; this is the main point of the idea of Helmholtz's assumption that each substance has a specific affinity for electricity which varies with the temperature: these forces will be assumed to be such that the typical electron will on the average have a potential energy $e\mu$ under the standard conditions relative to the metal molecules surrounding it. Thus if we now assume that the impressed field is derived from a potential ϕ we shall have

$$\mathbf{F} = -\frac{e}{m}\operatorname{grad}(\phi + \mu),$$

so that we have the currents of electricity and heat

$$\mathbf{C} = -\kappa\left[\operatorname{grad}(\phi + \mu) + \frac{R\theta}{Ae}\operatorname{grad} A + \frac{p+5}{2e}R\operatorname{grad}\theta\right],$$

and

$$\mathbf{H} = \frac{p+5}{2e}R\theta\,\mathbf{C} - \gamma\operatorname{grad}\theta.$$

From the first of these expressions we may deduce expressions for the rate of fall of potential at each point and for the difference of potential between the ends of the bar examined in the previous paragraph. It is however more interesting to make the calculations for a more general case. We therefore consider a circuit consisting of a thin curved wire, the dimensions of the normal section at any point of which are small compared with the radius of curvature of the curve of the wire at the point. We may then assume that the nature and temperature of the metal are very nearly the same at all points of a single cross section and we shall therefore only be concerned with the component fluxes tangentially along the wire, the other component being negligibly small. If we use s to denote a coordinate of distance along the wire, the equations for these fluxes are

$$\mathbf{C}_s = -\kappa\left[\frac{d}{ds}(\phi + \mu) + \frac{R\theta}{Ae}\frac{dA}{ds} + \frac{p+5}{2e}R\frac{d\theta}{ds}\right],$$

and

$$\mathbf{H}_s = \frac{p+5}{2e}R\theta\,\mathbf{C}_s - \gamma\frac{d\theta}{ds}.$$

We now examine special cases of these equations.

(1) In an open circuit in which no current is flowing there is a potential difference between the ends which may be regarded

as a measure of the electromotive force existing in the circuit when closed. In this case we have

$$\frac{d\phi}{ds} = -\frac{d\mu}{ds} - \frac{R\theta}{Ae}\frac{dA}{ds} - \frac{p+5}{2e}R\frac{d\theta}{ds},$$

so that on integration along the circuit from the point s_1 to the point s_2 we get

$$\phi_1 - \phi_2 = -(\mu_1 - \mu_2) - \frac{R}{e}\int_{s_1}^{s_2} \frac{\theta}{A}\frac{dA}{ds}\,ds - \frac{p+5}{2e}R\,(\theta_1 - \theta_2).$$

If the temperature is uniform along the circuit this gives

$$\phi_1 - \phi_2 = -(\mu_1 - \mu_2) - \frac{R\theta}{e}\log\frac{A_1}{A_2}.$$

This equation shows at once that the potential difference between any two points of the circuit will depend only on the nature of the metal at these two points and will be zero if these metals are the same, for then $\mu_1 = \mu_2$ and $A_1 = A_2$, these two quantities depending essentially on the character and conditions of the metal and nothing else. The explanation of the volta potential differences and the laws which it obeys is now obvious: the difference in the potential between the ends of a compound circuit may be due either to a difference in the "affinity" potential μ of the electron in the metals at the ends or to a difference in the concentration of the electrons at these ends.

(2) We now consider an open circuit consisting of one kind of metal only but in which the temperature is not uniform. In this case A and μ will be functions of the temperature only so that the expression for $d\phi$ will be an exact differential with respect to θ: again the potential difference between two points of the circuit will only be dependent on the temperature of these points and will be zero if these are the same:

$$\frac{d\phi}{ds} = -\frac{d\mu}{ds} - \frac{R\theta}{Ae}\frac{dA}{ds} - \frac{p+5}{2e}R\frac{d\theta}{ds},$$

$$\phi_1 - \phi_2 = -\int_1^2 \left(\frac{d\mu}{d\theta} + \frac{R\theta}{Ae}\frac{dA}{d\theta} - \frac{p+5}{2e}R\right)\frac{d\theta}{ds}\,ds.$$

334. We finally consider the more general case of a non-homogeneous circuit in which the temperature varies. We shall confine ourselves to the consideration of a circuit such as that described above in the text in which three pieces of metal are joined up in a circuit; the first and third pieces being however of the same material so that if they were joined up the circuit would in reality consist of only two pieces of metal. The junctions are P_1 and P_2 and we shall suppose that they are at temperatures θ_1 and θ_2 and also that the temperatures at the two ends of the circuit are both θ_0. Still retaining the coordinates s_1 and s_2 for these ends we have

$$\phi_1 - \phi_2 = -\int_1^2 \left(\frac{d\mu}{ds} + \frac{R\theta}{Ae}\frac{dA}{ds} + \frac{p+5R}{2e}\frac{d\theta}{ds}\right) ds.$$

Now $$\int_{s_1}^{s_2} \frac{\theta}{A}\frac{dA}{ds}\, ds = \Big|\,\theta \log A\,\Big|_1^2 - \frac{R}{e}\int_{s_1}^{s_2} \log A \frac{d\theta}{ds},$$

and the integrated term vanishes because the metal and temperature at the ends 1 and 2 are the same: thus

$$\begin{aligned}
\phi_1 - \phi_2 &= -\int_1^2 \left(\frac{d\mu}{ds} + \frac{R\theta}{Ae}\frac{dA}{ds} + \frac{p+5R}{2e}\frac{d\theta}{ds}\right) ds \\
&= -\int_1^2 \left(\frac{d\mu}{d\theta} - \frac{R}{e}\log A + \frac{p+5R}{2e}\right)\frac{d\theta}{ds}\, ds \\
&= \int_{\theta_0}^{\theta_1} \left(\frac{d\mu}{d\theta} - \frac{R}{e}\log A + \frac{p+5R}{2e}\right) d\theta \\
&\quad + \int_{\theta_1}^{\theta_2} \left(\frac{d\mu}{d\theta} - \frac{R}{e}\log A + \frac{p+5R}{2e}\right) d\theta \\
&\quad + \int_{\theta_2}^{\theta_0} \left(\frac{d\mu}{d\theta} - \frac{R}{e}\log A + \frac{p+5R}{2e}\right) d\theta \\
&= \int_{\theta_2}^{\theta_1} \left(\frac{d\mu}{d\theta} - \frac{R}{e}\log A + \frac{p+5R}{2e}\right) d\theta \\
&\quad - \int_{\theta_1}^{\theta_2} \left(\frac{d\mu}{d\theta} - \frac{R}{e}\log A + \frac{p+5R}{2e}\right) d\theta,
\end{aligned}$$

and the first integral now refers to the one type of metal and the second to the other so that the integrands are proper functions of the temperature. Using suffices a, b to denote quantities referring to the different metals we see that

$$\phi_{12} = \phi_1 - \phi_2 = \int_{\theta_1}^{\theta_2} \left(\frac{d\mu_a}{d\theta} - \frac{d\mu_b}{d\theta} - \frac{R}{e}\log\frac{A_a}{A_b}\right) d\theta.$$

The potential difference between the ends of the circuit depends therefore merely on the temperatures at the junctions of the two different metals and vanishes if these are equal.

335. Let us now examine the development of heat which takes place in the same circuit when the current is allowed to flow round it. To do this we shall assume that the conditions at each point of the circuit are stationary, the temperature being suitably maintained constant by conduction (if the circuit consists of a thin wire this may be done without appreciably altering the conditions under which we are treating these questions).

We now consider a small element of the circuit between the cross-sections at distances s and $s + ds$ from the origin on the circuit and we find the quantity of heat dH which must be extracted from such an element to maintain its temperature constant. This quantity will of course be equal to the difference between the amount of energy supplied to the electrons in the element on account of the extraneous forces and the amount which is brought into the element as a result of the electrons diffusing into it from other parts of the circuit. The state of the flow being assumed to be stationary the same number of electrons will flow into the element on one side as will flow out at the other and the amount of heat can be calculated from the expression for $\mathbf{H}_s$ which determines the amount of electronic energy, consisting in part of kinetic energy and in part of potential energy, which is transferred through unit area of a normal section of the circuit per unit time. If we take the area of the cross-section at s to be f, then we shall have

$$dH = -\left(f\mathbf{C}_s \frac{d\phi}{ds} + \frac{d}{ds}(f\mathbf{H}_s)\right) ds,$$

or introducing the total current flow J determined by

$$J = f\mathbf{C}_s = -f\kappa \frac{d\phi}{ds},$$

we get $$dH = \left[-\frac{J^2}{f\kappa} + J\left(\frac{d\mu}{ds} + \frac{R\theta}{eA}\frac{dA}{ds}\right) - \frac{d}{ds}\left(f\gamma \frac{d\theta}{ds}\right)\right] ds.$$

The first term in this expression, which is proportional to the square of the current strength, indicates of course the Joule's heat developed in the element. The last term, which is independent of the current, indicates the heat supply to the element on account of true thermal conduction in the circuit. The middle term is the important one: it denotes a development of heat in the circuit

$$dH' = J\left[\frac{R\theta}{eA}\frac{dA}{ds} + \frac{d\mu}{ds}\right] ds,$$

which is proportional to the strength of the current and which therefore changes sign when the direction of the current is reversed. It is this term which contains the expression of the Peltier and Thomson effects, as we see by examining two simple cases.

We first consider a part of the circuit in which the temperature is constant and in which a transition from the metal a to the metal b takes place; this transition is assumed to be gradual and on integration of the expression for dH' across it we find the corresponding amount of heat developed per unit of time by the passage of the current across the junction is equal to

$$J\left[\frac{R\theta}{e}\left(\log\frac{A_b}{A_a}\right) + (\mu_b - \mu_a)\right],$$

so that the Peltier effect as defined in the text is

$$\frac{R\theta}{e}\left(\log\frac{A_b}{A_a}\right) + (\mu_b - \mu_a).$$

Now consider a portion of the circuit within which the metal is the same but in which the temperature varies. We then find in a similar manner that the amount of heat developed per unit of time in a small part of the circuit in which the temperature changes uniformly from θ to $\theta + d\theta$ will be

$$J\left[\frac{R\theta}{Ae}\frac{dA}{d\theta} + \frac{d\mu}{d\theta}\right]d\theta,$$

so that the Thomson effect as defined is

$$-\left(\frac{R\theta}{Ae}\frac{dA}{d\theta} + \frac{d\mu}{d\theta}\right).$$

The expressions thus found for the quantities denoting the potential in a circuit of the type under consideration and the Peltier and Thomson effects are fully consistent with the relations between these quantities deduced by Kelvin from thermodynamical considerations and given above in the text.

This theory moreover effectively accounts for many features of the phenomena which it is rather difficult to explain on any other basis and in particular the phenomena associated with the Thomson effect, which on the present basis becomes almost self-evident.

It is however rather astonishing to find that in fact the order of magnitude of this effect is much smaller than the above argument and result would suggest. It is concluded therefore that the main part of the expression determining this effect must be zero or to a first approximation it is necessary that

$$\frac{R\vartheta}{Ae}\frac{dA}{d\vartheta}+\frac{d\mu}{d\vartheta}=0;$$

which means that $$\log A+\int\frac{e\,d\mu}{R\vartheta}=\text{const.}$$

If μ is neglected then $A\,(=Nq^{\frac{3}{2}}\pi^{-\frac{3}{2}})$ itself must be constant, a result which is often interpreted as implying that N varies as $\vartheta^{\frac{3}{2}}$; but it is doubtful whether the assumption $\mu=0$ can be satisfactorily justified and therefore the order of magnitude of this effect cannot be used as an argument in favour of any particular relation between N and ϑ.

INDEX

Aberration on the electromagnetic theory, 365–368

Abraham's electron, 354

Action, medium and distance, con trasted, 52; question of velocity of transmission of, 52; Faraday-Maxwell theory of transmission of electromagnetic, 52, 89–97

Action and Reaction, Newton's principle of, 194

Aether; displacement currents in, 75, 97–102; conception of, as a carrier of energy, 236; experimental evidence of fixity and rigidity of, 333, 362; position of, in the theory of relativity, 388

Ampère's electromagnetic theory, 163–165; the circuital equation of, 164; the position of magnetism in, 165; general dynamical formulation of, 244–251

Anion, 135

Anode, 135, 149

Availability of energy in electrostatic fields, 80–86; in magnetic fields, 184

Biot and Savart; law of, in electromagnetic theory, 192

Boundary problems; conditions for, in electrostatics, 24, 37; in dielectric theory, 72; in magnetic theory, 175–177; involving electromagnetic waves, 268; involving moving bodies, 337–342

Calamoids; Whittaker's 4-dimensional tubes of force, 380

Carnot's cycle, 40; application of, in thermoelectric problems, 133; to the voltaic phenomena, 140; in thermomagnetic problems, 395

Cathode, 135, 149

Cathode rays, 150; constitution of, 150; charge and mass of particles in, 151

Circuitality of electric displacement flux, 101; of total current flux, 220–223; of magnetic induction, 166–218

Condenser, discharge of, 290

Conduction of electricity, by metals, 110–117; electron theory of, in metals, 404–422; by gases and air under the action of Roentgen rays, 141–144; by liquids and electrolytes, 134–141

Conduction of heat, by metals, explained on electron theory, 411–416

Conductors, general properties of the static field of a system of, 33–37; ponderomotive forces acting on and between, 46–50; propagation of electric waves in, 267; reflection of waves by, 288; general relations of charged, in uniform motion, 334

Convection, field due to, of electrostatic system, 335; current due to, of charged and polarised media, 214

Convection potential, due to moving system, 335; characteristic equations for, 337; due to uniformly moving ellipsoidal shell, 339; due to spherical shell, 331; due to Lorentz's ellipsoidal shell, 342

Coulomb, the practical unit of capacity defined in terms of the theoretical units, 224; law of force in electro statics due to, 5; law of force in magnetostatics due to, 154

Crystals, analysis of structure by X rays, 303; pyro- and piezo-electric properties of, 105; magnetic pro· perties of, 398

Curie's law, in thermomagnetic effects, 391; its thermodynamic bearing, 395

D'Alembert's principle, in dynamics, its significance, 15

Dark space, Faraday's, 149; Crooke's, 149

Diamagnetism, 390

Dielectric constant, 53, 56; its relation to density, 102–105

Dielectrics, on Faraday-Maxwell theory, 52–59; the effect of, on capacity of a condenser, 53; the polar theory of, 60–64; specification of the field of polarised, 61–65; validity of expressions for the field functions in the theory of, 66–69; Poisson's transformation of the potential of polarised, 67; composition of the displacement current in the theory of polarised, 75; characteristic properties of the field with dielectrics, 70; the law of induction of polarisation in, 56; energy relations of, 78–85; intrinsic energy of, 82; energy restrictions on the law of induction in, 83; complete expression for the energy in the field of polarised, 85; mechanical relations of polarised, 87, 88; force and couple on the elements of, 87; the stress in polarised, 91; effect of, on the propagation of electric waves, 268

Dimensions, doctrine of, 223; of magnitudes in electric theory, 223–225

Discharge tube, phenomena in, 149

Displacement, *see* **Electric displacement**

Eichenwald's experiment on moving polarised dielectrics, 332

Electric currents, origin of, 102; general definition of, 112; special definition of, 113; resistance to flow of, 114, 115; hydrostatic analogy for, 117; energy relations of, 120–128; distribution of steady, in a network of conductors, 127; law of minimum dissipation for the flow of, 128; the thermal relations of, 128–134; thermodynamic discussion of these relations, 133; in liquids, 134–138; in gases, 141–152; magnetic field due to, 147–160; dimensions of, 224; complete expression for, on Maxwell's theory, 213, 214, 330; due to convexion of charged and polarised media, 214, 330; electronic aspects of the, in Maxwell's theory, 215, 216; statistical analysis of, in metallic conductors, 404–422; circuital property of, 220, 221; induction of, by varying magnetic fields, 195–197; mutual mechanical relations of linear conductors carrying, 199–208; interaction between magnets and circuits carrying, 191–194; energy in the field of linear conductors carrying, 182; dynamical theory of the interaction of linear conductors carrying, 199–208; in a discharging condenser, 290; in a network of conductors, 207; oscillating, in a conductor of finite cross-section, 274; in the 4-dimensional equations of the theory, 379

Electric displacement, 55; tube of, 56; characteristic properties of, 56–59; circuital property of, 57, 98–102; physical significance of, 60, 61; on the theory of polarised dielectrics, 60, 61; currents of, 70–75, 98–102; significance of, in the theory of electric waves, 265–269

Electricity, 1–4; constitution of, 3; atomic structure of, 3, 136; specific heat of, 132–134, 415; dissipation of charge of, by conduction, 267

Electro-caloric effects, 105–109

Electrodynamic potential, 205

Electrodynamics of linear circuits carrying currents, 199–208; Maxwell's general theory of, 244–251; of the single electron, 249, 382

Electrolysis, 134–140; Faraday's laws of, 136; electrolytic dissociation, 139; velocity of ions in, 139; thermodynamic relations of, 140

Electromagnetic field, Ampère's circuital equation for, 163–165; Faraday's circuital equation for, 195–197; definition of, in terms of scalar and vector potentials, 226; Maxwell's general theory of, 329; of the Hertzian oscillator, 293–296; of a given distribution of currents, 231; of given moving charges, 233–244, 335, 343; in radiation, plane waves, 284–289; distribution and flux of energy in, 236–244; distribution of force and momentum in, 258; Lorentz's transformation of the equations for, 372–375; the 4-dimensional form of the equations for, 378–390

Electromagnetic induction, Faraday's law of, 195; its connexion with the energy principle, 195

Electromagnetic momentum, in the general electromagnetic field, 258; in the field of a number of point charges, 259–263; of a single electron, 263; alternative expressions for, 263; of a general electrostatic system in motion, 346–359; of the general system, its connexion with the Lagrangian function, 351; of a moving ellipsoid, 353, 354; in the 4-dimensional analysis, 385

Electromagnetic potentials, *see also* **Scalar potential** *and* **Vector potential**; general definition of, 226; characteristic equations for, 227; the instantaneous, 227; the retarded, 228; in the 4-dimensional analysis, 381, 382

Electromagnetic waves, characteristic equations for the field in, 275–284; in conductors, 267; in dielectrics, 268; propagated along the surface of a conductor, 269; from an incandescent body, 258; from a group of electrons, 299–302; in Roentgen or X-rays, 302; mechanism of the generation of, 289–304; from a Hertzian oscillator, 295; boundary conditions at the wave front of, 305; energy carried by, 314–316; radiation pressure due to, 320–328; in moving media, 359–372

Electromotive force, definition of static, 7; connexion with the potential, 8, 37–40; complete expression for, on moving charge, 249, 329; in 4-dimensional analysis, 382
Electron, 3; mass of, 3, 151; charge on, 3, 147; electromotive force on, 249, 329; field of a moving, 231–236; momentum of, in quasi-stationary motion, 263, 352–355; mass of, in quasi-stationary motion, 354; Abraham's, 354; Lorentz's, 355, 356; discrimination of type by experiment, 357; relation between energy and mass of, in relativity mechanics, 383
Electrostatic field, definition of, 6, 7; of a point charge, 7; of a system of charges, 8; of continuous charge distributions, 9–18; surrounding charged conductors, 33–37; distribution of energy in, 40–46; transmission of force through, 47–50, 88–96; Green's analysis of, 18–26; Gauss' analysis of, 26
Electrostriction, 109
Energy, conservation of, 37–40; of electrostatic system, 40–46; distribution of, in the electrostatic field, 70–86; of the magneto-static system, 179–182; distribution of, in the magnetic field, 182–187; relations of an electric current, 124–128; of a voltaic cell, 140, 141; distinction between kinetic and potential, 187; general conception of, in the electromagnetic field, 236; flux of, in the electromagnetic field, 238–244; flux of, in radiation fields, 313–320; of an electron in relation to its mass, 383; relations of Lorentz electron, 355, 356
Equilibrium in the static distribution of charge, 43
Equilibrium theories in dynamics, their significance, 208
Equi-potential surfaces, 31

Farad, 224
Faraday-Maxwell theory, of the electrostatic field, 51–109; of the electromagnetic field, 212–264; of the field of radiation, 265–269; general dynamical formulation of, 244–251
Faraday's law of electromagnetic induction, 195; its differential form, 198; for circuits in general, 330; general relation of, to dynamical theory, 244–251
Ferromagnetism, 390–396; Weiss' theory of, 399–403
Fitzgerald-Lorentz contraction, 371
Force, lines of, in electrostatic field, 53, 54; in 4-dimensional analysis, 380, 381

Gases, conduction of electricity in, 140–152
Gauss' normal induction, theorem, 26; in Faraday-Maxwell theory, 57
Gauss' reciprocal energy theorem, 44
Green's analysis, of the electrostatic field, 18–24
Green's theorem, 18; 4-dimensional form equivalent to Maxwell's equations, 380

Henry, 226
Hertz's oscillator, 292; field of, 293–295
Huygen's principle, 276; Larmor's discussion of, 276–283
Hysteresis, curve of, in iron, 396; loss of energy due to, in magnetic cycles, 186

Ideal, electric distribution equivalent to dielectric polarisation, 68; magnetic distribution equivalent to magnetic polarisation, 156
Ignoration of coordinates, 187
Induced polarisation in dielectrics, 60; in magnetism, 172–176, 390
Induction, electromagnetic, Faraday's law of, 195; coefficients of self and mutual, for linear circuits, 177, 182; general dynamical theory of, as regards linear currents, 198–208
Induction, electrostatic, electrification by, 2
Induction, magnetic, vector of, as distinguished from the magnetic force, 157
Inverse square law, 4, 154

Joule, 224
Joule effect, in magnetism, 210

Kinetic energy, as distinct from potential energy, 187; in relativity mechanics, 383; in radiation fields, 288, 313–320, 384
Kirchhoff's equations for the flow of currents in a network of conductors, 127

Lagrange's equations for current cir cuits, 198–208
Lagrange's function, for current circuits, 205; for the general electrodynamic system, 247; for the uniformly convected electrical system, 350; for the moving electron, in relativity mechanics, 387

Laplace's equation in electrostatics, 24
Larmor-Lorentz transformation of the electromagnetic equations, 372–378
Least-action, principle of, 245; application in electromagnetic theory, 246–248; 4-dimensional form of, 387
Level surfaces, *see* **Equi-potential surfaces**
Leyden jar, discharge of, 290
Lines of force, in electrostatic field, 29; in 4-dimensional analysis, 380, 381

Magnet, poles of a long thin, 153; axis of a, 152; field of a finite, 156
Magnetic field, 152–166; reason for special choice of force at internal points, 157; mathematical relations of the, 165; boundary conditions in, 167; of the bi-pole, 155; of a finite magnet, 156; of a magnetic filament, 157; of a magnetic shell, 159; of a linear circuit current, 160–163; vector potential of, 168–175; boundary conditions in, in terms of the vector potential, 176; energy relations of, 174–187; energy of polarisations in, 179; energy of currents in, 180; degradation of energy in, by hysteresis, 186; forces on polarisation in, 189–191; forces on currents in, 191–194
Magnetic force, 157
Magnetic induction, new definition of, as mechanically effective force, 157; its connection with the magnetic potential, 157, 165
Magnetic matter, Poisson's ideal, 156
Magnetic shell, potential in field of, 159; equivalence of, and a linear current, 160–163; vector potential of, 171
Magnetic stress, 257
Magnetisation; specification of, in a magnet, 155; solenoidal, 160; lamellar, 160; Ampère's view of origin of, 165; law of induction of, 173; permanent, 172; hysteresis curve for, 396; energy of, 179; loss of energy of, due to hysteresis, 187; Curie's law of dependence of, on temperature, 391; thermodynamical relations of, 395
Magnetism, 153–177; unit of, 154; law of force in, 154; permanent and in duced, 172; law of induction in, 172; Ampère's theory of, 165; Langevin's theory of, 391–394; constitutional theories of, 397–399
Magnetostriction, 209
Mass, of cathode particle, 147; of electron, 3; of Abraham's electron, 355, 356; Kaufmann's experiments on, of electron, 357; electromagnetic, of a moving electrical system, 353; relation between, and kinetic energy, 383
Maxwell's equations, 330; 4-dimensional form of, 379, 380
Maxwell's stress in electrostatic field, 91; in electromagnetic field, 257; in the 4-dimensional relations, 385
Maxwell's theory of the electrostatic field, 51–109; of the electromagnetic field, 212–223; the vectors covered in, 213; the constitutional relations in, 219; dynamical aspects of, 244–251
Michelson-Morley experiment, 371
Momentum, electromagnetic, *see* **Electromagnetic momentum**
Mutual induction, 177

Ohm's Law, 114

Paramagnetism, 390
Peltier effect, 131, 421
Piezo-electricity, 105
Poisson's equation, 24
Polarisation, dielectric, 60–64; field of, 64, 65–71; contribution to displacement, 75; law of induction of, 56; energy relations of, 78–85; mechanical relations of, 87–91
Polarisation, magnetic, field of, 155; law of induction of, 173, 390–399; energy relations of, 179–183, 391–395; mechanical relations of, 189–191, 257
Potential, convection, in moving electrical systems, 335
Potential, electromotive, Volta's difference of, for substances in contact, 117–122; electron theory of contact, 418–422
Potential, electrostatic, definition of, 8; of point charge, 8; of continuous charge distributions, 9–18; physical nature of, 37–40; of polarised media, 61–65
Potential, magnetostatic, of magnetic bi-pole, 155; of finite magnets, 156; of a solenoid, 157; of a magnetic shell, 159; of a linear current, 160–163
Potential, scalar, *see* **Scalar potential**
Potential, vector, *see* **Vector potential**
Potential energy, *sec* **Energy**
Poynting's vector, for flux of energy, 241; relative to a moving framework, 367; position of, in the 4-dimensional analysis, 385
Pressure of radiation, 320–328; on moving conductor, 363–365

Pyro-electricity, 105

Quasi-stationary motions, significance of concept of, 347

Radiation, mechanism of, from a group of electrons, 298–304; plane fields of, 284–289; from a Hertzian oscillator, 292–298; pressure of, on any absorbing medium, 322; pressure of, on a mirror, 322; reaction from, on the source, 323; pressure of, on mirror in motion, 326; propagation of, in a moving medium, 359–362
Reaction, principle of action and, 194; Kelvin's kinetic, 206
Reflexion of electric waves by conductors, 288
Relativity, theory of, 368–378
Resistance of metals to flow of electricity, 114–117; mechanism of, in metals, 404–414
Retarded potentials, *see* **Scalar potential** *and* **Vector potential**
Roentgen rays, 302
Roentgen-Rowland experiment on moving polarised media, 332

Scalar potential, 198; in expression for electric force, 198, 226, 227; instantaneous, 226; retarded, 227; retarded, in field of given current distribution, 229; due to specified moving charges, 231–233; as multiplier in the variational form of Maxwell's theory, 247; in the 4-dimensional analysis, 381–382
Self-induction, 177, 182
Specific heat of electricity, 133, 134, 415–421
Stress, Maxwell's dielectric, 91; magnetic, 257; electromagnetic, 255, 258; in the 4-dimensional analysis, 385
Stress-energy tensor, 385
Surface distribution of charge, density of, 16; effect on continuity of force and potential, 24, 37

Thomson thermoelectric effect, 132
Tubes of force in electrostatic field, 30; in 4-dimensional field, 381

Vector potential of magnetic fields, 168–172; boundary conditions in terms of, 176; in expression for the electric force, 198; as a general electromagnetic potential, 226–228; instantaneous, 226; retarded, 227; retarded, in field of given current distribution, 231; due to specified moving charges, 231–233; as multiplier in the variational form of Maxwell's theory, 247; in the 4-dimensional analysis, 381
Velocity of propagation of electromagnetic waves, 208, 268; of propagation of electric waves in moving dielectric, 361; of electron and its mass, 354–357
Villari effect in magnetisation of iron, 210
Volta potential difference, 117–119
Voltaic cell, 138–141

Wave equation, significance of solution of, 275–284
Waves, electric, *see* **Electromagnetic waves**
Wiedemann effect in magnetisation, 210
Wiedemann-Franz law of metallic conduction, 415

Printed in the United States
By Bookmasters